Handbook of Ion Exchange Resins: Their Application to Inorganic Analytical Chemistry

Volume VI

Author

Johann Korkisch, Ph.D.
Professor of Analytical Chemistry
Institute of Analytical Chemistry
University of Vienna
Vienna, Austria

CRC Press, Inc.
Boca Raton, Florida

Library of Congress Cataloging-in-Publication Data

Korkisch, Johann.
  Handbook of ion exchange resins.

  Includes bibliographies and indexes.
  1. Ion exchange chromatography. 2. Ion exchange
resins. I. Title.
QD79.C453K67 1989   543′.0893   87-27829
ISBN 0-8493-3191-9 (v. 1)
ISBN 0-8493-3192-7 (v. 2)
ISBN 0-8493-3193-5 (v. 3)
ISBN 0-8493-3194-3 (v. 4)
ISBN 0-8493-3195-1 (v. 5)
ISBN 0-8493-3196-X (v. 6)

This book represents information obtained from authentic and highly regarded sources. Reprinted material is quoted with permission, and sources are indicated. A wide variety of references are listed. Every reasonable effort has been made to give reliable data and information, but the author and the publisher cannot assume responsibility for the validity of all materials or for the consequences of their use.

All rights reserved. This book, or any parts thereof, may not be reproduced in any form without written consent from the publisher.

Direct all inquiries to CRC Press, Inc., 2000 Corporate Blvd., N.W., Boca Raton, Florida, 33431.

© 1989 by CRC Press, Inc.

International Standard Book Number 0-8493-3191-9 (Volume I)
International Standard Book Number 0-8493-3192-7 (Volume II)
International Standard Book Number 0-8493-3193-5 (Volume III)
International Standard Book Number 0-8493-3194-3 (Volume IV)
International Standard Book Number 0-8493-3195-1 (Volume V)
International Standard Book Number 0-8493-3196-X (Volume VI)

Library of Congress Card Number 87-27829
Printed in the United States

PREFACE

In writing this handbook, the author's primary objective was to provide the reader with essentially all available information on the application of ion exchange resins to inorganic analytical chemistry. This information was extracted from more than 6000 original publications in many languages which included Japanese, Chinese, and Russian so that additional translation work was required in numerous cases. These publications comprised the entire relevant literature published until 1986, and the material as described in more than 1000 tables is presented in the six volumes of the book. In most of these tables comprehensive analytical methods involving the assay of virtually all elements of the Periodic Table are described in a concise manner and in a condensed and digested form. The ion exchange characteristics of the elements as well as other important information required by analysts using ion exchange resins are presented in separate tables which appear in the appropriate sections of the *Handbook*.

In the presentation of the analytical procedures special emphasis and attention has been paid to those methods which allow the multielement analysis of complex matrices such as geological, biological, and industrial materials. This is a treatment of the subject which is generally not adopted in books dealing with ion exchange resins. Therefore, the practical part of the *Handbook* is expected to find broad application in analytical laboratories of all types performing analyses which require the utilization of ion exchange resins (both for routine and research work).

The practical part describes analytical methods involving the rare earth elements and actinides, and then a systematic treatment is presented of the other elements which have been subdivided into transition metals and elements of the main groups of the Periodic Table.

The practical part in Volume I is preceded by a general description of the theoretical, instrumental, and other principles underlying the various applications of ion exchange resins in inorganic analytical chemistry. This general portion of the book was the basis of a very successful course on ion exchange separations in analytical chemistry held by the author at the Catholic University of Rio de Janeiro, Brazil, in 1986. Therefore, it is expected that this and other large portions of the *Handbook* can be used as a text at universities and colleges, and for reference and study by all who are familiar with inorganic analytical chemistry. These include research workers, graduate and undergraduate students of all branches of chemistry and of the other natural sciences, e.g., geology, environmental sciences, and medicine, as well as investigators engaged in analyses involving the quantitation of the elements in industrial products.

Since this book is devoted exclusively to ion exchange resins and their application to inorganic analytical chemistry, ion exchanging materials other than synthetic polymers containing ionogenic groups are not considered, except in a very few cases. Also not treated in this handbook are preparative procedures such as ion exchange methods used for the production of deionized water and the purification of inorganic and/or organic reagents. Included, however, are techniques for the isolation of radioisotopes which are required for radioanalytical purposes.

ACKNOWLEDGMENTS

While working on this handbook valuable help in many respects was provided by the following persons (listed in alphabetical order): J. J. Alberts (Director, Marine Institute, University of Georgia, Sapelo Island), P. Sanchez Batanero (Ph.D., Professor of Chemistry, Department of Analytical Chemistry, Faculty of Science, University of Valladolid, Spain), T. F. Cummings (Ph.D., Professor of Analytical Chemistry, Department of Chemistry, Bradley University, Peoria, Ill.), O. Egert (Typist, Vienna, Austria), R. Frache (Ph.D., Professor of Analytical Chemistry, Institute of General and Inorganic Chemistry, University of Genoa, Italy), J. M. Fresco (Ph.D., Professor of Chemistry, Department of Chemistry, McGill University, Montreal, Quebec, Canada), O. W. Lau (Ph.D., Professor of Chemistry, Department of Chemistry, University Science Centre, The Chinese University of Hong Kong, Shatin NT, Hong Kong), G. E. Janauer (Ph.D., Professor of Analytical Chemistry, Department of Chemistry, State University of New York at Binghamton, New York), R. Keil (Ph.D., Federal Institute for Reactor Research, Würenlingen, Switzerland), G. Korkisch (Typist, Vienna, Austria), G. Kurat (Ph. D., Professor of Geology, Museum of Natural History, Vienna, Austria), Chuen-Ying Liu (Ph.D., Professor of Chemistry, Department of Chemistry, National Taiwan University, Taipei, Taiwan), J. D. Navratil (Ph.D., Manager, Chemical Research, Rockwell International Rocky Flats Plants, Golden, Colo.), Kunio Ohzeki (Ph.D., Professor of Analytical Chemistry, Department of Chemistry, Faculty of Science, Hokkaido University, Sapporo, Japan), K. Orlandini (Ecological Sciences Section, Radiological and Environmental Research Division, Argonne National Laboratory, Argonne, Ill.), Isao Sanemasa (Ph.D., Professor of Analytical Chemistry, Department of Chemistry, Faculty of Science, Kurokami, Kumamoto, Japan), F. W. E. Strelow (Ph.D., National Chemical Research Laboratory, Pretoria, South Africa), Kyoji Tôei (Ph.D., Professor of Analytical Chemistry, Department of Chemistry, Faculty of Science, Okayama University, Tsushima, Okayama, Japan), N. E. Vanderborgh (Ph.D., Los Alamos Scientific Laboratory, Los Alamos, N.M.), Hirohiko Waki (Ph.D., Professor of Analytical Chemistry, Department of Chemistry, Faculty of Science, Kyushu University, Hakozaki, Higashiku, Fukuoka City, Japan), and H. Weiss (Ph.D., Department of the Navy, Naval Ocean Systems Center, San Diego, Calif.). The contributions of these 21 scientists and typists made it possible for me to finish the six volumes of the book in a 4-year period of most intensive and sometimes agonizing work. This assistance and additional help, with respect to typing and copying, received from some employees of the Institute of Analytical Chemistry, is most gratefully acknowledged.

INTRODUCTION

Ion exchange phenomena have been observed since about the middle of the 19th century, but the practical significance of ion exchange was not recognized immediately, and it was not until the first part of the 20th century that natural or synthetic ion exchangers were widely known. By then a number of relatively pure minerals such as zeolites and clays were found or synthesized that exhibited exchange characteristics. Although these exchangers were commonly used in water treatment and are still employed today, they are unstable in acid and alkaline solutions and can only be used satisfactorily under nearly neutral conditions. Consequently these aluminosilicates are of only little significance for the analytical separation of metal ions.

A more significant development took place in 1934, however, when it was discovered that some synthetic high-molecular weight organic polymers containing a large number of ionogenic groups, as an integral part, could be employed as ion exchange resins which can be considered as gel-like dispersed systems. The dispersed medium is usually water and the dispersed portion is the three-dimensional polymer of the ion exchange resin, which is of organic origin. Cross-linked bonds (e.g., divinylbenzene bridges) between the polymer chains form a three-dimensional matrix (network; skeleton), which hinders the motion of the polymer chains and formation of a solution in contact with the solvent (water). Only swelling of the matrix in contact with the solvent occurs. This swelling is governed especially by the character, number, and length of the cross-linked bonds.

An important feature differentiating the ion exchange resins from other types of gels is the presence of functional groups (also called ionogenic or exchangeable groups). The groups (e.g., $-SO_3H$, $-\overset{+}{N}R_3$, etc.) are attached to the matrix. The ion exchange process between the ions in the solution takes place on these functional groups.

The exchange of ions between the ion exchange resin and the solution is governed by the following two principles:

1. The process is reversible (only rare exceptions are known).
2. The exchange reactions take place on the basis of equivalency in accordance with the principle of electroneutrality. The number of millimoles of an ion sorbed by an exchanger should correspond to the number of millimoles of an equally charged ion that has been released from the ion exchanger.

Ion exchange resins, on account of their property to exchange ions in solutions, can be applied in various fields of chemistry. Concerning analytical chemistry, which is the topic of the present book, ion exchangers can be successfully applied not only in the quantitative separation of complex ionic mixtures (which is the main application of ion exchange resins) but also be used for other analytical purposes, e.g., for the microchemical detection of elements.

Ion exchange is an important and modern tool in every contemporary analytical laboratory.

THE AUTHOR

Johann Korkisch, Ph.D., is a Professor in the Institute of Analytical Chemistry at the University of Vienna, Austria.

Dr. Korkisch received his higher education at the University of Vienna. His undergraduate work was completed in Chemistry with a minor in Physics. His thesis research concerned the use of ion exchange resins in the microanalysis of gallium and uranium. After having obtained a Ph.D. in 1957 he was an Assistant Professor at the University of Vienna (Insitute of Analytical Chemistry) from 1957 to 1961.

In 1961—1962 Dr. Korkisch was a Visiting Professor at the Scripps Institution of Oceanography, University of California, La Jolla, where he performed research on the chemical analysis of marine sediments and seawater.

After his appointment in 1966 as an Associate Professor at the Institute of Analytical Chemistry (University of Vienna) Dr. Korkisch joined (from 1967 to 1968) the Argonne National Laboratory of the U.S. Atomic Energy Commission, Argonne, Ill. There, as a Visiting Scientist, he performed research work connected with the application of ion exchange resins to separations of uranium, plutonium, and trans-plutonium elements in mixed aqueous-organic solvent systems.

This work, as well as previous and later research on similar lines, was sponsored by the U.S. Atomic Energy Commission mainly acting through the International Atomic Energy Agency (IAEA) in Vienna. Research support was also provided by other sources such as the Fund for Peaceful Atomic Development USA, the American Petroleum Research Fund, as well as by the Austrian Fund for the Promotion of Scientific Research.

In 1970 Dr. Korkisch worked as a Visiting Scientist at the State University of New York at Binghamton, and in the following year was appointed by the IAEA to the Mexican Atomic Energy Commission to act as a Technical Assistance Expert in Mexico City. There, he introduced the use of ion exchange resins for separations involving the analysis of nuclear raw materials.

In 1973 he joined the University of Kumamoto, Japan as a Visiting Scientist where he performed research work connected with the application of ion exchange resins to the analysis of natural waters polluted with heavy metals.

In the same year Dr. Korkisch was promoted to the rank of Full Professor of Analytical Chemistry at the University of Vienna.

Since 1957 Dr. Korkisch has published over 200 scientific papers mainly relating to analytical applications of ion exchange resins. His numerous collaborators come from Austria, the U.S., Japan, Egypt, India, Brazil, Turkey, Bulgaria, Israel, Mexico, Romania, Italy, and Greece.

Dr. Korkisch is also the author and co-author, respectively, of the books entitled *Modern Methods for the Separation of Rarer Metal Ions,* Pergamon Press and *Handbook of the Analytical Chemistry of Uranium,* Springer-Verlag. Other publications are, in addition to numerous reports and two U.S. patents, three chapters in books published by CRC Press, Wiley-Interscience, and Pergamon Press.

At present his research in inorganic analytical chemistry is mainly concerned with practical applications of ion exchange resins to the analysis of natural materials (geological and biological samples) and also industrial products (uranium and thorium). Concerning natural materials Dr. Korkisch's special attention is directed towards the analysis of manganese nodules using ion exchange resins and ICP-emission spectroscopy.

At the University of Vienna, Dr. Korkisch is teaching courses in quantitative analysis and in the application of ion exchange resins to analytical chemistry. The latter course which is based on this 6-volume *Handbook* is also the topic of 3- to 4-week courses which he likes to present at foreign universities and institutions.

TABLE OF CONTENTS

Volume I

GENERAL DISCUSSION
Ion Exchange Resins and Fundamental Concepts of Ion Exchange 3

Special Analytical Techniques Using Ion Exchange Resins 53

PRACTICAL APPLICATIONS
Rare Earth Elements 115

Index 295

Volume II

PRACTICAL APPLICATIONS
Actinides 3

Index 311

Volume III

PRACTICAL APPLICATIONS
Noble Metals
    Platinum Metals 3
    Silver 67
    Gold 91

Copper 121

Index 281

Volume IV

PRACTICAL APPLICATIONS
Zinc, Cadmium, Mercury
    Zinc 3
    Cadmium 81
    Mercury 119

Titanium, Zirconium, Hafnium
    Titanium 147
    Zirconium and Hafnium 185

Vanadium, Niobium, Tantalum
    Vanadium 221

Niobium and Tantalum 257

Chromium, Molybdenum, Tungsten
   Chromium 275
   Molybdenum 301
   Tungsten 331

Index 345

Volume V

PRACTICAL APPLICATIONS
Manganese, Technetium, Rhenium
   Manganese 3
   Technetium 19
   Rhenium 27

Iron, Cobalt, Nickel
   Iron 43
   Cobalt 79
   Nickel 119

Alkali Metals
   Lithium 135
   Sodium 159
   Potassium 187
Rubidium, Cesium, Francium 193

Alkaline Earths
   Beryllium 215
   Magnesium 235
   Calcium 263
   Strontium 289
   Barium 321
   Radium 331

Index 343

Volume VI

PRACTICAL APPLICATIONS
Boron, Aluminum, Gallium, Indium, Thallium
   Boron 3
   Aluminum 21
   Gallium 39
   Indium 59
   Thallium 75

Silicon, Germanium, Tin, Lead
    Silicon 89
    Germanium 99
    Tin... 107
    Lead 119

Nitrogen, Phosphorus, Arsenic, Antimony, Bismuth
    Nitrogen 149
    Phosphorus.. 167
    Arsenic 189
    Antimony 205
    Bismuth.. 211

Sulfur, Selenium, Tellurium, Polonium
    Sulfur 225
    Selenium, Tellurium, and Polonium 255

The Halogens
    Fluorine.. 275
    Chlorine 293
    Bromine 305
    Iodine.. 311

Appendix: Ion Exchange Resins 323

Index 339

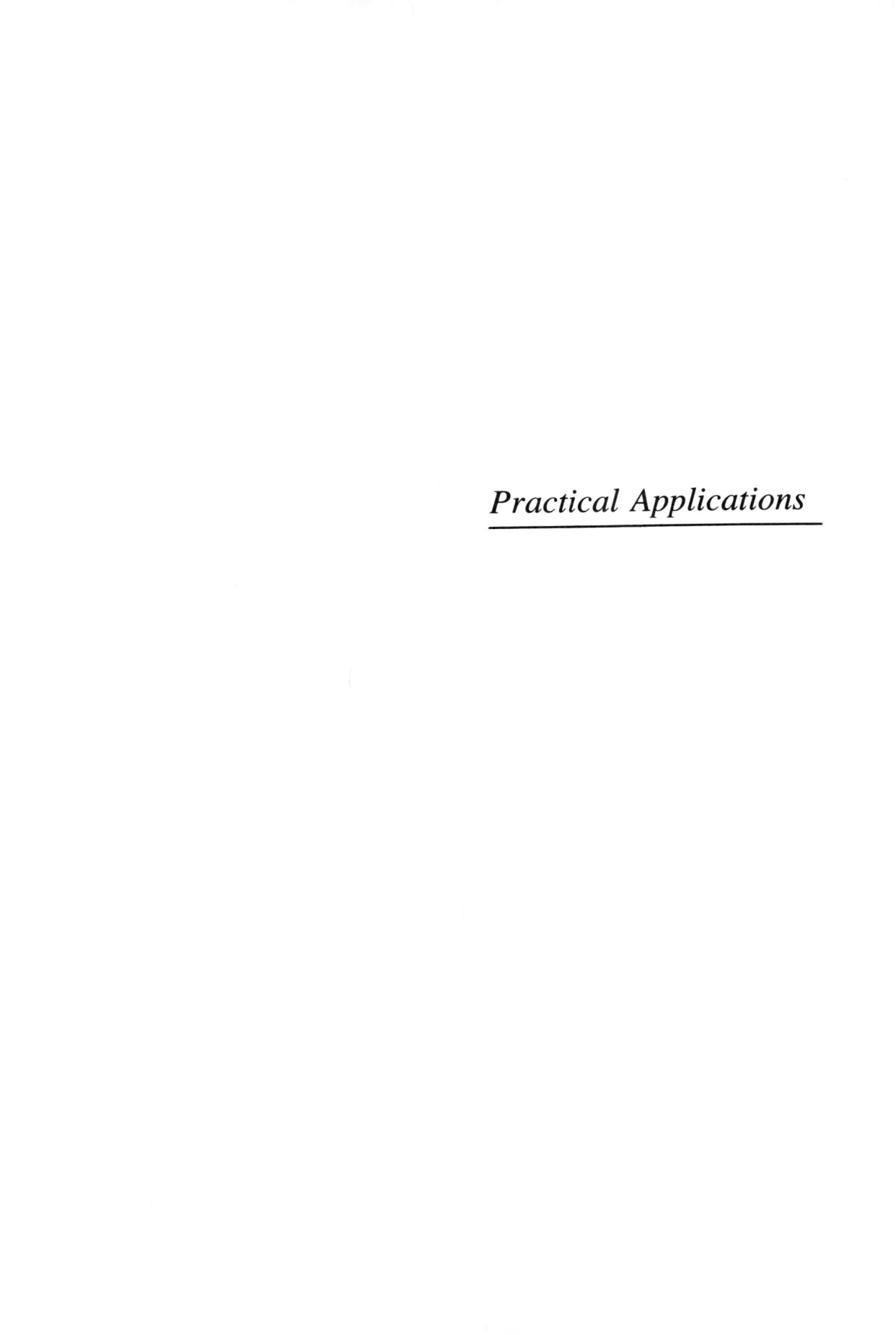

Practical Applications

BORON

Ion exchange methods based on the nonadsorbability of B, present as boric acid, on strongly acidic cation exchange resins are very frequently used to separate this element from accompanying metals interfering with its determination in geological, biological, and industrial materials. Removal of most cations and anions can be effected by the successive or simultaneous use of strongly acidic and weakly or strongly basic resins. Adsorption of borate on strongly basic resins in the hydroxyl forms has been employed for some separations.

CATION EXCHANGE RESINS

General

Cationic resins such as strongly acidic cation exchangers do not retain B (usually present as boric acid, i.e., H_3BO_3) from acid, neutral, and alkaline solutions. Consequently, this nonadsorbability of boric acid is the basis of most analytical methods so far employed for the separation of B from metal ions which interfere with its final determination. In order to prevent hydrolysis, especially of polyvalent ions such as Fe, Al, and Ti, this adsorption of metals is, in the majority of cases, performed from dilute mineral acid solutions (see Tables 1 to 12*).[1-37] Effective separations of B from many accompanying elements can also be achieved by their adsorption from neutral[5,6,38-41] and alkaline solutions,[42-45] as well as from untreated samples of natural waters, e.g., seawater.[46-51]

This removal of cationic constituents is usually performed on columns filled with cation exchangers in the H^+ form, although quite often the more rapid batch method is also utilized (see Tables 2, 3, 6, 9, and 11).[2,3,31,42,43,48,52-56]

An application of the slurry column technique has also been reported (see Table 10).[57]

Following adsorption of the cations, boric acid remaining in the resin is recovered by washing it with water[4,7,9-14,17,18,25,28-30,38-41,47,52,58,59] or very dilute mineral acid solutions, e.g., of HNO_3,[60,61] hydrochloric acid (HCl),[8] and H_2SO_4.[27]

By means of this cation exchange separation technique usually a strong acid eluate is obtained which, besides boric acid, also contains all anions of both strong and weak acids, as, for example, sulfate, chloride, nitrate, phosphate, carbonate, and silicate. In the presence of fluoride, fluoroborate, i.e., BF_4^-, is formed which is also not adsorbed on the cation exchanger (see Table 6).[59] Also not retained by the cation exchange resins used for B separation are metal ions forming stable anionic complexes under the conditions employed for the adsorption process. Thus, for example, when the adsorption takes place from very dilute HCl media, e.g., 0.01 to 0.1 M, the anionic chloro complexes of Pt metals, Au(III), and Hg(II) will pass into the effluent together with the boric acid. On neutralization of the eluate containing the above-mentioned anionic constituents, which in the case of the nonmetallic anions are present as the free acids, a high ionic strength medium is obtained. This may cause interferences in the subsequent determination of B by, for example, titrimetric or spectrophotometric procedures. Therefore, this cation exchange procedure is sometimes used in combination with an anion exchange process which removes these anionic interferences. From Tables 10 to 12 it is seen that effective deionization of weakly acid sample solutions[55,57,62] or water samples[63] can be effected by the use of coupled columns of which the upper column contains a strongly acid cation exchange resin in the H^+ form, while the lower column is filled with a weakly basic resin (e.g., Amberlite® IR-45 or Dowex® 3) in the free-base form. By using the downflow mode the cationic constituents are adsorbed in the upper resin bed and the liberated acids (e.g., H_2SO_4, HCl, HNO_3, and H_3PO_4) are then

---

* Tables for this chapter appear at the end of the text.

neutralized and adsorbed on the basic resin in the lower column. On the latter, boric acid and other weak electrolytes such as carbonic and silicic acids are not adsorbed and, thus, an effluent is obtained in which boric acid can be determined free from interferences. Any boric acid remaining in the coupled columns is readily eluted with water.[55,62-65] This separation of boric acid from accompanying cations and anions can also be effected simultaneously, i.e., by using a batch method in which a mixture of both resins is employed (see Table 11).[54]

For the removal of anions not only weakly, but also strongly basic anion exchange resins can be employed. However, the latter must not be used in the free-base, i.e., hydroxide form, but in other forms such as the chloride (see Table 10)[66] or formate form (see Table 12).[64,65] On strongly basic resins in the hydroxide form, boric acid is retained together with all other anions so that separations on coupled columns of the type described in Table 12 are used for special separations only.[67] If, however, three coupled columns are used of which the first and second column are filled with a strongly acid exchanger (H^+ form) and a weakly basic resin (free-base form), respectively, the adsorption of boric acid on the OH^--form strong base resin (third column) is not accompanied by the coadsorption of other anions, except very weak electrolytes (see Table 11).[68-71] After washing of the strong base resin with water[67-71] the adsorbed boric acid can be eluted with 5% NaOH solution,[68-71] 1 M NaOH,[67] or 1 M HCl (see Table 13).[72]

Applications

The separation principles discussed above have variously been used in connection with the determination of B in geological materials (see Tables 1 to 3, 10, and 11), biological samples (see Tables 4 and 11), and industrial products (see Tables 5 to 9 and 12).

ANION EXCHANGE RESINS

As mentioned in the preceding section, boric acid is not retained by basic resins except when using the hydroxide forms of resins of the quaternary ammonium type (see Tables 11 and 12).[66-71] An interesting example for this type of separations is also shown in Table 13. From this table it is further evident that boric acid is retained by Dowex® 1 from ethyl ether medium[73] and as an anionic mannitol complex.[74] Adsorption of the latter on anion exchange resin paper (Amberlite® IRA-410)[75,76] has been applied to the determination of B in plant materials.[76]

It has been observed that B as fluoroborate, i.e., BF_4^-, is adsorbed on anion exchange resins (fluoride form) from very dilute hydrofluoric acid (HF) solutions.[77,78] Maximum adsorption of B on the anionites AV-17-X2 and EDE-10P was found to occur from 0.48 M HF and it has been shown that this adsorption decreases with increasing concentration of HF. B is not adsorbed at acid concentrations >2 M. In the presence of an excess of Al, it is not the B which is retained by, e.g., Amberlite® IRA-400, but it is the Al which has been converted to adsorbable AlF_6^{3-}.[24]

From Table 13 it is seen that B is adsorbed on Amberlite® XE-243 containing N-methylglucamine groups (which complex the B) and also tertiary amine groups which allow B to be retained as BF_4^-.[79,80]

Adsorption of B as tetrafluoroborate has also been utilized in connection with the determination of B using ion chromatography (see the chapter "Special Analytical Techniques Using Ion Exchange Resins," Volume I).

Table 1
DETERMINATION OF B IN ROCKS, MINERALS, AND GLASSES AFTER SEPARATION BY CATION EXCHANGE USING THE COLUMN METHOD

Material	Ion exchange resin, separation conditions, and remarks	Ref.
Silicates	Amberlite® IR-120 (H⁺ form) Column of 2 cm ID containing the resin to a height of 10 in. a) Neutral sample solution (~50 mℓ) (adsorption of cations; boric acid passes into the effluent) b) Water (200 mℓ) (elution of remaining boric acid) B is determined titrimetrically using the mannitol method Before the separation, the sample (≯1 g; containing 10—20 mg of B_2O_3) is either dissolved in dilute HCl or by fusion with Na_2CO_3	38
Rocks	Dowex® 50W-X12 (50-100 mesh; H⁺ form) Column: 15 × 1.5 cm containing 30 mℓ of the resin The sample (0.1—0.5 g) is fused for 1 hr with Na_2CO_3 (2.5 g) and the melt is treated with water (100 mℓ) and conc HCl (8 mℓ); the resulting solution is passed through the column to adsorb Al and Fe and in the effluent B is determined spectrophotometrically using the azomethine-H method	1
Tourmaline and glass	Dowex® 50-X8 (H⁺ form) Column: 25 × 1.8 cm operated at a flow rate of 3—4 mℓ/min a) Neutral sample solution (200 mℓ) (adsorption of Na and other cations; B passes into the effluent) b) Water (50 mℓ) (elution of residual B) B is determined colorimetrically Before the separation, the sample (1 g) is decomposed by fusion with Na_2CO_3 (or Li_2CO_3), the melt is leached with water and HCl, the solution neutralized (methyl orange), and a precipitate of silica and basic salt is filtered off	39

Table 2
DETERMINATION OF B IN ROCKS AND MINERALS AFTER SEPARATION BY CATION EXCHANGE USING THE BATCH METHOD

Material	Ion exchange resin, separation conditions, and remarks	Ref.
Rocks and minerals	Dowex® 50W-X8 (H⁺ form) The sample (0.2 g, containing ⩾150 ppm of B) is fused with K_2CO_3 (1.25 g) for 1 hr and to the cooled melt are added in this order: 1% aqueous mannitol solution (2 mℓ), resin (20 mℓ), and 5% HCl (2 mℓ); then the melt is broken up, mixed with the above reagents, and the mixture is left overnight; subsequently, the resin (on which the cations are adsorbed) is removed by filtration and in the filtrate B is determined spectrophotometrically with carminic acid; the standard deviation was 1.43 ppm for 137 ppm of B (25 determinations)	3
Silicates	Amberlite® IR-120 (35-60 mesh; Na⁺ form) The sample (0.5 g) is sintered for 10 min at 900°C with ZnO (1 g) and Na_2CO_3 (2 g), dissolved in water, and the solution is filtered (sintered glass); the filtrate is made acid to methyl red and set aside in vacuo for 3 min; then the solution is stirred with the resin (2 mℓ) to adsorb metal ions and the resin is separated by filtration (paper filter); in the filtrate, B (present as borate) is determined titrimetrically (mannitol method) The coefficient of variation is 0.1% for 10%, 0.06% for 12.5%, and 0.05% for 15% of B, respectively; the time required is <2 hr; on sintering with the above reagents the silica in the samples is almost completely converted to a compound which is insoluble in water	2
Natural borosilicates (tourmaline and axinite)	Dowex® 50W-X8 (50-100 mesh) The weakly alkaline sample solution (50 mℓ) is kept for ≃12 hr in contact with the resin (10 g) to adsorb cations such as Al, Mn, Cr, U, and alkali metals Then the resin is separated and in the filtrate B is determined titrimetrically using the mannitol method Before the separation, the sample (0.3 g) is sintered with Na_2CO_3-ZnO, the melt is leached with water (100 mℓ), and the solution is filtered	42

Table 3
DETERMINATION OF B IN WATERS AFTER CATION EXCHANGE SEPARATION

Material	Ion exchange resin, separation conditions, and remarks	Ref.
Natural waters	Cationite KU-2 (H^+ form) Column containing 10 g of the resin The water sample (1—10 mℓ) is adjusted to pH 6 and the cations are adsorbed on the resin, while B passes into the effluent and water washings ($\not>$100 mℓ) in which it is determined spectrophotometrically This procedure can also be used for determining B in ores, soils, coals, and plant material; the coefficient of variation is 5%	4
Seawater	Amberlite® IR-120 (H^+ form) Column: 45 × 1.3 cm A dilute solution (100 mℓ) of seawater is passed through the resin bed to adsorb the cations and in the last 30 mℓ of the effluent, B is determined fluorimetrically with benzoin (with a precision of better than ±2%) Before the ion exchange separation step, the seawater (20 mℓ) is diluted to 200 mℓ with deionized water	46
Natural waters and effluents	Zeo-Karb® 225 (H^+ form) Column: 25 × 1.2 cm operated at a flow rate of ~2 drops/sec a) Sample solution (25 mℓ or less) (adsorption of cationic constituents; B passes into the effluent) b) Deionized water (25 mℓ) (elution of residual B) B is determined spectrophotometrically with curcumin	47
Tap-, sea-, and hot-spring water	Amberlite® IR-122 (Na^+ form) washed with hot water To remove large amounts of metal ions such as Ca and Mg at concentrations of 10^{-2} M, and Al and Fe at concentrations of 10^{-4} M, the sample solution (10 mℓ) is equilibrated for 5 min with 0.5 g of the resin; then the resin containing the interfering metals is removed and B determined fluorimetrically with chromotropic acid using a continuous-flow system	48

Table 4
DETERMINATION OF B IN BIOLOGICAL MATERIALS AFTER CATION EXCHANGE SEPARATION

Material	Ion exchange resin, separation conditions, and remarks	Ref.
Plants	The dried (at 60°C) sample (1 g) is ashed at 550°C for 3 hr and the ash is heated at 80°C for 1 hr with 0.5 M HCl (10 mℓ), with occasional stirring; then the pH is adjusted to 6—7 with Na_2CO_3 solution and the solution is diluted to 100 mℓ With use of an automatic analyzer, the sample solution is mixed with water, passed through a column (18 × 0.3 cm) of Amberlite® IR-120 (Na^+ form) to adsorb metal cations, and in the effluent B is determined fluorimetrically with carminic acid The limit of detection is 5 μg/ℓ; recoveries of B from pine needles range from 97—104%	5, 6

Table 5
TITRIMETRIC DETERMINATION OF B IN Ti AND Fe MATRICES AFTER SEPARATION BY CATION EXCHANGE[a]

Material	Ion exchange resin, separation conditions, and remarks	Ref.
Ti alloys	Dowex® 50-X8 (50-100 mesh; H⁺ form) Column: 25 × 3 cm containing 80 g of the resin a) Dilute HCl sample solution (200 mℓ) (adsorption of Ti and Fe; B passes into the effluent) (flow rate: 2—3 mℓ/min) b) Water (200 mℓ) (elution of residual B) (flow rate: 3—4 mℓ/min) c) 0.5 M oxalic acid (500 mℓ) (elution of Ti) B is determined titrimetrically Before the separation, the sample (2 g) is dissolved in 6 M HCl (40—45 mℓ), Ti(III) is oxidized with H_2O_2, and an excess of the oxidant is destroyed with Fe(II)-sulfate	7
	Dowex® 50 (H⁺ form) Column of 3 cm ID containing 80 g of the resin a) Sample solution (~150 mℓ) (adsorption of Ti[IV] and other cations such as Fe, Mn, Al, Co, Ni, and Cu; B passes into the effluent) b) (1:50) HCl (elution of residual B) B is determined titrimetrically using the mannitol method (after neutralization of eluate [b] with $CaCO_3$ which reacts with the HCl only); before the separation, the sample (1—2 g, containing from 0.025—1% of B) is dissolved in 6 M HCl (20 mℓ) under reflux To the hot solution (1 + 1) HNO_3 is added until the blue color (due to Ti[III]) disappears, and after the addition of an excess of this acid (1—2 drops) the solution is diluted with water to obtain solution (a)	8
Steel containing 0.2—5% of B	Amberlite® IR-120 (grain size; 0.3—0.5 mm; Na⁺ form) Column: 30—40 × 1.2 cm operated at a flow rate of 3 mℓ/min a) Sample solution of pH 2 (adsorption of Fe[III] and other cationic constituents; B passes into the effluent) b) Water (4 column volumes) (elution of remaining B) B is determined by potentiometric titration in the presence of mannitol or fructose Before the separation, the sample (1—2 g) is dissolved in HCl + H_2O_2	9
Cast iron containing >0.1% of B	Zeo-Karb® 225 (H⁺ form) in a column operated at a flow rate of 5 mℓ/min a) Very dilute acid sample solution (100-mℓ aliquot containing 1 g of the sample) (adsorption of Fe and other cationic constituents; B passes into the effluent) b) Water (200—250 mℓ) (elution of remaining B) B is determined titrimetrically using the mannitol method Before the separation, the sample (5 g) is dissolved in H_2SO_4 and H_2O_2 and any residue is decomposed by fusion with Na_2O_2 (1 g); a final solution of 500 mℓ is prepared	10
Steel	Cation exchange resin FN (Czech) Column: 30 × 1.5 cm operated at a flow rate of 2—3 mℓ/min a) Dilute HCl sample solution (25-mℓ aliquot equivalent to 0.25 g of the sample) (adsorption of Fe[III]; B passes into the effluent) b) Water (80 mℓ) (elution of residual B) B is determined titrimetrically; the relative error is 0.7% Before the separation, the sample (0.5 g) is dissolved in HCl	11

Table 5 (continued)
TITRIMETRIC DETERMINATION OF B IN Ti AND Fe MATRICES AFTER SEPARATION BY CATION EXCHANGE[a]

Material	Ion exchange resin, separation conditions, and remarks	Ref.
Steel and ferroboron	Amberlite® IR-100 or Dowex® 50 (H$^+$ form) Column: 30 × 1.9 cm operated at a flow rate of ~250 mℓ/15 min a) Very dilute acid sample solution (75 mℓ) (adsorption of Fe, Ni, and other cations; boric acid passes into the effluent) b) Water (200 mℓ) (elution of residual boric acid) B is determined titrimetrically using the invert sugar method Before the separation, the sample (1—1.5 g) is dissolved in HCl; the method can also be used for the determination of B in casting alloys and Ni-plating baths	12

[a] See also Table 7 in the chapter on Aluminum.

Table 6
TITRIMETRIC DETERMINATION OF B, S, AND F IN ELECTROLYTES AFTER CATION EXCHANGE SEPARATION

Material	Ion exchange resin, separation conditions, and remarks	Ref.
Ni electrolytes	Cationite SBS and Wofatit® KS (H$^+$ form) Column operated at a flow rate of 4—5 mℓ/min a) Sample solution (2 or 3 mℓ diluted to 50 mℓ) (adsorption of Ni; B and sulfate pass into the effluent) b) Water (150 mℓ) (elution of residual B and sulfate) Boric acid and sulfate are determined titrimetrically	58
	Cationite KB-4p The sample (3—5 mℓ, containing ⋟100 mg of Ni) is diluted with water (20 mℓ) and made just alkaline to methyl red by addition of 2% NaOH solution; then the air-dried resin (5 g) is introduced, the solution is diluted to 100 mℓ, and shaken for 10 min to adsorb Ni; in the filtrate H$_3$BO$_3$ is determined titrimetrically	43
Cr chloride electrolyte	Cationite KU-2 (H$^+$ form) Column: 40 × 1.5 cm operated at a flow rate of 4 mℓ/min a) Sample solution (1 mℓ ⋟2 M in CrCl$_3$, ≈2 M in NH$_4$F, and ⋟1 M in H$_3$BO$_3$) diluted with water (30 mℓ) and adjusted to pH ⋞3 with 5% NaOH solution (adsorption of Cr[III]; B passes into the effluent) b) Water (3 × 30 mℓ) (elution of residual B) B is determined by potentiometric titration; the error of the method is ±10%	13
Pb fluoroborate electrolyte	Cationite SBS or Wofatit® II (H$^+$ form) Column operated at a flow rate of 8—10 mℓ/min a) Sample solution (50 mℓ) (adsorption of Pb; fluoroborate anion BF$_4^-$ passes into the effluent) b) Water (75 mℓ) (elution of residual BF$_4^-$) B and F are determined titrimetrically Before the separation, the electrolyte (20 mℓ) is diluted to 1 ℓ and to a 25-mℓ aliquot an equal volume of water is added to prepare solution (a)	59

Table 7
TITRIMETRIC DETERMINATION OF B IN VARIOUS INDUSTRIAL PRODUCTS AFTER CATION EXCHANGE SEPARATION

Material	Ion exchange resin, separation conditions, and remarks	Ref.
Borosilicate glass	Amberlite® IR-120 Column: 40 × 2 cm operated at a flow rate of 6 mℓ/min a) Sample solution of pH 2—4 (adsorption of cations; B passes into the effluent) b) Water (3 × 50 mℓ) (elution of residual B) In the combined effluents, B is determined by amperometric titration Before the separation, the sample (1 g) is fused with Na_2CO_3 (4 g); the method is applicable to Pb borosilicate glass containing ≃10% of B_2O_3	14
High-purity La boride	Strongly acidic cation exchange resin (100-200 mesh) The dilute acid sample solution is passed through a column (20 × 1 cm) of the resin to adsorb the La; H_3BO_3 passes into the effluent and is determined titrimetrically (with NaOH in the presence of mannitol) after removal of excess mineral acid Before the separation, the sample (100 mg) is dissolved in HCl, plus a small amount of HNO_3 to convert B into H_3BO_3	15
U and its oxides and other materials	Dowex® 50 or Amberlite® IR-100 (H⁺ form) Column: 30 × 1.9 or 33 × 2.6 cm operated at a flow rate allowing eluents (a) and (b) to pass through in ~15 min a) Aqueous sample solution (50 mℓ) (adsorption of U and other cations; B passes into the effluent) b) Water (200 mℓ) (elution of residual B) B is determined titrimetrically	17, 18, 25
Th-B and U-B alloys	Zeo-Karb® 225 (H⁺ form) On percolating an aliquot of the slightly acid sample solution through a column of the resin adsorption of Th and U occurs, while B passes into the effluent in which it is determined titrimetrically using the mannitol method; before the separation, the sample (1 g, containing 0.5—2.9% of B) is dissolved in conc HCl (30 mℓ) and a final sample solution of 250 mℓ is prepared; the dissolution of the alloy is effected by heating under reflux; there is no loss of B during refluxing with HCl	26

Table 8
SPECTROPHOTOMETRIC DETERMINATION OF B IN Ti, Fe AND Al MATRICES AFTER CATION EXCHANGE SEPARATION

Material	Ion exchange resin, separation conditions, and remarks	Ref.
Ti and its alloys	Dowex® 50-X8 (20-50 mesh; H⁺ form) Column: 50 × 1 cm conditioned with eluent (b) (100 mℓ) and operated at a flow rate of 1 drop/2 sec a) Sample solution (10-mℓ aliquot equivalent to 200 mg of sample) diluted with 30% H_2O_2 (2 mℓ) and water (18 mℓ) (adsorption of Ti as cationic peroxy complex; B passes into the effluent) b) Solution (50 mℓ) which is 2% in H_2O_2 (30%) and 1% in dilute H_2SO_4 (1 + 4) (elution of residual B) Boron is determined spectrophotometrically with carminic acid; the accuracy of the method is ±0.002% for samples containing 0.1—0.5% of B Before the separation, the sample (1 g) is dissolved in 10 mℓ of dilute H_2SO_4 (1 + 4) (in a CO_2 atmosphere to prevent the formation of easily hydrolyzable Ti[IV]) and then 30% H_2O_2 (5 mℓ) and water are added to obtain the sample solution (50 mℓ)	27
Steel	Dowex® 50-X4 (100-200 mesh; H⁺ form) Column of 1.9 cm ID containing 150 mℓ of the resin and operated at a flow rate of 2—3 mℓ/min a) Sample solution (250 mℓ) (adsorption of Fe, Al, Ti, Zr, and other cations; boric acid passes into the effluent) b) Distilled water (250 mℓ) (elution of remaining boric acid) B is determined colorimetrically with azomethine-H Before the separation, the sample (1 g) is dissolved in H_2SO_4 (d = 1.83) (5 mℓ) and the solution is diluted to 250 mℓ; insoluble B is fused with Na_2CO_3 and treated in a similar manner	28
Al	Dowex® 50 (50-100 mesh; H⁺ form) Column: 6 × 4.5 cm containing 30 mℓ of the resin The 1.1 M HCl sample solution (50 mℓ) is passed through the resin bed to adsorb Al and in the effluent B is determined spectrophotometrically Before the separation, the sample is dissolved in HCl-HNO_3-H_2SO_4	29
Al and Si Al alloys	Zeo-Karb® 225 (H⁺ form) a) Sample solution (~10 mℓ) adjusted to pH ≃2 with 2 N H_2SO_4 (adsorption of Al, Cr, and other cations; B passes into the effluent) b) Water (250 mℓ) (Elution of remaining B) B is determined spectrophotometrically Before the separation, the sample (0.1 g) is decomposed by fusion with Na_2O_2	30

Table 9
DETERMINATION OF B IN NUCLEAR AND VARIOUS OTHER MATERIALS AFTER CATION EXCHANGE SEPARATION

Material	Ion exchange resin, separation conditions, and remarks	Ref.
Pu and U-nitrate solutions	Dowex® 50W-X10 (200—400 mesh; H⁺ form) Columns 8 × 1 and 14 × 1 cm for samples containing up to 0.5 and 2 g of U (or Pu), respectively; the column is conditioned with 0.2 M HNO₃ and operated at a flow rate of ~0.3—0.5 mℓ/min a) 0.2 M HNO₃ sample solution containing 0.1—20 μg of B per milliliter (adsorption of U[VI] and Pu[III]; B passes into the effluent) b) 0.2 M HNO₃ (50 mℓ) (elution of residual B) B is determined by emission spectrography; the lower limit of detection ranges from 0.007—0.001 μg of B; the coefficient of variation is <9% for 0.1—0.3 ppm of B in 0.1—2 g of U, or 3—30 ppm of B in 0.1—0.4 g of Pu If Pu is present in solution (a) it is first reduced to Pu(III) with NH₂OH In effluents (a) and (b) Si can also be determined spectrographically[61]	60, 61
High-purity Th sulfate	Strong acid cation exchange resin Nalcite® HCR (H⁺ form) An aqueous suspension of the sample (1.8 g) is shaken with the resin (10 mℓ) to adsorb the Th; then the resin is filtered off and washed with water and in the filtrate + washings B is determined spectrophotometrically From 0.2—10 ppm of B in the sample can be determined to within ±10%	52
Nuclear graphite	Amberlite® IR-120 (H⁺ form) The sample (1 g) is mixed with CaO (60 mg) and ignited for 2 hr at 800°C The ash is dissolved in ~6 M HCl (10 mℓ) and the solution is stirred with a batch (10 g) of the resin to remove Ca from the solution; subsequently, the resin (containing the adsorbed Ca) is filtered off and in the filtrate borate is determined spectrophotometrically using the curcumin method	31
Organoboron compounds	Amberlite® IR-120 (H⁺ form) Column with a 18-cm bed of the resin a) Aqueous sample solution (5 mℓ) (adsorption of Na; B passes into the effluent) b) Deionized water (~90 mℓ) (elution of residual B) B is determined flame photometrically; the error is ≥ ±0.1% Before the separation, the sample (5—10 mg) is fused with Na metal and the melt is dissolved in water	40
Fertilizer	Amberlite® IR-120 (H⁺ form) The sample (1 g) is stirred with the resin (10 g) and water (30 mℓ) for 3 min to adsorb cationic constituents; after filtration of the mixture the B is determined inthe filtrate using Ar plasma emission spectrometry	53
B in binary mixtures with Ni, Co, Fe(II, III), Cu, and Ag	Zeo-Karb® 225 (H⁺ form) Column: 3 × 1 in. operated at a flow rate of ~0.65—0.7 mℓ/min a) Test solution containing the metal (milligram amounts) and H₃BO₃ (milligram amount) (adsorption of metal cation; H₃BO₃ passes into the effluent) b) 2 M HCl (elution of Fe, Cu, and Co) c) 4 M HCl (elution of Ni) d) 20% Ca(NO₃)₂ solution or 2 M HNO₃ (elution of Ag)	81

Table 10
DETERMINATION OF B IN MINERALS AND GLASSES AFTER CATION AND ANION EXCHANGE SEPARATION

Material	Ion exchange resin, separation conditions, and remarks	Ref.
B minerals (tourmaline, kotoite, and paigeite)	The sample (~0.2—0.4 g) is decomposed by heating under reflux with a slight excess of 6 M HCl or by fusion with Na_2CO_3 and dissolution of the melt in 6 M HCl After dilution to 50—100 mℓ with water, the weakly acid solution (pH 1—5) is passed through successive columns of Amberlite® IR-120 (H^+ form; which adsorbs the cations) and Amberlite® IRA-400 (chloride form; which retains anions but not B) In the effluent B is determined titrimetrically or colorimetrically with curcumin The error ranges from ±0.01—0.1 μg on 1—11 μg of B	66
Glasses	Mixed resin consisting of Dowex® 50-X8 (50-100 mesh; H^+ form) and Amberlite® IR-45 (40—60 mesh; weakly basic resin in the free-base form) (combined on the basis of equivalent exchange capacity using the capacity ratings of the manufacturer) The glass sample (0.2—0.3 g) is fused with K_2CO_3 (1.2—1.8 g), the melt taken up in a minimum of hot water and 30—40 mℓ of the mixed resin are added to adsorb cations and to destroy carbonate; the mixture is swirled and the contents allowed to settle until no further sign of CO_2-evolution is apparent; then the resin solution mixture is transferred to a column of 1.1 cm ID already containing about 20 mℓ of the mixed resin; the B passes into the effluent and residual amounts are eluted by washing the resin bed with water using a flow rate of 2.5 mℓ/min (total volume of effluent = 250 mℓ) B is determined by potentiometric titration A similar procedure can be used for the assay of B in steels	57

Table 11
DETERMINATION OF B IN NATURAL WATERS AND BIOLOGICAL MATERIALS AFTER CATION AND ANION EXCHANGE SEPARATION

Material	Ion exchange resin, separation conditions, and remarks	Ref.
Tap-, river-, and seawater and biological samples	Dowex® 50W-X8 (H⁺ form) and Amberlite® IR4B (weakly basic resin in the OH⁻ form) The sample solution (1 mℓ, containing from parts per billion to parts per million amounts of B) is shaken with a mixed ion exchange resin slurry (2 mℓ of cation exchanger and 5 mℓ of the anion exchanger) and after centrifugation, the B is determined in 0.1-mℓ aliquots using emission spectrometry Interfering cations and anions are retained by the resins Before the separation, the sample of biological material (hair, orchard leaves, and bovine liver) is subjected to RF low-temperature dry ashing and the ash is dissolved in HCl	54
Natural waters	The sample (e.g., 100 mℓ) is passed successively (at a rate of 5 mℓ/min) through 3 columns (20 × 1.6 cm) containing 10 g each of the cationite KU-2 (H⁺ form), anionite AN-2F (weakly basic resin), and anionite AV-16 (strongly basic resin) (both in OH⁻ form), respectively; the first adsorbs the cations, the second adsorbs anions of strong acids, and the third retains H_3BO_3; subsequently, the set of 3 columns is washed with water (75 mℓ), and after their disconnection the third column is treated with 5% NaOH (50—75 mℓ) to elute B which is determined photometrically or titrimetrically If there is no need to concentrate the B, the sample is passed through the first two columns only; in place of anionite AN-2F the anionite AN-1 can be used	68—71
Mineral waters	Composite column (25 × 0.8 cm) containing 4 mℓ each of Dowex® 50W-X8 (50-100 mesh; H⁺ form) (upper resin bed) and Dowex® 3 (weakly basic anion exchange resin; 20—50 mesh; OH⁻ form) operated at a flow rate of 0.3 ± 0.1 mℓ/min a) Sample (1—10 g of mineral water) (adsorption of interfering strong electrolytes; into the effluent pass B and other weak electrolytes such as silicic and carbonic acids) b) Distilled water (40 mℓ) (elution of residual B) B is determined by mass spectrometry	63
Water from salt lakes	The sample (2—3 mℓ, containing a few milligrams of B as borate) is acidified with a few drops of 6 M HCl and shaken with a mixture of 15 mℓ each of strongly acidic cation exchange resin (<100 mesh), previously treated with 1 M HCl, and a weakly basic anion exchange resin (grain size: 0.4—0.5 mm), previously treated with 10% $NaHCO_3$ solution; when all CO_2 has been evolved, the B is eluted with water (at a rate of 1—1.5 mℓ/min); during passage of 60 mℓ of water ≈98% of the B and negligible amounts of other salts and impurities are collected	55

Table 12
DETERMINATION OF B IN INDUSTRIAL PRODUCTS AFTER CATION AND ANION EXCHANGE SEPARATION

Material	Ion exchange resin, separation conditions, and remarks	Ref.
Fertilizers	Amberlite® IR 120 (H$^+$ form) and Amberlite® IRA 400 (formate form) in coupled columns with the cation exchanger in the upper column; the columns are operated at a flow rate of ~3—4 mℓ/min a) Aqueous sample solution (50-mℓ aliquot containing 1 g of the fertilizer) (adsorption of all cations in the upper column and of anions such as phosphate by Amberlite® IRA-400; not retained are H$_3$BO$_3$ and carbonate) b) Water (190 mℓ) (elution of H$_3$BO$_3$) The first 80—100 mℓ of the effluents (a) and (b) can be discarded because they do not contain H$_3$BO$_3$ B is determined by flame photometry Before the ion exchange separation step, the sample (20 g) is shaken for 30 min with water (950 mℓ), then the pH is adjusted to 4, and after further 30 min shaking the solution is diluted to 1000 mℓ and filtered	64, 65
Na metal and NaOH	Column (18 × 3.5 cm) of Amberlite® CG-120 (100-200 mesh; H$^+$ form) connected in series with a column (2.5 × 1.8 cm) of Amberlite® CG-400 (100-200 mesh; OH$^-$ form) Through this set of columns the following solutions are passed a) Sample solution prepared by dissolving Na (5 g) or NaOH (10 g) in water (50 mℓ) (adsorption of Na on the cation exchanger and of borate on the anion exchange resin) (flow rate: 5—8 mℓ/min) b) Water (220 mℓ) (as a rinse) c) 1 M NaOH (10 mℓ) (elution of B from the anion exchange resin column after disconnecting it from the large cation exchange column) (flow rate: 1 mℓ/min) B is determined spectrophotometrically with dianthimide (for 0.5—5 µg of B) or curcumin (for 0.1—0.5 µg) The method can also be used for the determination of B (~0.01 ppb) in water There is no interference from the following species (amounts in milligrams): sulfate (<60), phosphate (<0.3), arsenate (<1), silicate (<0.22) and Al (<1)	67
Steels	Mixture of the resins Nalcite® HCR-X8 (60—100 mesh; H$^+$ form) and Amberlite® IR-45 (weak-base resin in the free-base form) Column: 50-mℓ buret containing ≈50 mℓ of the resin mixture and operated at a flow rate of 2.5—2.8 mℓ/min a) Weakly acid sample solution (1 mℓ of conc HCl/250 mℓ) (adsorption of Fe[III] and other cations on the cationic resin and of anions such as Cl$^-$ on the anion exchanger which also neutralizes the acid present; into the effluent passes boric acid together with other weak acids such as silicic acid) b) Distilled water (elution of remaining boric acid until the eluate + effluent from [a] reaches a volume of 250 mℓ) B is determined colorimetrically with carminic acid; before the separation, the sample (2—5 g) is dissolved under reflux with aqua regia and H$_2$O$_2$	62

Table 13
DETERMINATION OF B IN INDUSTRIAL PRODUCTS AND NATURAL WATERS AFTER ANION EXCHANGE SEPARATION[a]

Material	Ion exchange resin, separation conditions, and remarks	Ref.
Enamels and glazes	Dowex® 2 (20-50 mesh; OH⁻ form) Column: 10 × 2.2 cm connected to a condensor of a distillation apparatus a) Methanol-distillate containing methyl borate (reaction of methyl borate with OH⁻-form resin, i.e., saponification of the ester producing H_3BO_3 and CH_3OH; the former is neutralized by the OH⁻ and borate anion is retained by the resin; methanol passes into the effluent and back into the distillation flask) b) Water (50 mℓ) (removal of methanol) c) 1 M HCl (100 mℓ) (elution of boric acid) B is determined titrimetrically Before the separation, the sample (0.5—1 g) is decomposed by heating with NaOH (5 g) and then the B is distilled directly into the resin column from a mixture of conc H_2SO_4 (20 mℓ) and methanol (75 mℓ) (distillation time: ~2 hr)	72
U	Dowex® 1-X8 (100-200 mesh; chloride form) Column: 10 × 1 cm conditioned with ethyl ether and operated at a flow rate of 5 mℓ/min Column: 4 × 0.8 cm is used for the separation of smaller quantities of B to be determined by emission spectrography a) Solution of uranyl nitrate in ethyl ether (~10 g/100 mℓ of ether) (adsorption of H_3BO_3; U passes into the effluent) b) Ethyl ether (100 mℓ) (elution of residual U) c) 5 M HCl (24 mℓ) (elution of H_3BO_3) B is determined by flame photometry Before the separation the U metal is dissolved in conc HNO_3; with eluent (c) traces of U that might still be present are not coeluted with B, but remain adsorbed (as anionic chloride complex)	73
Detergents	Strongly basic anion exchange resin, e.g., Dowex® 1-X4 (grain size: 150—300 μm; Cl⁻ form) Column: 22 × 1.5 cm operated at a flow rate of 1.0—1.5 mℓ/min The filtered sample solution (25 mℓ; comprising water plus 20 mℓ of an aqueous 1% solution of the detergent adjusted to pH 3) is passed through the resin bed to adsorb PO_4^{3-} and in the effluent B is determined titrimetrically using the mannitol method	82
Boric acid in mixtures with alkali metals	Anionite AV-17 (chloride or acetate form) Column of 1.5—1.8 cm ID containing 8—10 g of the resin and operated at a flow rate of 2 mℓ/min a) Sample solution (e.g., 200 mℓ of a neutral 1% solution of an alkali metal salt containing ≤5 ppm of B and which is 0.5 M in mannitol) (adsorption of B as anionic mannitol complex) b) 0.5 M mannitol (75 mℓ) (as a rinse) c) 2 M HCl (50 mℓ) or 2 M acetic acid (50 mℓ) followed by water (30 mℓ) (elution of B) B is determined spectrophotometrically	74
Natural waters	Amberlite® XE-243 boron-specific anionic resin (40—80 mesh in a column: 5 cm × 3/16 in. ID operated at a flow rate of ~1 mℓ/min) and Dowex® 50W-X8 (50-100 mesh; H⁺ form in a column: 10 cm × 1/4 in ID)	79

Table 13 (continued)
DETERMINATION OF B IN INDUSTRIAL PRODUCTS AND NATURAL WATERS AFTER ANION EXCHANGE SEPARATION[a]

Material	Ion exchange resin, separation conditions, and remarks	Ref.
	A) First column operation (Amberlite® resin column) a) Water sample (adsorption of B) b) 10% HF (1 mℓ) (formation of BF_4^- which is retained by the resin) c) After 10 min the excess of HF is removed with water (2 mℓ) d) 3 M NaOH (10 mℓ) (elution of BF_4^-) B) Second column operation (Dowex® 50 resin column) The eluate (d) obtained by (A) is passed directly through the resin bed (adsorption of Na$^+$) and in the effluent (25 mℓ, containing HBF_4 + HF) the B is determined potentiometrically The method is suitable for the determination of 2.6—520 µg of B in tapwater and can also be used for the analysis of borosilicate glasses[80]	

[a] See also Table 1 in the chapter on Germanium; Table 4 in the chapter on Aluminum; and Tables 16 and 27 in the chapter on Copper, Volume III.

REFERENCES

1. **Yoshimura, K., Kariya, R., and Tarutani, T.,** Spectrophotometric determination of boron in natural waters and rocks after specific adsorption on Sephadex gel, *Anal. Chim. Acta,* 109, 115, 1979.
2. **Moriya, Y.,** Separation and determination of boron in silicate by sintering with zinc oxide and sodium carbonate, *Jpn. Analyst,* 8(10), 667, 1959.
3. **Fleet, M. E.,** Spectrophotometric method for determining trace amounts of boron in rocks and minerals, *Anal. Chem.,* 39, 253, 1967.
4. **Vasilevskaya, A. E. and Lenskaya, L. K.,** Determination of boron in some natural materials by means of salicylic acid, *Zh. Anal. Khim.,* 20(6), 747, 1965.
5. **Ogner, G.,** Automatic determination of boron in plants, *Analyst (London),* 105, 916, 1980.
6. **Ogner, G.,** Automatic determination of boron in water samples and soil extracts, *Commun. Soil Sci. Plant Anal.,* 11(12), 1209, 1980.
7. **Newstead, E. G. and Gulbierz, J. E.,** Determination of boron in titanium alloys by an ion-exchange method, *Anal. Chem.,* 29, 1673, 1957.
8. **Norwitz, G. and Codell, M.,** Determination of boron in titanium alloys by means of ion exchange, *Anal. Chim. Acta,* 11, 233, 1954.
9. **Kysil, B. and Vobora, J.,** Determination of boron in steel, *Collect. Czech. Chem. Commun.,* 24(12), 3893, 1959.
10. **Clarke, W. E. and Norbury, R.,** Determination of boron contents greater than 0.1% in cast iron, *B.C.I.R.A.J.,* 12(6), 787, 1964.
11. **Janoušek, J. and Študlar, K.,** Determination of one to five per cent of boron in steel, *Hutn. Listy,* 14(5), 458, 1959.
12. **Martin, J. R. and Hayes, J. R.,** Application of ion-exchange to determination of boron, *Anal. Chem.,* 24, 182, 1952.
13. **Yakovlev, P. Ya. and Kozina, G. V.,** Determination of boron in a chromium chloride electrolyte in the presence of fluorine, *Sb. Tr. Tsentr. Nauchno Issled. Inst. Chem. Metall.,* No. 31, 173, 1963.
14. **Ishizuka, T., Sunahara, H., and Kato, E.,** Amperometric titration of boron in borosilicate glass, *Jpn. Analyst,* 20(7), 818, 1971.
15. **Kobayashi, M., Ichinose, A., and Einaga, H.,** Determination of boron and lanthanum in lanthanum boride, *Yogyo Kyokai Shi,* 84(10), 513, 1976.
16. **Khan, S. A., Stone, M., and Bawden, M. G.,** Rapid improved colorimetric method for micro-determination of boron in silicate rocks and minerals by use of carminic acid, *Rev. Roum. Chim.,* 25(9-10), 1433, 1980.
17. **Eberle, A. R., Lerner, M. W., and Kramer, H.,** Spectrophotometric determination of boron in uranium oxide with 1,1′-dianthrimide, Report NBL-143, U.S. Atomic Energy Commission, 1958.
18. **Yoshimori, T. and Takeuchi, K.,** Colorimetric micro-determination of boron with curcumin. Determination of boron in metallic uranium, *Jpn. Analyst,* 9(4), 354, 1960.
19. **Grob, R. L., Cogan, J., Mathias, J. J., Mazza, S. M., and Piechowski, A. P.,** 1,1′-Iminodi-(6-chloroanthraquinone) as a reagent for trace amounts of boron, *Anal. Chim. Acta,* 39, 115, 1967.
20. **Tkacz, W. and Pszonicki, L.,** Fluorimetric determination of boron by means of the boric acid-morin-oxalic acid complex in anhydrous acetic acid medium, *Anal. Chim. Acta,* 90, 339, 1977.
21. **Danielsson, L.,** Determination of boron in iron and low-alloy steels with 1,1′-dianthrimide: a colorimetric method that does not require preliminary separations, *Talanta,* 3, 203, 1959.
22. **Saavedra Alonso, J. and Garcia Sanchez, A.,** Determination of boron in geological materials, *An. Quim.,* 68(3), 335, 1972.
23. **Gattorta, G. and Servello, V.,** Determination of boron in fertilizers, *Ann. Stn. Chim. Agrar. Sper. Roma,* 3(204), 1962.
24. **Ryabchikov, D. I. and Kuril'chikova, G. E.,** Determination of small amounts of boron in the presence of fluorine and silicon, *Zh. Anal. Khim.,* 19(12), 1495, 1964.
25. **Martin, J. R. and Hayes, J. R.,** Application of ion exchange to determination of boron, *Anal. Chem.,* 24, 182, 1952.
26. **Barnett, G. A. and Milner, G. W. C.,** The analysis of thorium-boron and uranium-boron alloys, Report C/R 2307, A.E.R.E., Harwell, England, 1958.
27. **Calkins, R. C. and Stenger, V. A.,** Photometric determination of boron in titanium and its alloys, *Anal. Chem.,* 28, 399, 1956.
28. **Capelle, R.,** Colorimetric micro-determination of boron in steel using azomethine-H., *Anal. Chim. Acta,* 25, 59, 1961.
29. **Ichiryu, A. and Hashimoto, S.,** Spectrophotometric determination of traces of boron in aluminum with 1,1′dianthrimide, *Jpn. Analyst,* 10(10), 1137, 1961.
30. **Towndrow, E. G. and Webb, H. W.,** Determination of boron in aluminum and silicon-aluminum alloys, *Analyst (London),* 85, 850, 1960.

31. **Valentini, G., Conti, M. L., and Moro, A. L.,** Extension of the determination of boron as a boron-curcumin complex to materials for nuclear energy use, *Ann. Chim. (Roma),* 49(5-6), 1039, 1959.
32. **Pogorelova, M. G.,** Analysis of ludwigite by a rapid complexometric method, with use of ion-exchange resins, *Zh. Anal. Khim.,* 21(2), 235, 1966.
33. **British Iron and Steel Research Association Methods for Analysis Committee,** Determination of boron in carbon and low-alloy steels, *J. Iron Steel Inst. London,* 189(3), 227, 1958.
34. **British Iron and Steel Research Association Metallurgy Division,** Determination of boron in ferroboron, B.I.S.R.A. Open Report MG/D/341/63, 1967.
35. **Tkacz, W. and Pszonicki, L.,** Fluorimetric determination of nanogram amounts of boron, *Chem. Anal. (Warsaw),* 22(4), 801, 1977.
36. **Peshev, P., Leyarovska, L., and Bliznakov, G.,** Boro-thermic preparation of some vanadium, niobium and tantalum borides. Analysis of tantalum and niobium borides, *J. Less Common Met.,* 15(3), 259, 1968.
37. **DeParade, D., Doerffel, K., and Voigt, U.,** Solution-spectrometric determination of boron in steel by using a stabilised d.c. arc, *Z. Chem.,* 22(9), 339, 1982.
38. **Kramer, H.,** Determination of boron in silicates after ion-exchange separation, *Anal. Chem.,* 27, 144, 1955.
39. **Patrovský, V.,** Photometric determination of boron in aqueous medium using phthalein violet, *Talanta,* 10, 175, 1963.
40. **Shah, R. A., Qadri, A. A., and Rehana, R.,** Flame-photometric micro-determination of boron in organoboron compounds by ion-exchange, *Pak. J. Sci. Ind. Res.,* 8(1), 282, 1965.
41. **Dobizha, E. V., Shteinberg, G. A., and Ekgol'm, E. A.,** Determination of trace impurities in boric acid, *Khim. Promst. Ser. Reakt. Osobo Chist. Veshchestva,* No. 4, 39, 1979.
42. **Povondra, P. and Hejl, V.,** Volumetric determination of boron in natural borosilicates, *Collect. Czech. Chem. Commun.,* 41(5), 1343, 1976.
43. **Lazarev, A. I. and Lazareva, V. I.,** Analysis of nickel electrolyte by means of static ion-exchange chromatography, *Zavod. Lab.,* 25(11), 1301, 1959.
44. **Schuele, W. J., Hazel, J. F., and McNabb, W. M.,** Analysis of boron tribromide and its addition compounds, *Anal. Chem.,* 28, 505, 1956.
45. **Babko, A. K., Chalaya, Z. I., and Voronova, E. D.,** Luminescence determination of boron in alkaline media by means or ion-exchange resins, *Zavod. Lab.,* 31(2), 157, 1965.
46. **Parker, C. A. and Barnes, W. J.,** Fluorimetric determination of boron. Application to silicon, sea-water and steel, *Analyst (London),* 85, 828, 1960.
47. **Bunton, N. G. and Tait, B. H.,** Determination of boron in water and effluents using curcumin, *J. Am. Water Works Assoc.,* 61(7), 357, 1969.
48. **Motomizu, S., Oshima, M., and Toei, K.,** Fluorimetric determination of boron with chromotropic acid by using a continuous-flow system, *Jpn. Analyst,* 32(8), 458, 1983.
49. **Tsaikov, S., Boichinova, E. S., and Brynzova, E. D.,** Coulometric determination of boron in coastal seawater, *Zh. Anal. Khim.,* 37(7), 1329, 1982.
50. **Tsaikov, S.,** Coulometric determination of boron in coastal seawaters, *Dokl. Bolg. Akad. Nauk,* 35(1), 61, 1982.
51. **Sato, S. and Uchikawa, S.,** Spectrophotometric determination of boron by solvent extraction with hydrobenzoin and crystal violet, *Jpn. Analyst,* 31(9), 479, 1982.
52. **Federgruen, L. and Abrão, A.,** Spectrophotometric determination of boron in thorium sulphate, Report IEA, No. 420, Instituto de Energia Atomica São Paulo, 1976.
53. **Woodis, T. C., Hunter, G. B., Holmes, J. H., and Johnson, F. J.,** Argon-plasma emission spectrometry of boron in fertilizers: comparison with spectrometric and distillation procedures, *J. Assoc. Off. Anal. Chem.,* 63(1), 5, 1980.
54. **Daughtrey, E. H. and Harrison, W. W.,** Analysis for trace levels of boron by ion-exchange-hollow cathode emission, *Anal. Chim. Acta,* 72, 225, 1974.
55. **Xiao, Y., Wang, Y., and Cao, H.,** Ion exchange separation of boron for mass spectrometric measurement of boron isotopic abundances in salt lakes, *Fenxi Huaxue,* 11(8), 604, 1983.
56. **Inlow, R. O. and Scarborough, J. M.,** Isotope-dilution assay of boron-10-enriched elemental boron, *Anal. Chem.,* 48, 1357, 1976.
57. **Wolszon, J. D., Hayes, J. R., and Hill, W. H.,** Application of anion-exchange resins to determination of boron, *Anal. Chem.,* 29, 829, 1957.
58. **Abesgauz, D. M. and Kheifets, Z. I.,** Determination of boric acid and sulfate ion in nickel electrolytes by ion-exchange, *Tr. Leningr. Tekhnol. Inst. im. Lensoveta,* No. 35, 183, 1956.
59. **Degtyarenko, Ya. A.,** Rapid determination of fluorine and boron in lead fluoroborate electrolytes by ion exchange, *Ukr. Khim. Zh.,* 22(6), 813, 1956.
60. **Wenzel, A. W. and Pietri, C. E.,** Emission-spectrographic determination of boron in plutonium and uranium nitrate solutions after cation-exchange separation, *Anal. Chem.,* 36, 2083, 1964.

61. **Wenzel, A. W. and Pietri, C. E.**, The emission spectrochemical determination of impurities in plutonium, Report NBL-215, U.S. Atomic Energy Commission, 1964.
62. **Callicoat, D. L., Wolszon, J. D., and Hayes, J. R.**, Separation of microgram quantities of boron in mixed-resin-bed ion-exchange, *Anal. Chem.*, 31, 1437, 1959.
63. **Gorenc, B., Marsel, J., and Tramsek, G.**, Application of ion exchange to mass-spectrometric determination of boron in mineral waters, *Mikrochim. Acta*, No. 1, 24, 1970.
64. **Bovay, E. and Cossy, A.**, Determination of boron in fertilizers by ion-exchange and flame spectrophotometry, *Mitt. Lebensmittelunters. Hyg. (Bern)*, 48(2), 59, 1957.
65. **Schütz, E.**, Contribution to the titrimetric determination of boron in fertilizers utilizing ion-exchange resins, *Trav. Chim. Alim. Hyg.*, 44(2), 213, 1953.
66. **Muto, S.**, Analytical studies of boron. VI. Determination of boron using ion-exchange resin, *Bull. Chem. Soc. Jpn.*, 30(8), 881, 1957.
67. **Fukasawa, T., Yajima, K., Kano, S., and Mizuike, A.**, Ion-exchange in basic media and its analytical applications. II. Determination of traces of boron in sodium metal and sodium hydroxide, *Jpn. Analyst*, 20(2), 193, 1971.
68. **Eristavi, D. I. and Brouchek, F. I.**, Ion-exchange separation and concentration of boron in its determination in natural water, *Soobshch. Akad. Nauk Gruz. SSR*, 30(5), 565, 1963.
69. **Eristavi, D. I., Brouchek, F. I., and Cheishvili, L. I.**, Use of ion-exchange resins for the determination of boron in natural water. I, *Tr. Gruz. Politekh. Inst.*, 5(85), 3, 1962.
70. **Eristavi, D. I. and Brouchek, F. I.**, Use of ion-exchange resins for the determination of boron in natural water. II, *Tr. Gruz. Politekh. Inst.*, 5(85), 17, 1962.
71. **Brouchek, F. I.**, Use of ion-exchange resins for the determination of boron in natural water. III, *Tr. Gruz. Politekh. Inst.*, 5(85), 27, 1962.
72. **Trentelman, J. and van Velthuijsen, A.**, Estimation of boron in enamels and glazes, *Chem. Weekbl.*, 53(11), 117, 1957.
73. **Muzzarelli, R. A. A.**, Separation of boric acid from uranyl nitrate by anion-exchange chromatography, *Anal. Chem.*, 39, 365, 1967.
74. **Vinkovetskaya, S. Ya. and Nazarenko, V. A.**, Determination of traces of boron after ion-exchange separation as the mannitol complex, *Zavod. Lab.*, 32(10), 1202, 1966.
75. **Madrowa, M.**, Application of anion-exchange paper to the determination of microgram amounts of boron, *Chem. Anal. (Warsaw)*, 10(3), 345, 1965.
76. **Madrowa, M.**, Direct determination of boron in plant material by using anionite paper, *Chem. Anal. (Warsaw)*, 10(5), 895, 1955.
77. **Otmakhova, Z. I., Chashchina, O. V., and Kataev, G. A.**, Use of ion-exchange resins in spectrochemical analysis, *Izv. Sib. Otd. Akad. Nauk SSSR*, No. 9, 1967; *Ser. Khim. Nauk*, No. 4, 84.
78. **Otmakhova, Z. I. and Chashchina, O. V.**, Spectrochemical determination of boron and silicon in gallium arsenide by ion-exchange, *Tr. Tomsk. Gos. Univ.*, 204, 140, 1971.
79. **Carlson, R. M. and Paul, J. L.**, Potentiometric determination of boron as tetrafluoroborate, *Anal. Chem.*, 40, 1292, 1968.
80. **Kochen, R. L.**, Potentiometric determination of boron in borosilicate glasses, *Anal. Chim. Acta*, 71, 451, 1974.
81. **Huq, A. K. M. A., Deb, S. K., and Khundkar, M. H.**, Separation of heavy-metal ions from mixtures with boric acid by ion-exchange resin, *J. Indian Chem. Soc. Ind. News Ed.*, 20(3-4), 127, 1958.
82. British Standards Institution, Analysis of formulated detergents. III. Quantitative test methods. Section 3.13. Method for determination of total boron content, British Standard, BS 3762: Section 3.13: 1983 (ISO 6835-1981), December 30, 1983.

ALUMINUM

For the quantitative separation of Al by ion exchange both anionic and cationic resins can be used with about equal effectiveness. Separation methods utilizing strongly basic anion exchange resins are mainly based on the nonadsorbability of Al from hydrochloric acid (HCl) media. On cation exchange resins of the sulfonic acid type selective separations of Al from some accompanying elements can be achieved in the presence of organic solvents or complexing agents. Chelating resins have found only limited application for separations involving Al.

ANION EXCHANGE RESINS

General

Al(III) is not adsorbed on anion exchange resins, e.g., Dowex® 1, at any concentration of HCl. This fact has variously been used to separate Al from accompanying metal ions which can be adsorbed as anionic chloro complexes. The HCl systems most frequently employed for this type of separation are 9[1-12] and 12 M HCl.[10,12-15] At these concentrations of HCl, Fe(III), UO_2(II), Co(II), Mo(VI), Cu(II), Sn, Pu(IV), Zn, Ga, Te, and several other elements are adsorbed, while Al is eluted together with a large number of other elements which include the alkali metals, alkaline earth elements, rare earths, Ni, and Th. Similar separations can be achieved in media which are 6,[16] 7,[17,18] 8,[19,20] and 10 M[21-23] in HCl or 40% in ethanol and 5 M in HCl.[24] In 3 M HCl Al can be separated from Sn and Cu(I) (in the presence of ascorbic acid).[25] Separation procedures based on this nonadsorbability of Al from HCl media are illustrated in Tables 1 to 3*.

Also, no adsorption of Al on anionic resins of the quaternary ammonium type is observed in media containing nitric or sulfuric acids. This fact has been utilized for the separations of Al from Pu(IV)[26] and Th[27] outlined in Table 3. Al and other nonadsorbed elements can also be separated from Pu in 8 M nitric acid medium.[28] To separate Al from Zr, acetone solutions containing dilute sulfuric acid may be employed.[29]

In hydrobromic and perchloric acid systems Al is not adsorbed on basic resins, while, however, weak to moderate adsorption was observed from very dilute hydroiodic and hydrofluoric (HF) acid media, respectively. In the latter acid, distribution coefficients of the order of 10 to ~50 were measured at HF concentrations of <2 M,[30] suggesting the presence of adsorbable anionic species such as AlF_4^- or AlF_6^{3-}. At acid molarities >10 the adsorption of Al is negligible, as is also the case when nitric acid (or HCl) is present simultaneously. Thus, for example, 0.5 M HF - 0.1 M HNO_3 can be used for the elution of Al that has been adsorbed from 1 M HF.[31]

The nonadsorbability of Al from aqueous solutions of the type indicated in Table 4 has been used to separate it from chromate[32] and V,[33] as well as from chloride on a resin in the nitrate form.[34] To separate Al from U the adsorption of the latter element from a Na acetate solution of pH 4 can be employed.[35]

In oxalic acid media Al forms an anionic oxalate complex which is retained by strongly basic resins such as Bio-Rad® AG1 from, for example, 0.05 M oxalic acid - 0.1 M HCl.[36] Coadsorbed with Al, under these conditions, are Fe(III), Zr, Ti, V, and Mo, while, however, many other elements including the alkali metals, alkaline earth elements, and Mn are not retained. This fact has been utilized for the two-column separation technique illustrated in Table 1. At higher concentrations of HCl the Al oxalate complex is destroyed so that 0.05 M oxalic acid - 0.5 M HCl can be used as an eluent for the Al which is separated from coeluted V by means of a subsequent cation exchange separation step (see Table 1).

---

* Tables for this chapter appear at the end of the text.

Al can also be adsorbed from solutions containing other organic complexing agents as, for example, from lactate solutions of pH 4[37] or a tiron solution of pH 9 (see Table 4).[38] Coadsorbed elements include Fe(III), Ga, In, and Cu. The adsorbed Al may be eluted with NaOH solution[37] or with the mineral acids from which no adsorption of this element occurs (see preceding paragraphs).

On a resin in the hydroxide form Al was found to be adsorbed as $Al(OH)_4^-$.[39]

Applications

From Tables 1 to 4 it is seen that anion exchange separation procedures have been used in connection with the determination of Al in geological materials (Table 1), biological samples (see Table 9 in the chapter on Copper, Volume III), and industrial products such as steel, alloys, metals, and radioactive materials (see Tables 2 to 4).

CATION EXCHANGE RESINS

General

From the distribution data presented in Table 125 (see in the chapter on Actinides, Volume II) it is seen that Al is very strongly adsorbed on strongly acidic cation exchange resins from dilute hydrochloric, nitric, and sulfuric acid media. In these systems the distribution coefficient of Al shows values of $>10^3$ to $>10^4$ in the 0.1 to 0.2 N acids and then decreases with increasing acid concentration to reach values of $>10^2$, $<10^2$, ~ 10, and <10 in the 0.5, 1, 2, and 3 to 4 N mineral acids, respectively. A very similar behavior is shown by many other elements such as the alkaline earths, alkali metals, rare earths, and Fe(III), so that these elements will be coadsorbed with the Al when adsorbing this element from very dilute acid solutions of pH 1 to 7.[22,23,40-49] Not adsorbed together with Al from these media as well as from more acid systems as, for example, from 0.5 M HNO_3,[50,51] 0.75[52] and 1 M HCl,[53] and 1 M HNO_3[54,55] are anions such as phosphate, sulfate, silicate, borate, etc. (see Table 5), and very stable anionic complexes of metals such as those of Pt metals[46] and Au.

A higher selectivity of separations of Al from accompanying elements is achieved when carrying out the adsorption from HCl media containing CIESE-active components such as tetrahydrofuran (THF)[56] or acetone.[57] In the presence of high concentrations of these solvents Al can be separated from large amounts of Fe which is not retained by the cation exchange resins (see Table 6). From this table it is seen that from acetic acid[58] and ethanol media[59] of similar HCl concentrations, the Fe is adsorbed together with the Al, but can be removed by elution with 5% ammonium thiocyanate solution.[58] Eluents of this type have also been used to separate Al from Cr(III) which, like Fe(III), is eluted as an anionic thiocyanate complex.[60] Coadsorbed with Al from the organic solvent HCl media are Mg, Ca, Ni, and all other elements not forming anionic chloro complexes under the conditions selected.

In the presence of organic complexing agents such as oxalic acid (see Tables 1 and 8),[36,61] ethylenediaminetetracetic acid (EDTA) (see Table 8),[62] and glycerol,[63] Al can be separated from Fe(III), Ti, Zr, and other elements, Co(III), and borate, respectively. With these chelating ligands anionic complexes are formed which prevent the adsorption of the elements on the cation exchange resins.

To separate Al from coadsorbed elements (listed in parentheses) the following eluents have been used: 0.01 M HNO_3 - 0.02% H_2O_2 (V as anionic peroxy complex),[36] 0.1 N H_2SO_4 - 1% H_2O_2 - 0.1 M HNO_3 (Mo as anionic peroxy complex; see Table 7),[49] 0.5 M HCl (V),[36] 0.5 N H_2SO_4 - 13.2% $(NH_4)_2SO_4$ solution (Ti),[64] 0.7 M HCl (Ni),[56] 0.75 M HCl (Ti),[52] and 1.9 M HCl (Fe[III]).[65] With the latter eluent, Fe(III) is eluted chromatographically before the Al (in this medium the separation factor for this pair of ions is $\simeq 2$).

For the elution of Al from cation exchange resins usually 3 M HCl is employed.[18,36,48,52,58,59,64,66] Other eluents include: 2[47,60,67] and 4 M HCl,[62] 4 M HNO_3,[46] 5 M

HClO$_4$,[44] 6[56] and 10 M HCl,[22,23] 1 M NH$_4$NO$_3$,[41] 0.1 M HF,[50] 0.2 M oxalic acid of pH 4.0,[61] and 1 M NaOH.[45] With the last three eluents Al is eluted in the form of anionic complexes.

No elution of Al is required when this element can be determined directly on the resin using radiometry (see Tables 5 and 7),[49,51,54,55] emission spectrography,[43,63,68] and X-ray spectrometry.

Separations based on the nonadsorbability of Al on cation exchange resins are shown in Tables 6 and 8. The media from which Al is not adsorbed are sulfosalicylic acid solution at pH 6 to 6.2 (Table 8),[69] neutral EDTA solution (see Table 8),[70] and 50% ethanol - 50% 6.5 M HCl (see Table 6).[71]

Applications

The cation exchange separation methods presented in Tables 5 to 7 have been employed in connection with the determination of Al in a variety of biological matrices (see Table 5) and industrial products which include Fe, steel, ceramics, alloys, and glass (see Table 7). To separate Al from synthetic mixtures with other elements the procedures outlined in Table 8 can be used. Cation exchange methods that have been used in connection with the determination of Al in silicate rocks, ores, meteorites, and natural waters[82] are described in Table 1 (see in the chapters on Titanium [Volume IV], Magnesium [Volume V], and Sodium [Volume V], Table 2 (see in the chapters on Sodium [Volume V] and Cobalt [Volume V]), Tables 4 and 87 (see in the chapter on Rare Earth Elements [Volume I]), Table 5 (see in the chapters on Lithium [Volume V] and Sodium [Volume V]), Table 9 (see in the chapter on Chromium [Volume IV]), Tables 17 and 18 (see in the chapter on Titanium [Volume IV]), and Table 42 (see in chapter on Copper [Volume III]).

CHELATING RESINS AND ION EXCHANGE RESINS USED FOR MICROCHEMICAL DETECTION

In Table 9 three methods based on ion exchange separation with chelating resins are presented. Besides Chelex® 100 (see Table 9) other exchangers containing iminodiacetate groupings, as, for example, Wofatit® MC 50, have been used for separations of Al from Fe(III), and Ti(IV).[72,73] With a chelating resin prepared by treating Wofatit® L150 with chloroacetic acid, Al and other elements were separated from a Li matrix.[74] In connection with the determination of Al in biological materials by ICP-atomic emission spectrometry, the Al was preconcentrated on a poly(acrylamidoxime) resin.[75] Resin spot tests that can be used for the microchemical detection of trace amounts of Al are described in Table 10.

Table 1
DETERMINATION OF Al IN GEOLOGICAL MATERIALS AFTER ANION EXCHANGE SEPARATION[a]

Material	Ion exchange resin, separation conditions, and remarks	Ref.
Silicate rocks	A) First column operation (Bio-Rad® AG1-X8, 200—400 mesh; oxalate-chloride form)	36
	Column: 14 × 2.0 cm containing 46 mℓ of the resin conditioned with eluent (b) (100 mℓ) and operated at a flow rate of 3.0 ± 0.3 mℓ	
	a) Dilute HCl sample solution (250 mℓ) containing 5% oxalic acid (80 mℓ), 0.5% H_2O_2 (10 mℓ), and boric acid (10—12 g) (adsorption of the anionic oxalate complexes of Al, Fe[III], Zr, Ti, V, and Mo; into the effluent pass Mn[II], Mg, Ca, K, and Na)	
	b) 0.05 M oxalic acid - 0.1 M HCl - 0.02% H_2O_2 (200 mℓ) (elution of residual nonadsorbed elements)	
	c) The column is now connected to a cation exchange resin bed (lower column; see below under [B])	
	B) Second column operation (Bio-Rad® AG50-X8, 200—400 mesh; H^+ form)	
	Column 17 × 2.5 cm containing 75 mℓ of the resin	
	a) 0.05 M oxalic acid - 0.5 M HCl - 0.02% H_2O_2 (550 mℓ) (this eluent is passed through the connected columns causing selective elution of Al and V from the anion exchanger and adsorption on the cation exchange resin)	
	b) Disconnection of the columns and passage of the following eluents through the cation exchanger bed	
	c) 0.5 M HCl (150 mℓ) (removal of oxalic acid and V[V])	
	d) 0.01 M HNO_3 - 0.02% H_2O_2 (100 mℓ) (oxidative elution of V[IV])	
	e) 3 M HCl (elution of Al which is determined complexometrically)	
	Before the first ion exchange separation step, the sample (~1 or 0.5 g when Al_2O_3 >20%) is decomposed with HF, HCl, and H_2SO_4; any insoluble material is dissolved with H_3PO_4-$HClO_4$-HF; the presence of H_3BO_3 in solution (a) (A) permits up to ≃150 mg of Ca to be tolerated; the concentration of the elements coadsorbed with the Al should be <10 mmol	
Fe ores	Dowex® 1-X8 (200 mesh; chloride form)	1
	Column: 10 × 1 cm conditioned with 9 M HCl and operated at a flow rate of 1 mℓ/min	
	a) 9 M HCl sample solution (5-mℓ aliquot equivalent to 0.1-g sample) (adsorption of Fe[III]; Al passes into the effluent)	
	b) 9 M HCl (50 mℓ) (elution of Al)	
	Al is determined spectrophotometrically with oxine	
	Before the separation, the sample (2.5 g) is decomposed with HCl and HF and the residue is fused with Na_2CO_3	

[a] See also Table 1 in the chapter on Iron, Volume V; Table 1 in the chapter on Titanium, Volume IV; Table 2 in the chapter on Cobalt, Volume V; Table 2 in the chapter on Copper, Volume III; Table 2 in the chapter on Zinc, Volume IV; Table 11 in the chapter on Magnesium, Volume V; and Table 86 in the chapter on Rare Earth Elements, Volume I and Reference 81.

Table 2
DETERMINATION OF Al IN INDUSTRIAL PRODUCTS AFTER ANION EXCHANGE SEPARATION IN HCl MEDIA[a]

Material	Ion exchange resin, separation conditions, and remarks	Ref.
Steel	Dowex® 1-X8 (100-200 mesh; chloride form) Column: 15 × 1 cm conditioned with conc HCl a) 12 M HCl sample solution (5-mℓ aliquot containing 250 mg of the original sample) (adsorption of Fe[III] and Ti; Al passes into the effluent) b) 12 M HCl (25 mℓ) (elution of Al) Al is determined spectrophotometrically Before the separation, the sample (1 g) is dissolved in 6 M HCl and Fe is oxidized to Fe(III) with HNO$_3$; any residue is decomposed by fusion with Na$_2$CO$_3$	13
	Amberlite® IRA-400 (50-100 mesh; chloride form) Column: 13 × 0.8 cm conditioned with 9 M HCl and operated at a flow rate of 1 mℓ/min a) ~9 M HCl sample solution (5-mℓ aliquot containing 5 mg of the original sample) (adsorption of Fe[III]; Al passes into the effluent) b) 9 M HCl (15 mℓ) (elution of residual Al) Al is determined spectrophotometrically Before the separation, the sample (0.2 g) is dissolved in 6 M HCl and HClO$_4$	2
Steel and cast iron	Anionite AV17-X8 (grain size: 0.1—0.25 mm; chloride form) a) 40% Ethanol - 5 M HCl sample solution (adsorption of Fe[III]) b) 40% Ethanol - 5 M HCl (elution of Al) Al is determined spectrophotometrically; the coefficient of variation was <9% (three results); before the separation, the sample (0.1—0.5 g) is decomposed by HCl-HNO$_3$	24
Alloys and steel	Dowex® 1-X10 (50-100 mesh; chloride form) Column: portion of a 10-mℓ graduated pipet containing 4.5 mℓ of the resin conditioned with 9 M HCl a) 9 M HCl sample solution (adsorption of Fe[III] and other elements; Al and Pb pass into the effluent) b) 9 M HCl (15 mℓ) (elution of Al and Pb) To separate Al from Pb the combined effluents (a) and (b) are diluted to be 2 M in HCl and the resulting solution is passed through a column of the same resin to adsorb Pb; in the effluent and 2 M HCl washings (15 mℓ), Al is determined spectrophotometrically	3
Al-Fe alloys	Dowex® 1-X8 (200-400 mesh; chloride form) Column: ~25 × 2.5 cm conditioned with 8 M HCl and operated at a flow rate of 1.5—2 mℓ/min a) 8 M HCl sample solution (20 mℓ) (adsorption of Fe[III]; Al passes into the effluent) b) 8 M HCl (450—500 mℓ) (elution of Al) Al is determined gravimetrically as Al$_2$O$_3$ Before the separation, the sample (0.5 g) is dissolved in aqua regia	19
Bronze	Amberlite® CG-400 (100-200 mesh; Cl$^-$ form) Column of 15 cm length conditioned with 10 M HCl (100 mℓ) a) 10 M HCl sample solution (10 mℓ) (adsorption of accompanying elements, e.g., Cu and Sn; Al passes into the effluent) b) 10 M HCl (~110 mℓ) (elution of remaining Al)	21

Table 2 (continued)
DETERMINATION OF Al IN INDUSTRIAL PRODUCTS AFTER ANION EXCHANGE SEPARATION IN HCl MEDIA[a]

Material	Ion exchange resin, separation conditions, and remarks	Ref.
Al-Sb-Ga alloys	Al is determined fluorimetrically Before the separation, the alloy sample (usually 10 mg) is dissolved in conc HNO$_3$ (2—3 drops) Anionite EDE-10P (chloride form) Column containing 25 g of the resin conditioned with 7 M HCl a) 7 M HCl sample solution (50 mℓ) (adsorption of Ga; Al passes into the effluent) b) 7 M HCl (150 mℓ) (elution of remaining Al) c) Water (200 mℓ) (elution of Ga) Al and Ga are determined complexometrically Before the separation, the sample (0.15 g) is dissolved in H$_2$SO$_4$ and Sb is separated from Al and Ga by precipitation	17
Zn-base die-casting alloys	Dowex® 1-X8 (20-50 mesh; Cl$^-$ form) Column: 36 × 1.5 cm operated at a flow rate of 5 mℓ/min a) 6 M HCl sample solution (adsorption of Zn [up to 1 g], Fe[III], and Cu[II]; Al passes into the effluent) b) 6 M HCl (elution of remaining Al) Al is determined titrimetrically In the absence of Fe and Cu the Zn can be removed by adsorption from ≮2 M HCl	16

[a] See also Table 5 in the chapter on Iron, Volume V; Table 11 in the chapter on Beryllium, Volume V; Tables 16, 21, and 24 in the chapter on Copper, Volume III; Table 7 in the chapter on Titanium, Volume IV; and Table 4 in the chapter on Gallium.

Table 3
DETERMINATION OF Al IN NUCLEAR MATERIALS AFTER ANION EXCHANGE SEPARATION[a]

Material	Ion exchange resin, separation conditions, and remarks	Ref.
Pu-Al alloys	Dowex® AG1-X8 Column containing the resin to a height of 5 in. washed with 12 M HCl - 0.12 M HNO$_3$ and operated at a flow rate of 40 ± 5 drops/min a) ~12 M HCl sample solution (5 mℓ containing 100—500 µg of Al) (adsorption of Pu[IV]; Al passes into the effluent) b) 12 M HCl - 0.12 M NHO$_3$ (3 × 5 mℓ) (elution of residual Al) c) 0.5 M HCl (elution of Pu) Al is determined spectrophotometrically; from 100—600 µg of Al can be determined with a coefficient of variation of <2% and a reproducibility within 2—200-µg level Before the separation, the sample (0.5 g) is dissolved in HCl and Pu is oxidized to adsorbable Pu(IV) by evaporation with conc HNO$_3$	14
	Dowex® 1-X4 Column: 25 × 2.2 cm conditioned with 8 M HCl (250 mℓ) and operated at a flow rate of 2 mℓ/min a) Sample solution (4 mℓ) consisting of 2 mℓ each of 8 M HCl and conc HNO$_3$ (adsorption of Pu[IV]; Al passes into the effluent) b) 8 M HCl (60 mℓ) (elution of Al) Al is determined titrimetrically Before the separation, the sample (containing ≃5 mg of Al) is dissolved in eluent (a) Alloys containing ≃1% of Al gave a standard deviation of ±0.012% (22 samples)	20
Pu metal	Dowex® 1-X2 (nitrate form); flow rate through the column: 10 mℓ/min a) The solution of the sample in a minimum volume of 7.2 M HNO$_3$ is mixed with the resin (10 g) and the mixture is stirred for ~10 min to adsorb Pu(IV); then the slurry is transferred to a column containing 10 g of the resin b) 7.2 M HNO$_3$ (100 mℓ) (elution of Al) Al is determined polarographically; the average recovery of Al is 99.3% with a coefficient of variation of 7.2% Before the separation, the sample (containing 1—50 µg of Al) is dissolved in 3 M HCl and Pu is oxidized to adsorbable Pu(IV) first by evaporation with conc HNO$_3$ and then with 30% H$_2$O$_2$ in 7.2 M HNO$_3$	26
Homogeneous reactor fuel	Dowex® 1-X10 (chloride form) Column containing 4 mℓ of the resin a) 9 M HCl sample solution (adsorption of U and other elements; Al passes into the effluent) b) 9 M HCl (2 column volumes) (elution of residual Al) Al is determined spectrophotometrically with aluminon The same procedure is used for the isolation of Ni, Mn(II), and Zr	4
High-purity Th compounds	Strong base anion exchange resin Merck III (sulfate form) Column: 20 × 1.5 cm conditioned with dilute H$_2$SO$_4$ of pH 2.1 ± 0.1 and operated at a flow rate of 2 mℓ/min a) Dilute H$_2$SO$_4$ (pH 2.1) sample solution (30 mℓ) containing	27

Table 3 (continued)
DETERMINATION OF Al IN NUCLEAR MATERIALS AFTER ANION EXCHANGE SEPARATION[a]

Material	Ion exchange resin, separation conditions, and remarks	Ref.
	6—30 μg of Al in Th-nitrate (corresponding to 0.5 g of Th) (adsorption of Th as anionic sulfate complex; Al passes into the effluent) b) Dilute H_2SO_4 of pH 2.1 (120 mℓ) (elution of residual Al) c) 2 M HCl (200 mℓ) (elution of Th) Al is determined spectrophotometrically with oxine	

[a] See also Table 6 in the chapter on Chromium, Volume IV; Table 9 in the chapter on Zirconium and Hafnium, Volume IV; Tables 27 and 28 in the chapter on Copper, Volume III; Table 28 in the chapter on Actinides, Volume II; and Tables 29 and 30 in the chapter on Zinc, Volume IV.

Table 4
DETERMINATION OF Al IN INDUSTRIAL PRODUCTS AFTER ANION EXCHANGE SEPARATION[a]

Material	Ion exchange resin, separation conditions, and remarks	Ref.
High-purity Bi	Anionite AV-17-X8 (sulfate form) Column containing 50 mg of the resin a) Sample solution (20 mℓ; pH 9) containing tiron (1.13 g) and 1 M NaF (1 mℓ) (adsorption of Al and Fe as tiron complexes and of B and Si as anionic fluoride complexes; Bi passes into the effluent) b) Water (2 mℓ) (elution of residual Bi) c) The resin is taken from the column and Al and the other impurity elements are determined directly on the resin using a spectrographic method The lower limits of determination (parts per million) are Al (0.1), Fe (0.5), B (0.25), and Si (0.35); the coefficients of variation are <5.2% (ten results) Before the separation, the sample (0.5 g) is dissolved in conc HNO$_3$ (2 mℓ)	38
Al corrosion products	Dowex® 1-X8 (50-100 mesh; Cl⁻ form) Column: 2 in. × 1.4 cm a) Sample aliquot (<70 mℓ) containing 0.1—40 mg of Al (adsorption of chromate; Al and H$_3$PO$_4$ pass into the effluent) b) Water (30 mℓ) (elution of remaining Al) Al is determined spectrophotometrically with aluminon Before the separation, the corrosion products are stripped from the Al with 2% chromic - 5% phosphoric acid and a sample solution is prepared which contains ≤170 ppm of Al	32
Ferro-vanadium	Anionite Wofatit® L-150 To a 5-mℓ aliquot of the sample solution, 15% H$_2$O$_2$ solution (2 mℓ) and 3 g of the resin + 10 mℓ of water are added to adsorb V as anionic peroxy complex; after 10 min the resin is filtered off and in the filtrate Al is determined spectrophotometrically with aluminon (this filtrate also contains the Fe which, however, does not interfere if reduced to Fe[II], e.g., with ascorbic acid) Before the separation, the sample (0.5 g) is first treated with HNO$_3$ and then fused with KHSO$_4$; the melt is dissolved in 1:1 H$_2$SO$_4$ (1 mℓ) and diluted to 500 mℓ with water to prepare the sample solution	33

[a] See also Tables 7 and 10 in the chapter on Titanium, Volume IV; Table 7 in the chapter on Iron, Volume V; and Table 25 in the chapter on Copper, Volume III.
[b] See also Reference 81.

Table 5
DETERMINATION OF Al IN BIOLOGICAL MATERIALS AFTER CATION EXCHANGE SEPARATION[a]

Material	Ion exchange resin, separation conditions, and remarks	Ref.
Animal tissue (e.g., beef liver)	Dowex® 50W-X12 (200-400 mesh; H⁺ form) Column containing 0.5 mℓ of the resin to a height of ~5 cm a) 0.5 M HNO₃ sample solution (adsorption of Al; phosphate and silicate pass into the effluent) b) 0.5 M HNO₃ (2.5 mℓ) (elution of residual phosphate and silicate) c) Deionized water (2 mℓ) (as a rinse) d) 0.1 M HF (2.4 mℓ) (elution of Al) Eluate (d) is subjected to neutron irradiation and Al is determined radiometrically; the precision of the procedure is ±0.1 ppm for samples containing ≃1 ppm of Al Before the ion exchange separation step, the sample (≃0.3 g) is dry ashed and the ash dissolved in HNO₃	50
Archeological bone	Bio-Rad® AG50W-X8 (H⁺ form) Column: 4 × 0.7 cm conditioned with 1 M HNO₃ (5 mℓ) and operated at a flow rate of 3 mℓ/hr a) 1 M HNO₃ sample solution (3 mℓ) (adsorption of Al; phosphate passes into the effluent) b) 1 M HNO₃ (16 mℓ) (as a wash) c) The resin is taken from the column, irradiated with neutrons, and Al is determined by γ-ray counting Before the separation, the sample (≥0.5 g) is decomposed by digestion with conc HNO₃ for 24 hr	54
Bone	Amberlite® CG-120 (H⁺ form) a) 0.5 M HNO₃ sample solution (adsorption of Al; phosphate passes into the effluent) b) 0.5 M HNO₃ solution (elution of phosphate) c) Neutron irradiation of the resin ²⁸Al is determined radiometrically	51
Urine	Bio-Rad® 50W-X8 (200-400 mesh; H⁺ form) Column of 0.7 cm ID containing 1.5 g of the resin conditioned with 1 M HNO₃ (5 mℓ) a) 1 M HNO₃ sample solution (3 mℓ) (adsorption of Al; into the effluent pass Na, Cl, Si, and PO₄³⁻) b) 1 M HNO₃ (16 mℓ) (elution of residual nonadsorbed elements) c) After drying, the resin is irradiated with neutrons and ²⁸Al is determined radiometrically Recoveries of Al are ≃80—90% with a coefficient of variation of <5% on samples with 5 µg/mℓ of added Al; in the range of 0.05—0.3 µg/mℓ of Al, the coefficient of variation is ≃20%; below these amounts it increases to 100% Before the separation, the sample (10 mℓ) is wet digested with an equal volume of conc HNO₃ at 80°C	55
Blood serum	A) First column operation (Dowex® 50-X8; H⁺ form) Column containing 5 mℓ of the resin and operated at a flow rate of 5—10 drops/min a) Aqueous sample solution (5 mℓ) acidified with 10 M HCl (2 drops) (adsorption of Al and other metals; into the effluent pass phosphate and other anions) b) Water (5 × 5 mℓ) (elution of residual anions) B) Second column operation (Dowex® 1-X10, chloride form) Column containing 5 mℓ of the resin conditioned with 10 M HCl (75 mℓ) and operated at a flow rate of 5—10 drops/min	22, 23

Table 5 (continued)
DETERMINATION OF Al IN BIOLOGICAL MATERIALS AFTER CATION EXCHANGE SEPARATION[a]

Material	Ion exchange resin, separation conditions, and remarks	Ref.
	The column is connected to the Dowex® 50 column (upper column) and 10 M HCl (6 × 5 mℓ) is passed through both resin beds (elution of Al and other metals from the cation exchanger and adsorption of Fe[III] on Dowex® 1); Al passes into the effluent and is determined spectrophotometrically; the relative precision is 13—5.2% for 1—2 μg of Al	
	Before the cation exchange separation step, the sample of dried serum is ashed at 550°C, the residue is dissolved in HCl, dried, and reashed	

[a] See also Table 8 in the chapter on Sodium, Volume V.

Table 6
DETERMINATION OF Al IN INDUSTRIAL PRODUCTS AFTER CATION EXCHANGE SEPARATION IN ORGANIC SOLVENT HCl MEDIA[a]

Material	Ion exchange resin, separation conditions, and remarks	Ref.
Fe and steel	Dowex® 50-X8 (100-200 mesh; H⁺ form) Column: 5 × 1 cm conditioned with eluent (b) a) Sample solution (containing 1 g of Fe per 80 mℓ) consisting of 90% THF and 10% 6 M HCl (adsorption of Al and Ni; Fe passes into the effluent) b) 90% THF - 10% 6 M HCl (50 mℓ) (elution of residual Fe) c) 0.7 M HCl (150 mℓ) (elution of Ni) d) 6 M HCl (35 mℓ) (elution of Al) Al is determined spectrophotometrically with oxine Before the separation, the sample (<5 g) is dissolved in HCl + H_2O_2 The method is also applicable to high-purity Zn	56
	Dowex® 50W-X12 (H⁺ form) Column containing the resin to a height of 5 cm a) 12 M HCl sample solution (40 mℓ containing 1 g of Fe) which is 90% in acetone (adsorption of ^{27}Mg and Al; into the effluent passes the Fe[III] matrix as well as ^{56}Mn and ^{53}Fe) b) Acetone saturated with HCl-gas (elution of ^{27}Mg) ^{27}Mg is determined radiometrically; prior to the separation, the sample (containing Al at levels of 0.00115%) is converted into Fe_2O_3 and Al_2O_3, respectively, which are then irradiated with neutrons (producing ^{27}Mg from Al) before dissolution in 12 M HCl	57
Steel	Dowex® 50W-X8 (H⁺ form) a) 90% Acetic acid - 10% 12 M HCl (adsorption of Al) b) 5% NH_4SCN solution (elution of Fe[III] as anionic thiocyanate complex) c) 3 M HCl (elution of Al) Al is determined spectrophotometrically In the presence of Ti or Zr, the resin column is washed with 1 N H_2SO_4 to elute these elements as anionic sulfate complexes	58
Zr ceramics	Dowex® 50W-X8 (100-200 mesh; H⁺ form) Column: ~15 × 1 cm conditioned with eluent (b) (50 mℓ) and operated at a flow rate of ~0.5 mℓ/min a) 1:1 Mixture (40 mℓ) of 6.5 M HCl and ethanol (adsorption of Zr and rare earth elements; Al passes into the effluent) b) 50% Ethanol - 3 M HCl (150 mℓ) (elution of residual Al) Al is determined spectrophotometrically Before the separation, the sample (50 mg) is fused with borax-H_3BO_3-LiOH and the melt is dissolved in 6.5 M HCl	71
Zn alloys	Strongly acid resin Catex® S-X8 (H⁺ form) a) 75% Ethanol - 0.5 M HCl (adsorption of Al, Cu, Mg, and Fe; Zn passes into the effluent) b) 3 M HCl (elution of Al and coadsorbed elements) Al and Mg are determined by EDTA titration, while Cu and Fe are assayed using electrogravimetry and spectrophotometry, respectively	59

[a] See also Table 8 in the chapter on Silver, Volume III and Table 56 in the chapter on Copper, Volume III.

Table 7
DETERMINATION OF Al IN INDUSTRIAL PRODUCTS AFTER CATION EXCHANGE SEPARATION[a]

Material	Ion exchange resin, separation conditions, and remarks	Ref.
C steels	Diaion® SK1 (40-60 mesh; H⁺ form) Column of 1 cm ID containing 1.5—2.0 g of the resin which is converted to the Ca form by passing a neutral solution of $CaCl_2$ (9.75 mg/mℓ) labeled with ^{45}Ca (specific activity of 5 × 10³ cpm/mℓ) followed by water to remove nonadsorbed Ca a) Sample solution (50 mℓ) adjusted to pH 3.5 with ammonia solution (the NH_4^+ concentration must not exceed 3 M) (adsorption of Al; a quantity of Ca equivalent to the adsorbed amount of Al passes into the effluent) b) Water (freed of CO_2 by boiling) (elution of residual displaced Ca) In the combined eluates (a) and (b) the radioactivity of the eluted Ca is exactly equivalent to the amount of Al present in the sample solution, and before measurement of the ^{45}Ca activity, the Ca is precipitated as Ca-oxalate dihydrate (if necessary in the presence of Ca-carrier [1.5 mg]) Before the ion exchange separation, the sample (1 g) is dissolved in 6 M HCl and most of the Fe(III) is removed by ether extraction; residual Fe is reduced with SO_2 to Fe(II) which does not interfere when present in amounts <5 mg; also, no interference is caused in the presence of Mg By selecting a Ca standard solution of suitable specific activity, Al can be determined over a range from 0.001—10 mg	40
Glass (containing P)	Dowex® 50-X4 (grain size: 0.25—0.5 mm; H⁺ form) Column containing 5 g of the resin a) ~Neutral sample solution (~100 mℓ) (adsorption of Al; H_3PO_4 passes into the effluent) b) 0.1 M HCl (250 mℓ) (elution of residual H_3PO_4) c) 5 M $HClO_4$ (80 mℓ) (elution of Al) Al is determined titrimetrically Before the separation, the sample (0.1 g) is dissolved in 25% NaOH solution (5—8 mℓ)	44
Al boride	Dowex® 50W-X8 (100-200 mesh; H⁺ form) Column: 9 × 1.1 cm containing 8 mℓ of the resin a) Dilute HNO_3 sample solution (adsorption of Al; B passes into the effluent) b) 2 M HCl (50 mℓ) (elution of Al) Al and B are determined titrimetrically Before the separation, the sample (AlB_{12}; ~23—26 mg) is decomposed by heating at 150°C for 3 hr with 50% HNO_3	47
Mo	Dowex® 50-X8 (200-400 mesh; H⁺ form) Column: 3 × 1.8 cm a) ~0.1 N H_2SO_4 sample solution (80 mℓ) which is 1% in H_2O_2 (adsorption of Al; Mo passes into the effluent as anionic peroxy complex) b) 0.1 N H_2SO_4 - 1% H_2O_2 - 0.1 M HNO_3 (2 × 50 mℓ) (the first 50 mℓ also contains 0.1 mg Mo per milliliter) (elution of residual Mo) c) Al is determined radiometrically directly on the resin Before the separation, the sample (50—500 mg) is irradiated with neutrons and then dissolved in conc H_2SO_4 (2.4 mℓ) and conc HNO_3 (0.6 mℓ)	49

[a] See also Tables 5, 19, and 20 in the chapter on Titanium, Volume IV; Tables 48, 49, 53, and 54 in the chapter on Copper, Volume III; and Table 11 in the chapter on Sodium, Volume V.

Table 8
CATION EXCHANGE SEPARATION OF Al FROM OTHER ELEMENTS

Elements separated	Ion exchange resin, separation conditions, and remarks	Ref.
Al from binary mixtures with Zn, Mn(II), Mg, Ni, Pb, and rare earths	Dowex® 50W-X8 (100-200 mesh; Na⁺ form) Column: 4 × 2.2 cm conditioned with eluent (b) (75—140 mℓ) and operated at a flow rate of 8—13 mℓ/min a) Sample solution of pH 6.0—6.2 containing 0.05—0.5 mmol of each metal ion, 0.15 M sulfosalicylic acid (equivalent to three times the molar concentration of total metal ion), and 0.05 M acetate buffer (30 mℓ) (adsorption of all elements except Al which passes into the effluent as anionic sulfosalicylate complex) b) 0.05 M acetate buffer solution(3 × 30 mℓ) (as a rinse) c) 3 M HCl (150 mℓ) (elution of Zn, Mn, Mg, and Ni) d) 4 M HCl (350—450 mℓ) (elution of La and Y) e) 4 M HNO$_3$ (150 mℓ) (elution of Pb)	69
Al from binary mixtures with Ti, Zr, and Fe	Amberlite® IR-120 (60-100 mesh; H⁺ form) Column: 10 × 0.9 cm After adsorption of the elements the following eluents are used a) 0.05 M oxalic acid - 0.4 M HCl (50 mℓ) (elution of Ti, Zr, and Fe[III]) b) 0.2 M oxalic acid of pH 4.0 (50 mℓ) (elution of Al)	61
Al from Mn, Ca, Mg, and Ti	Amberlite® IR-120 (NH$_4^+$ form) Column: 30 × 2 cm operated at a flow rate of ~8 mℓ/min a) Test solution adjusted to a pH corresponding to the orange of methyl orange and containing enough EDTA to chelate Al (and if present also Fe and Cu) (adsorption of the other metals; Al together with any Cu and Fe[III] passes into the effluent as anionic EDTA complex) b) Water (until the total volume of effluent amounts to 200—230 mℓ) (elution of residual Al)	70
Al from Co(III)	Amberlite® IR-120 Column: 20 × 1 cm operated at a flow rate of ~2 mℓ/min To the sample solution 0.1 M EDTA is added (corresponding to a 50% excess over the amount of Co present) and the pH is adjusted to 1.5—2; then the Co is oxidized with 6% H$_2$O$_2$ solution (5—10 mℓ) (while heating the solution for 2 min) and the pH is adjusted to ≃1 with 0.5 M HCl This solution is passed through the resin column on which Al is adsorbed, while the anionic Co(III)-EDTA complex passes into the effluent; Al is eluted with 4 M HCl Al and Co are determined titrimetrically	62

Table 9
DETERMINATION OF Al IN VARIOUS MATERIALS AFTER SEPARATION ON CHELATING RESINS

Material	Ion exchange resin, separation conditions, and remarks	Ref.
Tap-, rain-, and lake water	Chelex®-100 (-400 mesh; NH_4^+ form) The pH of the sample (250 mℓ) is adjusted to 4 with aq NH_3 and acetic acid, 5 mℓ of 0.2 M acetic acid - 0.05 M Na acetate buffer of pH 4.0 and the resin (0.1 g) are added, and the mixture is stirred for 10 min to adsorb Al; after filtration, the resin is suspended in water to give 5 mℓ of suspension, a 10-$\mu\ell$ portion of which is injected into a graphite furnace for atomic absorption spectrometry The coefficient of variation for 2 ppb Al was 4.1% (ten determinations)	76
Industrial water	Weak acid cation exchange resin Amberlite® IRC-50 (20-50 mesh) The aqueous sample solution (~80 mℓ) to which 2 mℓ of 0.1 M HCl and 4 mℓ of NaF solution (1 mg F^-/mℓ) had been added is mixed with the resin (20 mℓ), and after addition of buffer reagent (0.5 mℓ) (238 mℓ conc NH_3 solution + 111 mℓ glacial acetic acid in 1 ℓ of solution) the mixture is shaken for 20 ± 3 min (adsorption of cations not forming stable anionic fluoride complexes; Al is not retained) Then the resin is removed by decantation and after extraction of residual interfering ions with 8-quinolinol-chloroform, the Al is determined fluorimetrically or spectrophotometrically Less than 0.5 ppm of Al can be determined with a standard deviation of ±0.02 ppm	77
Al in mixtures with Sr	Chelating resin based on the reaction product of 4-chloropyridine-2,6-dicarboxylic acid and poly(ethyleneimine) After adsorption of the elements on a column of the resin the following eluents are used a) Ammonium acetate buffer solution (pH 5.5) (elution of Sr) b) 1 M HNO_3 (elution of Al)	78

Table 10
MICROCHEMICAL DETECTION OF Al BY RESIN SPOT TESTS[a]

Ion exchange resin	Experimental conditions and remarks	Ref.
Dowex® 1-X1	On resin beads (impregnated with 0.1% alizarin S solution) the color changes gradually (3—6 hr) from blue-violet to strong purple in an ammoniacal solution containing >0.1 μg of Al; by measuring the time taken for the development of the color, the amount of Al can be approximately estimated A similar coloration results from the presence of Zr, Be, WO_4^{2-}, MoO_4^{2-}, Cu, and Zn; Ca and F^- suppress the coloration; metals forming insoluble hydrated oxides under the given conditions cover the surface of the beads and interfere with the detection of Al	79
Dowex® 1-X1 (chloride form)	The sample solution (1 drop) is adjusted to pH 6 with saturated Na-acetate solution (1 drop) and a few beads of impregnated resin are added (for impregnation, the resin is treated with lumogallion, catechol violet, or stilbazo; 0.6 mℓ of 0.01% aqueous solution are added to 0.03 mℓ of the resin) In the presence of >0.8, >7, and >7 ng, respectively, of Al, the impregnated resin beads give a purple fluorescence, a blue color, or a dark-purple color, respectively	80

[a] See also Table 80 in the chapter on Copper, Volume III.

REFERENCES

1. **Sen Sarma, R. and Majumdar, M. K.**, Determination of aluminum in iron ores, *J. Indian Chem. Soc.*, 59(6), 790, 1982.
2. **Foglino, M. L.**, Photometric determination of aluminum in steel, after separation with ion-exchange resins, *Metall. Ital.*, 50(8), 372, 1958.
3. **Horton, A. D. and Thomason, P. F.**, Ion-exchange spectrophotometric determination of aluminum, *Anal. Chem.*, 28, 1326, 1956.
4. **Horton, A. D., Thomason, P. F., and Kelley, M. T.**, Remote control determination of corrosion products and additives in homogeneous reactor fuel. Application of ion exchange, *Anal. Chem.*, 29, 388, 1957.
5. **U.K.A.E.A.**, Volumetric determination of aluminum in uranium metal, Report PG 293 (S), U.K. Atomic Energy Authority, 1962.
6. **U.K.A.E.A.**, Spectrographic determination of aluminum in uranium alloys, Report PG 390 (S), U.K. Atomic Energy Authority, 1964.
7. **Zagorchev, B., Doicheva, R., Dodova, L., Koeva, M., and Ruseva, N.**, Determination of aluminum in low-alloy steel by using Eriochrome Cyanine, *Khim. Ind.*, 42(7), 303, 1970.
8. **U.K.A.E.A.**, Volumetric determination of aluminum in uranium-aluminum alloys, Report PG 271(S), U.K. Atomic Energy Authority, 1962.
9. **Liao Chen-Chiang**, Ion-exchange separation and colorimetric determination of aluminum, *Acta Chim. Sin.*, 25(3), 152, 1959.
10. **Michaelis, C., Tarlano, N. S., Clune, J., and Yolles, R.**, Complete separation of a mixture of iron(III), cobalt(II), molybdenum(VI), aluminum(III), and nickel(II) by ion exchange chromatography, *Anal. Chem.*, 34, 1425, 1962.
11. **Wilkins, D. H. and Hibbs, L. E.**, Determination of aluminum, nickel, cobalt, copper, and iron in Alnico, *Anal. Chim. Acta*, 18, 372, 1958.
12. **Strel'nikova, N. P. and Pavlova, V. N.**, Determination of aluminum in tellurium by means of an anionite, *Zavod. Lab.*, 26(4), 425, 1960.
13. **Wetlesen, C. U.**, Spectrophotometric determination of aluminum in steel with stilbazo, *Anal. Chim. Acta*, 26, 191, 1962.
14. **Evans, H. B. and Hashitani, H.**, Separation and spectrophotometric determination of sub-milligram amounts of aluminum in aluminum-plutonium alloys, *Anal. Chem.*, 36, 2032, 1964.
15. **Bok, L. D. C. and Schuler, V. C. O.**, Analytical application of the separation of ferric iron from aluminum employing anionic exchange resin, *J. S. Afr. Chem. Inst.*, 11(1), 1, 1958.
16. **Khasnobis, S. K. and Das, H. B.**, Estimation of aluminum and magnesium in zinc-base die-casting alloys. Rapid complexometric method, *Indian J. Appl. Chem.*, 27(5-6), 187, 1964.
17. **Denisova, N. E. and Tsvetkova, E. V.**, Analysis of aluminum-antimony-gallium alloys, *Zavod. Lab.*, 27(6), 656, 1961.
18. **deGelis, P.**, Ion exchange separation of iron, chromium, vanadium, titanium, nickel, and aluminum. Application to the analysis of steel, *Chim. Anal.*, 49(1), 30, 1967.
19. **Gilfrich, J. V.**, Determination of aluminum in aluminum-iron alloys, *Anal. Chem.*, 29, 978, 1957.
20. **Miner, F. J., Degrazio, R. P., Forrey, C. R., and Jones, T. C.**, Separation and determination of aluminum in plutonium-aluminum alloys, *Anal. Chim. Acta*, 22, 214, 1960.
21. **Haddad, P. R., Alexander, P. W., and Smythe, L. E.**, Effect of nitric acid, hydrochloric acid and hydrobromic acid on the fluorescence of the aluminum complex of 1-(2-pyridylazo)-2-naphthol, and an improved fluorimetric procedure for determination of aluminum, *Talanta*, 21, 123, 1974.
22. **Seibold, M.**, Determination of aluminum in blood serum with ion-exchange resins and Eriochrome Cyanine R, *Klin. Wochenschr.*, 38(3), 117, 1960.
23. **Seibold, M.**, Combined use of ion exchangers for the pre-treatment of serum for the determination of aluminum with Eriochrome Cyanine R, *Fresenius' Z. Anal. Chem.*, 173, 388, 1960.
24. **Belyavskaya, T. A., Ivanova, N. Yu., and Brykina, G. D.**, Chromatographic separation of iron(III) from aluminum on AV-17 X8-Cl anion exchanger in mixed solvents, *Zh. Anal. Khim.*, 34(11), 2124, 1979.
25. **Shevchuk, I. A. and Alemasova, A. S.**, Atomic-absorption determination of aluminum in bronze, *Zavod. Lab.*, 45(12), 1101, 1979.
26. **Plock, C. E. and Vasquez, J.**, Determination of aluminum in plutonium metal by differential linear-sweep oscillographic polarography, *Talanta*, 15, 1391, 1968.
27. **Athavale, V. T. and Subramanian, A. R.**, Estimation of aluminum in high-purity thorium compounds. An ion-exchange separation method, *J. Sci. Ind. Res. India B*, 19(11), 431, 1960.
28. **Wenzel, A. W. and Pietri, C. E.**, Emission-spectrochemical determination of impurities in plutonium, Report NBL-215, U.S. Atomic Energy Commission, 1964.
29. **Plyusnin, A. V.**, Separation and concentration of elements on anionite AV-17 from sulfuric acid-acetone solution, *Tr. Ural. Politekh. Inst.*, No. 226, 123, 1975.

30. **Faris, J. P.**, Adsorption of the elements from hydrofluoric acid by anion exchange, *Anal. Chem.*, 32, 520, 1960.
31. **Neirinckx, R., Adams, F., and Hoste, J.**, Determination of impurities in titanium and titanium dioxide by neutron activation analysis. III. Determination of vanadium and aluminum in titanium and titania by pre-separation, *Anal. Chim. Acta,* 47, 173, 1969.
32. **Groot, C., Peekema, R. M., and Troutner, V. H.**, Determination of aluminum in chromic-phosphoric acid solutions. Determination in aluminum corrosion products, *Anal. Chem.*, 28, 1571, 1956.
33. **Musil, J.**, Spectrophotometric determination of aluminum in ferro-vanadium, *Hutn. Listy*, 23(9), 649, 1968.
34. **Misra, U. K., Graham, E. R., Upchurch, W. J., and McKown, D. M.**, Activation analysis of aluminum from water extract of soil, *Proc. Soil Sci. Soc. Am.*, 37(2), 193, 1973.
35. **Riley, J. P. and Williams, H. P.**, Micro-analysis of silicate and carbonate minerals. IV. Determination of aluminum in the presence of interfering elements, *Mikrochim. Acta*, No. 6, 825, 1959.
36. **Strelow, F. W. E., Liebenberg, C. J., and Toerien, F. von S.**, Separation of aluminum from other elements by anion-exchange chromatography in oxalic-hydrochloric acid mixtures and its application to silicate analysis, *Anal. Chem.*, 41, 2058, 1969.
37. **Pyatnitskii, I. V. and Kolomiets, L. L.**, Anion-exchange separation of aluminum and gallium from indium, iron and copper with use of lactic acid, *Zh. Anal. Khim.*, 25(3), 479, 1970.
38. **Slezko, N. I., Dyadik, L. S., Demesheva, E. P., Chashchina, O. V., and Otmakhova, Z. I.**, Analysis of high purity bismuth, with ion-exchange separation, *Zh. Anal. Khim.*, 33(1), 102, 1978.
39. **Kulprathipanja, S., Hnatowich, D. J., and Brownell, G. L.**, Novel method for removing aluminum from spallation-produced strontium-82, *J. Radioanal. Chem.*, 43(2), 439, 1978.
40. **Amano, H.**, Studies of analytical methods for trace elements in metals with radioactive isotopes. IV. Determination of aluminum with calcium-45 type cation exchanger, *Sci. Rep. Res. Inst. Tohoku Univ.*, 11(5), 367, 1959.
41. **Kudryavskii, Yu. P., Kazantsev, E. I., and Spiridonov, E. N.**, Sorption and separation of aluminum and gallium on cation exchange resins, *Izv. Vyssh. Uchebn. Zaved. Tsvetn. Metall.*, 17(2), 62, 1974.
42. **Driscoll, C. T.**, Procedure for the fractionation of aqueous aluminum in dilute acidic waters, *Int. J. Environ. Anal. Chem.*, 16(4), 267, 1984.
43. **Dobizha, E. V., Krasil'shchik, V. Z., and Ekgol'm, E. A.**, Determination of trace impurities in aluminum oxide, oxide monohydrate and hydroxide, *Khim. Promst. Ser. Reakt. Osobo Chist. Veshchestva*, No. 4, 31, 1979.
44. **Luginin, V. A. and Tserkovnitskaya, I. A.**, Determination of aluminum in glass containing phosphorus, *Zavod. Lab.*, 37(3), 287, 1971.
45. **Lazarev, A. I.**, The determination of aluminum in alloys using cationites, in *Sovrem. Metody Anal. Metall.*, M., Metallurgizdat, 1955, 182.
46. **Chwastowska, J., Dybczyński, R., and Kucharzewski, B.**, Determination of impurities in platinum metal and platinum-rhodium alloys, *Chem. Anal. (Warsaw)*, 13(4), 721, 1968.
47. **Takahashi, Y.**, Analysis of aluminum boride and zirconium boride, *Jpn. Analyst*, 31(9), T74, 1982.
48. **Babachev, G. N.**, Determination of iron, aluminum and chromium in refractory materials by means of EDTA and ion-exchange resins, *Zh. Anal. Khim.*, 21(7), 881, 1966.
49. **Fedoroff, M.**, Determination of aluminum in molybdenum by neutron-activation, *C.R. Sci. (Paris)*, 267, 1227, 1968.
50. **Fritze, K. and Robertson, R.**, Neutron-activation analysis for aluminum in animal tissue, *J. Radioanal. Chem.*, 7(2), 213, 1971.
51. **Gilmore, G. R. and Goodwin, B. L.**, Neutron-activation analysis for aluminum in bone samples, *Radiochem. Radioanal. Lett.*, 10(4), 217, 1972.
52. **Tsyvina, B. S. and Kon'kova, O. V.**, Determination of aluminum in titanium and its alloys by ion exchange chromatography, *Zavod. Lab.*, 25(4), 403, 1959.
53. **Kolobova, K. K.**, Complexometric determination of aluminum in zirconium concentrates after chromatographic separation, *Tr. Vses. Gos. Inst. Nauchno Issled. proektn. Rab. Ogneup. Promsti.*, No. 37, 74, 1965.
54. **Blotcky, A. J., Rack, E. P., Recker, R. R., Leffler, J. A., and Teitelbaum, S.**, Trace aluminum determination and sampling problems of archaeological bone employing destructive neutron-activation analysis, *J. Radioanal. Chem.*, 43(2), 381, 1978.
55. **Blotcky, A. J., Hobson, D., Leffler, J. A., Rack, E. P., and Recker, R. R.**, Determination of trace aluminum in urine by neutron-activation analysis, *Anal. Chem.*, 48, 1084, 1976.
56. **Okochi, H.**, Determination of traces of aluminum in iron and steel by a cation exchanger and extraction-spectrophotometry with use of 8-hydroxyquinoline, *Jpn. Analyst*, 20(11), 1381, 1971.
57. **Peters, J. M. and Del Fiore, G.**, Rapid separation of magnesium-27 from a matrix of activated iron, *Radiochem. Radioanal. Lett.*, 23(1), 5, 1975.
58. **Oberhauser, R.**, Photometric determination of aluminum in steel after separation on a cation exchanger, *Materialpruefung*, 15(3), 85, 1973.

59. **Hanak, J. and Simek, M.**, Determination of aluminum, copper, magnesium, and iron in alloys of zinc-aluminum-copper type after the separation of zinc on a cation exchanger, *Hutn. Listy*, 26(5), 356, 1971.
60. **Panchev, N. P. and Evtimova, B.**, Separation of aluminum from chromium by ion-exchange chromatography, *C.R. Acad. Bulg. Sci.*, 18(12), 1127, 1965.
61. **Nozaki, T., Takeuchi, K., Yamauchi, T., and Nomura, T.**, Cation-exchange separation of metals in mixed oxalic acid-hydrochloric acid media, *Jpn. Analyst*, 27, 454, 1978.
62. **Giuffre, L. and Capizzi, F. M.**, Separation and determination of aluminum and cobalt, *Ann. Chim. (Rome)*, 51(6-7), 563, 1961.
63. **Otmakhova, Z. I., Cashchina, O. V., and Kashkan, G. V.**, Spectrochemical determination of trace impurities in boric anhydride, *Zavod. Lab.*, 42(2), 146, 1976.
64. **Dymov, A. M., Kozel', L. Z., and Iskandaryan, R. D.**, Determination of traces of aluminum and titanium in steel, *Izv. Vyssh. Uchebn. Zaved. Chern. Metall.*, No. 9, 184, 1969.
65. **Osborn, W. O.**, Separate determination of aluminum and of lithium, sodium and potassium in fire-clay and silica raw materials and refractories by cation-exchange chromatography, *J. Am. Ceram. Soc.*, 44(11), 527, 1961.
66. **Babachev, G. N.**, The ion exchange separation of iron and titanium from aluminum and the determination with EDTA, *Chim. Anal.*, 48(5), 258, 1966.
67. **Váchová, J. and Palaček, M.**, Analysis of refractory materials containing zirconium dioxide, *Sklar Keram.*, 13, 73, 1963.
68. **Govindaraju, K.**, New scheme of silicate analysis (for sixteen major, minor and trace elements) based mainly on ion-exchange dissolution and emission-spectrometric methods, *Analusis*, 2(5), 367, 1973.
69. **Fritz, J. S. and Palmer, T. A.**, Ion-exchange separation using sulfosalicylic acid, *Talanta*, 9, 393, 1962.
70. **Cimerman, C., Alon, A., and Marshall, J.**, Titrimetric determination of aluminum with ethylenediamine-tetra-acetic acid, in the presence of iron, copper, titanium, manganese, calcium, magnesium, and phosphate, *Talanta*, 1, 314, 1958.
71. **Kruidhof, H.**, Determination of small amounts of aluminum and silicon in stabilized-zirconia ceramics, *Anal. Chim. Acta*, 99, 193, 1978.
72. **Hering, R.**, Ion exchange resins with complex-forming groups. XLVI. Separation of aluminum from iron(III) and titanium(IV) by means of complex-forming ion-exchange resins, *Z. Chem. (Leipzig)*, 12(7), 272, 1972.
73. **Hering, R. and Ertel, W.**, Ion-exchange resins with complex-forming groups. XLVII. Quantitative separation of aluminum from iron(III) and titanium(IV) on Chelonite Wofatit MC 50, *Z. Chem. (Leipzig)*, 12(9), 345, 1972.
74. **Wieteska, E. and Witkiewicz, Z.**, Separation of microgram quantities of elements from a lithium matrix by ion-exchange chromatography, *Biul. Wojsk. Akad. Tech.*, 29(4), 83, 1980.
75. **Kagan, L. M. and Yankovski, A. A.**, Determination of aluminum and silicon in biological materials by inductively coupled plasma atomic-emission spectrometry with electrothermal vaporization, *Spectrochim. Acta Part B*, 39(7), 891, 1984.
76. **Isozaki, A., Kawakami, T., and Utsumi, S.**, Electrothermal atomic absorption spectrometry for aluminum by direct heating of aluminum adsorbed chelating resin, *Jpn. Analyst*, 31(10), E311, 1982.
77. **Noll, C. A. and Stefanelli, L. J.**, Fluorimetric and spectrophotometric determination of aluminum in industrial water, *Anal. Chem.*, 35, 1914, 1963.
78. **Wilkins, D. H. and Hibbs, L. E.**, Determination of aluminum, nickel, cobalt, copper, and iron in Alnico, *Anal. Chim. Acta*, 18, 372, 1958.
79. **Katou, K., Murase, T., and Kakihana, H.**, Detection of a micro amount of aluminum with an ion-exchange resin, *J. Chem. Soc. Jpn. Pure Chem. Sect.*, 77(8), 1233, 1956.
80. **Ichikawa, T., Kato, K., and Kakihana, H.**, Detection of aluminum with ion-exchange resin and lumogallion, catechol violet or stilbazo, *J. Chem. Soc. Jpn. Pure Chem. Sect.*, 88(10), 1064, 1967.
81. **Sarzanini, C., Mentasti, E., Gennaro, M. C., and Marengo, E.**, Enrichment of aluminum traces in liquid samples, *Anal. Chem.*, 57, 1960, 1985.
82. **Roegeberg, E. J. S. and Henriksen, A.**, Automatic method for fractionation and determination of aluminum species in fresh waters, *Vatten*, 41(1), 48, 1985.

GALLIUM

For the isolation of Ga from complex matrices, such as geological and biological materials as well as industrial products, mainly anion exchange in hydrochloric acid (HCl) media, utilizing strongly basic resins, has been employed. With cation exchange resins, especially when used in acetone-mineral acid media, selective separations of Ga from almost all elements are obtained.

ANION EXCHANGE RESINS

General

From the distribution data presented in Table 1* it is seen that Ga(III) shows very strong adsorption on basic resins of the quaternary ammonium type in 3 to 12 M HCl media and the species retained is $GaCl_4^-$. This anionic chloro complex is the predominant species in the 6 to 12 M acid, while at lower HCl concentrations it is the cation Ga^{3+} which predominates in solution. This exceedingly high adsorption of Ga from systems such as 4,[1] 6,[2-7] 7,[6,8,9] and 12 M HCl[10] has variously been utilized for separations from elements not forming adsorbable anionic complexes under these conditions. Thus, in these media Ga can be separated from alkali metals, alkaline earth elements, rare earths, Al, Ni, Th, Fe(II), and other elements, but not from Fe(III), UO_2(II), Pu(IV), Cu(II), Bi, Sn, Zn, Cd, Hg, In, and Tl(III) (see Table 1 in chapter on Actinides, Volume II), and some other metals. Examples for separations of this type are presented in Tables 2 to 5. To prevent coadsorption of most of the Fe present, it has been suggested to dissolve the sample in an atmosphere of N (see Table 2) or to reduce Fe(III) with Ag, prior to passage of the 6 M HCl sample solution through the resin bed (see Table 3).[3,4] More effective reductions of Fe(III) to nonadsorbed Fe(II) are achieved in the presence of holding reductants such as Ti trichloride (see Tables 2 and 3),[10,11] ascorbic acid (see Table 4),[12] and hydroxylamine hydrochloride (see Table 6). By means of these reducing agents Pu(IV) and Cu(II) are also reduced to lower, less strongly adsorbed valency states (see Tables 3 and 4).[11,12] The reduction with ascorbic acid can also be performed in dilute hydrobromic acid (HBr) solutions containing high concentrations of organic solvents such as in 90% methanol - 10% 4.5 M HBr. In this system highly selective separations of Ga from most other elements can be achieved (see separation method in Table 2).[13] This method is also suitable for the quantitative isolation of In and Pb. In this methanol medium of 0.45 M overall acidity, distribution coefficients for Ga, In, and Pb of 60, 165, and >10^3, respectively, have been measured. Comparison of the coefficients of Ga and In with those determined in pure aqueous hydrobromic acid systems of comparable acidity (see Table 1) makes it evident that in the methanol system the two elements show considerably higher adsorption.

Ga is also retained by anionic resins from 90% methyl glycol - 10% 6 M HCl (see Table 5),[14] 44% acetone - 2 M HCl,[15] 6 M Li chloride,[16] 2 M NH_4SCN - 0.05 M HCl (see Table 7),[17] 4 M KSCN - 0.1 M HCl,[18] and from phosphoric acid solutions such as 0.1 M H_3PO_4.[19,20] From the thiocyanate and phosphoric acid solutions Ga is adsorbed as anionic thiocyanate and phosphate complexes of the probable formulas $Ga(SCN)_4^-$ and $(Ga[HPO_4]_2)^-$, respectively. Coadsorbed from the thiocyanate media are Fe(III), Co, UO_2(II), Mo, Zn, Cd, and several other elements, while Al is not retained (a similar separation of Ga from Al can also be effected by the use of 0.5 M hydrazoic acid, i.e., HN_3).[21] From phosphoric acid medium In is coadsorbed as $(In[HPO_4]_3)^{3-}$.

Adsorption of Ga on strongly basic resins also takes place from carbonate media[22-25] as

---

* Tables for this chapter appear at the end of the text.

well as from systems containing organic complexing agents, as, for instance, tiron at pH 2.0,[26] oxalic acid at pH 4.0,[27,28] ammoniacal citrate medium,[28] 4-aminosalicylic acid solution on a resin in the hydroxide form,[28] and on a malate-form resin.[28] Examples of separations that can be carried out in media of this type are illustrated in Tables 5, 6, and 8.

After adsorption of the Ga from the media discussed in the preceding paragraphs, coadsorbed elements (listed in parentheses) can be removed using the following eluents: 12 M HCl - 1% $TiCl_3$ (Cu),[11] 4 M HCl - 1% $TiCl_3$ (Fe),[10] 10% HCl - 15% $TiCl_3$ (Fe),[11] 3 M HCl (Cu),[8] 4 M HCl (Cu),[5] 90% methyl glycol - 10% 6 M HCl (Al),[14] 3 M ammonia solution,[22] and ammoniacal 2 M ammonium carbonate solution.[25] With the ammoniacal eluents Zn, Cd, Cu, and Ni are eluted as cationic amine complexes, while Al is desorbed as anionic hydroxide complex.

Elution of the adsorbed Ga can be effected with water,[9] but much more frequently dilute HCl solutions are used (in which the distribution coefficients of Ga have very low values), as, for example, with 0.1,[1,6,15] 0.5,[8,12] 0.6,[4] 1,[2,3,14] and 1.2 M HCl.[10,28] With the 1 M acid sequential elution of Ga and In in this order can be effected (see Table 5),[14] because in this medium Ga and In show a separation factor α of $\simeq 10$ (see Table 1). Other eluents which do not elute the In together with the Ga are 80% acetone - 20% 3 M HCl (see Table 6)[14] and 0.5 M KI of pH 1 (see Table 7).[29,30] Following elution of the Ga with these systems, the In can be desorbed by means of 90% acetone - 10% 6 M HCl and 1 M ammonium chloride at pH 1 (or 0.5 M ammonium sulfate at the same pH), respectively. Other eluents that have been recommended for the elution of Ga include 0.5[31] and 1 M H_2SO_4,[21] 2 M HNO_3,[11] 2[18] and 2.5 M HCl,[23-25] as well as alkaline media. To the latter belong 0.03[26] and 1 M NaOH,[27] 2[28] and 4 M ammonia solutions,[5] and 3 M Na carbonate.[22] With these eluents Ga is eluted as the anionic gallate complex, i.e., $Ga(OH)_4^-$. With the dilute NaOH eluent and Na carbonate solution In is not coeluted and can subsequently be desorbed with 0.2 M sulfuric acid[26] and 2.5 M ammonium carbonate,[22] respectively.

From HBr media Ga is less strongly adsorbed than from the corresponding HCl systems,[32] and its adsorption from hydroiodic acid solutions is negligible (see Table 1). Therefore, the nonadsorption of Ga, e.g., at pH 1 from iodide solution (see Table 7), can be employed for its effective separation from In which is strongly adsorbed under these conditions (see Table 1).

In hydrofluoric acid (HF) solutions of the concentrations 0.5, 1, 2.5, 5, 10, and 15 M distribution coefficients for Ga of 1500, 260, 55, 36, ~ 7, and ~ 3, respectively, have been measured on Dowex® 1.[33] This is in contradiction with the findings of other investigators who claim that Ga is not adsorbed on strongly basic resins from HF media.[34,35]

No adsorption of Ga on anionic resins is observed in sulfuric, nitric, and perchloric acid media of any concentration. Therefore, in dilute sulfuric and strong nitric acid media possibilities exist for separating Ga from U, Th, Mo, Zr, Hf, and Fe(III) and from Th, Np(IV), Pu(IV), and Bi(III), respectively (see in chapters of the elements mentioned).

Applications

In Tables 2 to 5 methods based on anion exchange separations are presented, which have been employed in connection with the determination of Ga in rocks, sediments, meteorites, bauxite, biological tissue, bone, Al, and alloys. To separate Ga from synthetic mixtures with other elements the procedures illustrated in Tables 6 to 8 can be used, utilizing systems containing HCl (Table 6), thiocyanate and iodide (Table 7), and carbonate (Table 8).

CATION EXCHANGE RESINS

General

As is evident from the distribution coefficients shown in Table 9, Ga(III) is very strongly retained by resins of the sulfonic acid type from HCl and HBr solutions containing very low and very high concentrations of these acids. Thus, this behavior of Ga resembles that observed for Fe(III) in HCl media under comparable conditions (see in the chapter on Iron, Volume V). When adsorbing Ga from dilute solutions of the two acids or from dilute nitric acid media (from which Ga is also strongly adsorbed; see Table 9) a very large number of elements are coadsorbed with the Ga. This includes alkali metals, alkaline earth elements, rare earths, and virtually all other metals which do not form stable anionic complexes in these acids. However, effective separations of Ga are obtained from the common anions such as phosphate, sulfate, nitrate, chloride, and fluoride. In the presence of dilute HF, e.g., when using 0.05 M HNO$_3$ - 0.01 M HF or 0.1 M HNO$_3$ - HF[7] as sorption media (see Table 10), elements forming anionic fluoride complexes such as Ti, Zr, Hf, Fe(III), Al, Pa, W, and Mo are not adsorbed together with the Ga. This element is strongly retained on cationic resins from very dilute HF solutions.[36] Coadsorbed with Ga from such media are, among others, the alkali metals, Zn, Cu, Mn, Ni, and rare earth elements.

Very selective separations of Ga from accompanying metals can be achieved by the use of acetone-mineral acid systems of the type described in Table 11.[15,37-39] In the 80% acetone - 0.5 M HBr medium the distribution coefficient of Ga has a value of 65 and the separation factor α is ≃8 for the separation of Ga from Fe(III) and Cu(II). A similar high selectivity of separations is attained when adsorbing Ga from 8 M HCl (see Table 12)[40] in which the distribution coefficient of this element has a very high value (see Table 9). Under these conditions separation of Ga from Fe is achieved in the presence of Ti trichloride as a holding reductant and coadsorbed Sc is removed by elution with 7 M HCl.

After adsorption of the Ga from very dilute mineral acid media (see Tables 12 and 13)[41-43] coadsorbed elements (listed in parentheses) can be removed by means of the following eluents: 0.2 M HCl (Cd),[43] 0.4 M HCl (In, Sb, Pb, Cu, Zn, and Fe),[41] 1 M NH$_4$NO$_3$(Al),[42] and saturated ammonium acetate solution (Pb).[43]

Usually, the adsorbed Ga is eluted by application of HCl solutions in which its distribution coefficients have low values such as in the following media: 4,[6] 3.5,[7] 3,[37,38] 2.5,[39] 2.0,[42] and 1.5 M HCl[41] (see Tables 10 to 12). Other eluents that can be used for the quantitative elution of Ga include 2 M NaCl[44] and 1 M HF[6] as well as solutions of complexing agents (see Tables 13 and 14). To the latter belong media containing oxalic acid,[45,46] tartrate,[46] sulfosalicylate,[46] ethylenediaminetetraacetic acid (EDTA),[46] and 4 M KSCN - 0.1 M HCl.[18] In all cases Ga is eluted as an anionic complex. With ammoniacal eluents of the type described in Table 14 the Ga is eluted as an anionic hydroxide complex, i.e., Ga(OH)$_4^-$, while elements forming cationic amine complexes, e.g., Cu, Ni, Co, and Zn, are retained by the cation exchanger.[23,43,47]

Applications

From Tables 10 to 14 it is seen that procedures based on the cation exchange separation of Ga have been used in connection with the determination of this element in industrial products (see Table 10) as well as for the isolation of this element from synthetic mixtures with numerous other elements (see Tables 11 to 14).

CHELATING RESINS

Applications of chelating resins for the isolation of Ga are described in Tables 19 and 78 in the chapters on Lead and Copper (Volume III), respectively, and in Reference 53.

MICROCHEMICAL DETECTION

Trace amounts of Ga can be detected by means of the resin spot tests described in Table 15.[48,49]

Table 1
DISTRIBUTION COEFFICIENTS OF Ga, In, AND Tl IN HYDROHALIC ACID MEDIA ON STRONGLY BASIC ANION EXCHANGE RESINS[50-52]

Acid molarity (M)	Ga(III) HCl	Ga(III) HBr	Ga(III) HI	In(III) HCl	In(III) HBr	In(III) HI	Tl(III) HCl	Tl(III) HBr	Tl(III) HI
0.1	<1	—	—	~1	~2	st.a	>10^2	~17×10^3	st.a
0.5	<1	n.a.	—	~5	~7	st.a	~2×10^2	st.a	st.a
1	~1	—	n.a	~10	~20	st.a	>10^5	st.a	s.a
2	~20	—	n.a	~20	s.a	st.a	~10^5	st.a	s.a
3	~2×10^2	st.a	n.a	~20	s.a	st.a	~10^5	st.a	n.a
4	~7×10^3	st.a	n.a	~20	s.a	st.a	~7×10^4	st.a	n.a
6	>10^5	st.a	n.a	~20	s.a	s.a	~2×10^4	st.a	n.a
8	~10^5	st.a	n.a	>10	s.a	s.a	~6×10^3	st.a	s.a
10	~6×10^4	—	—	~10	—	—	~2×10^3	—	—
12	3×10^4	—	—	~7	—	—	~6×10^2	—	—

Note: St.a = strong adsorption; s.a = slight adsorption; n.a = negligible adsorption.

Table 2
DETERMINATION OF Ga AND OTHER ELEMENTS IN GEOLOGICAL MATERIALS AFTER ANION EXCHANGE SEPARATION[a]

Material	Ion exchange resin, separation conditions, and remarks	Ref.
Rocks and sediments (Mn nodules)	Bio-Rad® AG1-X8 (100-200 mesh; chloride form) Column of 0.8 cm ID containing 2 g of the resin pretreated successively with 6 M HCl (50 mℓ), 4 M HCl (10 mℓ), and TiCl$_3$ solution (10 mℓ; 15% solution of TiCl$_3$ in 10% HCl); the column is operated at a flow rate of ~0.5 mℓ/min a) 6 M HCl sample solution (20-mℓ aliquot equivalent to 0.4 g of sample) to which sufficient TiCl$_3$ solution is added until a faintly violet color persists, and then a 10-mℓ excess of this reductant is added (adsorption of Ga; Fe[II] and other matrix elements such as Mn, Al, alkaline earths, alkalies, etc. pass into the effluent) b) TiCl$_3$ solution (25 mℓ) (elution of residual Fe and other nonadsorbed elements) c) 6 M HCl (20 mℓ) (removal of Ti) d) 2 M HNO$_3$ (50 mℓ) (elution of Ga) Ga is determined by atomic absorption spectrophotometry Before the separation, the dried sample (1 g containing parts per million amounts of Ga) is dissolved in mineral acids	11
Fe meteorites	A) First column operation (Dowex® AG1-X10, 100-200 mesh; chloride form) Column: 15 × 1.5 cm conditioned with 6 M HCl a) 6—7 M HCl sample solution (adsorption of Ga and Fe[III]; Fe[II] passes into the effluent) b) 6 M HCl (several column volumes) (elution of residual Fe[II] and other nonadsorbed elements such as Ni and Co) c) 0.1 M HCl (2 column volumes) (elution of Ga and Fe[III]) B) Second column operation (separation of Ga from Fe contained in eluate [c] obtained by [A]) (Dowex® AG50-X8, 200-400 mesh; H$^+$ form) Column: 5 × 0.5 cm conditioned with 4 M HCl a) 4 M HCl solution (minimum volume) of the residue obtained after evaporation of eluate (c) (elution of Fe[III] just ahead of Ga)	6

Table 2 (continued)
DETERMINATION OF Ga AND OTHER ELEMENTS IN GEOLOGICAL MATERIALS AFTER ANION EXCHANGE SEPARATION[a]

Material	Ion exchange resin, separation conditions, and remarks	Ref.

Meteorites	b) 4 M HCl (2 column volumes) (elution of Ga)
	Ga is determined by mass spectrometry
	Before the first ion exchange separation step (A) the sample (1—2 g) is dissolved in HCl in an atmosphere of N_2 so that most of the Fe can be kept in the ferrous state, thus minimizing the amount of Fe(III) accompanying the Ga
	Dowex® 1-X8 (100-200 mesh; Cl⁻ form)
	Column containing 5 mℓ of the resin conditioned with 7 M HCl
	A) Stony meteorites
	  a) 7 M HCl sample solution (adsorption of Ga, Fe[III], Au[III], and Cu; Mn, Cr, and Na pass into the effluent)
	  b) 7 M HCl (3 column volumes) (elution of Mn, Cr, Na, and part of the Cu)
	  c) 3 M HCl (3 column volumes) (elution of residual Cu)
	  d) 0.5 M HCl (4 column volumes) (elution of Ga and Fe)
	B) Fe meteorites
	  a) 7 M HCl sample solution (adsorption of Ga, Fe[III], Au[III], and Cu)
	  b) 7 M HCl (2.5 column volumes) (elution of Mn and Na)
	  c) 0.5 M HCl (4 column volumes) (elution of Ga, Cu, and Fe[III])
	Ga and other elements are determined radiometrically
	Before the separation, the sample (30—100 mg) is irradiated with neutrons and then dissolved in HCl + H_2O_2 (Fe meteorites) or aqua regia + HNO_3-HF followed by treatment with H_2SO_4 (stony meteorites)
		8

Bauxites	Dowex® 1-X8 (100-200 mesh; Br⁻ form) Column: 80 × 1 cm pretreated with eluent (b) and operated at a flow rate of ≃0.4 mℓ/min a) Sorption solution (≯100 mℓ) consisting of 90% methanol and 10% 4.5 M HBr and containing 100—200 mg of ascorbic acid per 20 mℓ (adsorption of Ga, In, Bi, Pb, Zn, Cd, and Cu; Al, Fe, and other elements pass into the effluent) b) 90% Methanol - 10% 4.5 M HBr containing ascorbic acid (elution of remaining nonadsorbed elements) c) 0.45 M HBr (100 mℓ) (elution of Ga, In, and Zn) Ga is determined titrimetrically; before the separation, the sample (2 g) is decomposed with HBr and HF	13

[a] See also Table 2 in the chapter on Copper, Volume III; Table 6 in the chapter on Zinc, Volume IV; Table 11 in the chapter on Actinides, Volume II; and Table 86 in the chapter on Rare Earth Elements, Volume I.

Table 3
DETERMINATION OF Ga IN BIOLOGICAL MATERIALS AFTER ANION EXCHANGE SEPARATION[a]

Material	Ion exchange resin, separation conditions, and remarks	Ref.
Animal and plant tissues	Bio-Rad® AG1-X8 Column of 0.8 cm ID containing 2 mℓ of the resin conditioned with eluent (b) a) 12 M HCl sample solution (10 mℓ) (adsorption of Ga and Fe; Cu passes into the effluent) b) 12 M HCl - 1% TiCl$_3$ solution (30 mℓ) (reductive elution of residual Cu) c) 4 M HCl - 1% TiCl$_3$ solution (40 mℓ) (reductive elution of Fe) d) 1.2 M HCl (10 mℓ) (elution of 98% of Ga) ^{72}Ga is determined radiometrically; the limit of detection in blood serum was ≃50 parts per 10^{12} and the reproducibility is ±4% Before the separation, the dried sample (~250 mg) is irradiated with neutrons and then wet ashed with H$_2$SO$_4$-HNO$_3$ in the presence of a mixed carrier solution and HBr and ^{67}Ga; a solution of the residue in 12 M HCl is passed through a column of Sb$_2$O$_5$ to remove Na	10
Animal tissue and bone	Dowex® 1-X8 (50-100 mesh; chloride form) Column: 7 × 0.25 cm conditioned with 6 M HCl a) 6 M HCl sample solution (~10 mℓ) (adsorption of Ga and Fe[III]; alkalies, alkaline earths, phosphate, etc. pass into the effluent) b) 6 M HCl (20 mℓ) (elution of residual nonadsorbed elements) c) 1 M HCl (10 mℓ) (elution of Ga and Fe) Ga is determined fluorometrically with lumogallion; the useful range of the method is 5 ng—5 μg of Ga, and the limit of detection is ≃4 ng; the procedure was devised for use in connection with the study of Ga compounds in the detection and therapy of cancer Before the separation, the sample (≯1 g) is wet ashed with H$_2$SO$_4$, HNO$_3$, and HClO$_4$	2

[a] See also Table 12 in the chapter on Zinc, Volume IV.

Table 4
DETERMINATION OF Ga IN INDUSTRIAL PRODUCTS AFTER ANION EXCHANGE SEPARATION[a]

Material	Ion exchange resin, separation conditions, and remarks	Ref.
Pu-Ga alloys	Dowex® 1-X4 (50-100 mesh; chloride form) Column: ~7 × ~1.2 cm conditioned with eluent (b) (10 mℓ) and operated at a flow rate of 4 mℓ/min a) 5 M HCl - 1% ascorbic acid sample solution (25 mℓ containing ≈10 mg of Ga) (adsorption of Ga and Cu; Pu[III] and Fe[II] pass into the effluent) b) 5 M HCl - 1% ascorbic acid (~20 mℓ) (elution of residual Pu[III]) c) 0.5 M HCl (50 mℓ) (elution of Ga and Cu) Ga is determined by EDTA titration; interference by Cu (>250 ppm) is removed by precipitation of CuS from eluate (c) at pH 12 in the presence of Fe(III) (carrier) and sulfosalicylic acid to prevent hydrolysis of Ga	12
Al-Sb-Ga alloys	Anionite EDE-10P (chloride form) Column containing 25 g of the resin conditioned with 7 M HCl a) 7 M HCl sample solution (50 mℓ) (adsorption of Ga; Al passes into the effluent) b) 7 M HCl (150 mℓ) (elution of residual Al) c) Water (200 mℓ) (elution of Ga) Ga and Al are determined titrimetrically; before the separation, the sample (0.15 g) is dissolved in H_2SO_4 and Ga + Al are preconcentrated by hydroxide precipitation; the precipitate is dissolved in 7 M HCl	9
Industrial Al chloride	Dowex® 1-X8 (chloride form) a) 4 M HCl sample solution (adsorption of Ga; Al passes into the effluent) b) 4 M HCl (elution of residual Al) c) 0.1 M HCl (elution of Ga) Ga is determined spectrophotometrically	1
Bayer liquor	Dowex® 1-X4 (100-200 mesh; chloride form) Column of 1 cm ID containing the resin to a height of 4 cm a) 6 M HCl sample solution (~40 mℓ) (adsorption of Ga; Fe[II] passes into the effluent) b) 6 M HCl (60 mℓ) (elution of residual Fe[II]) c) 1 M HCl (20 mℓ) (elution of Ga) Ga is determined by EDTA titration Before the separation, the sample (10 mℓ of Bayer liquor containing ~40 g Al per liter) is diluted to 20 mℓ with water and adjusted to be 6 M in HCl; the resulting solution is passed through a Ag-reductor column to reduce Fe to Fe(II)	3

[a] See also Table 2 in the chapter on Aluminum; Tables 8 and 9 in the chapter on Gold, Volume III; Table 29 in the chapter on Zinc, Volume IV; Tables 16, 17, 19, and 26 in the chapter on Copper, Volume III; and Table 96 in the chapter on Rare Earth Elements, Volume I.

Table 5
DETERMINATION OF Ga AND In IN INDUSTRIAL PRODUCTS AFTER ANION EXCHANGE SEPARATION[a]

Material	Ion exchange resin, separation conditions, and remarks	Ref.
Al	Dowex® 1-X8 (100-200 mesh; chloride form) Column: 75 × 1 cm conditioned with eluent (b) (100 mℓ) and operated at a flow rate of 0.2—0.3 mℓ/min a) Sorption solution (20 mℓ) consisting of 6 M HCl (2-mℓ aliquot equivalent to 100 mg of sample) and methyl glycol (2-methoxyethanol-1) (18 mℓ) (adsorption of Ga and In) b) 90% Methyl glycol - 10% 6 M HCl (200 mℓ) (elution of Al) c) 70% Methyl glycol - 30% 2 M HCl (100 mℓ) (to further develop the adsorption bands of Ga and In) d) 1 M HCl (240 mℓ) (the first 110 mℓ of eluate contains only Ga, whereas the next 70 mℓ contains neither Ga nor In; in the last fraction, from 180—240 mℓ, all the In is present) Ga and In are determined by EDTA titration Before the separation, the sample (1 g) is dissolved in HCl containing HNO_3 (few drops)	14
Ga-In alloys	Amberlyst® A-26 (macroreticular anion exchanger, chloride form; exchange capacity: 3.46 meq/g impregnated with xylenol orange [Xo]) The resin (2 g) impregnated by shaking with aqueous Xo solution of pH 6.5 is used in a column of 0.8 cm ID which is operated at a flow rate of 1 mℓ/min a) Aqueous sample solution of pH 2.0 (10-mℓ aliquot equivalent to 16 mg of the sample) (adsorption of Ga and In as Xo complexes) b) 0.03 M NaOH (elution of Ga) c) 0.2 M H_2SO_4 (elution of In) Ga and In are determined by atomic absorption spectrophotometry Before the separation, the sample (160 mg) is dissolved in 1 + 1 H_2SO_4 (10 mℓ)	26

[a] See also Tables 9 and 10 in the chapter on Gold, Volume III and Table 2 in the chapter on Tungsten, Volume IV.

Table 6
ANION EXCHANGE SEPARATION OF Ga IN HCl MEDIA

Elements separated	Ion exchange resin, separation conditions, and remarks	Ref.
Ga, In, and Al	Dowex® 1-X8 (100-200 mesh; chloride form) Column: 75 × 1 cm conditioned with eluent (a) (100 mℓ) and operated at a flow rate of 0.2—0.3 mℓ/min After adsorption of the elements from eluent (a) (20 mℓ) the following solutions are passed through the column a) 80% Acetone - 20% 3 M HCl (200 mℓ) (elution of Ga) b) 90% Acetone - 10% 6 M HCl (140 mℓ) (elution of In) c) 70% Acetone - 30% 2 M HCl (100 mℓ) (elution of Al) The elements (present in milligram amounts) are determined by EDTA titration	14
Ga from Fe	Permutit® ES (nitrate form) Column of 1.6 cm ID containing 30 mℓ of the resin and operated at a flow rate of 4—5 mℓ/min a) Solution (150 mℓ) containing oxalic acid (1 g) and adjusted to pH 4.0 with NH$_4$-acetate (adsorption of Ga and Fe as anionic oxalate complexes) b) 0.001 M oxalic acid (50 mℓ) (as a rinse) c) 1 M NaOH (300 mℓ) (elution of Ga) d) Water (50 mℓ) (as a rinse) e) 2 M HNO$_3$ (300 mℓ) (elution of Fe)	27
Ga from Fe, Al, Mn, Ni, and Co	Anionite AV-17 a) 6 M HCl sample solution containing NH$_2$OH·HCl (adsorption of Ga; into the effluent pass Fe[II], Mn[II], Ni, and Co) b) 4 M HCl (elution of residual nonadsorbed elements) c) 0.5 M H$_2$SO$_4$ (elution of Ga together with any Cu, Cd, Bi, Sn, Pb, V, Cr, and Ti) Before passage through the column, solution (a) is heated for 15 min (or until the color changes from yellowish-brown [Fe{III}] to pale green [Fe{II}])	31
Ga from Fe	Anionite AV-17 (chloride form) a) 6 M HCl (adsorption of Ga; Fe[II] passes into the effluent) b) 0.6 M HCl (elution of Ga) Before the separation, the Fe is reduced to Fe(II) with Ag-powder in 6 M HCl medium	4

Table 7
ANION EXCHANGE SEPARATION OF Ga IN THIOCYANATE AND IODIDE MEDIA

Elements separated	Ion exchange resin, separation conditions, and remarks	Ref.
Ga from Fe(III)	Amberlite® IRA-400 (grain size: 0.1—0.3 mm; thiocyanate form) Column: 12 × 0.6 cm conditioned with eluent (b) (20 mℓ) and operated at a flow rate of 1 mℓ/min a) Sorption solution (100 mℓ) consisting of 50 mℓ each of 4 M KSCN and 0.1 M HCl (adsorption of Ga and Fe as anionic thiocyanate complexes) b) 2 M KSCN - 0.05 M HCl (20—50 mℓ) (elution of any Al that may be present) c) 0.1 M HCl (50—100 mℓ) (elution of Ga) Ga is determined fluorometrically Separation of 15 µg of Ga from a maximum of 1.25 mg of Fe can be accomplished with recoveries of 96—98% of Ga; if the content of Fe is ≯0.5 mg, up to 100 µg of Ga can be separated	17
Ga from Al	Wofatit® L-150 (chloride form) Column conditioned with 2 M KSCN - 0.05 M HCl a) 4 M KSCN - 0.1 M HCl (adsorption of Ga as anionic SCN$^-$ complex; Al passes into the effluent) b) Same as eluent (a) (elution of residual Al) c) 2 M HCl (elution of Ga) From a column containing the cationite KPS-200, the Ga is eluted first with the KSCN-HCl solution and then the Al is desorbed with 2 M HCl; Ga and Al mixed in ratios from 1:2—1:20 can be separated and determined with errors of <±1.4% and <±0.3%, respectively	18
Ga from In	Dowex® 1-X8 (50-100 mesh; chloride form) Column: 12 × 0.8 cm conditioned with 20 mℓ of 0.5 M KI of pH 1 and operated at a flow rate of 0.5 mℓ/min a) 0.5 M KI of pH 1 (30 mℓ) (adsorption of In as anionic iodide complex; Ga passes into the effluent) b) 0.5 M KI of pH 1 (40 mℓ) (elution of Ga) c) 1 M NH$_4$Cl of pH 1 (110 mℓ) (elution of In) Ga and In are determined by EDTA titration; the separation can also be effected by the use of 0.5 M (NH$_4$)$_2$SO$_4$ of pH 1 in place of eluent (c)	29, 30

Table 8
ANION EXCHANGE SEPARATION OF Ga IN CARBONATE MEDIA

Elements separated	Ion exchange resin, separation conditions, and remarks	Ref.
Ga from Cd, Zn, Tl, Al, In, Fe(III), and Pb	Anionite EDE-10P (carbonate form) Column: 9 × 1.6 cm operated at a flow rate of 1 mℓ/min a) Test solution (25 mℓ) of pH 2.5 (retention of Ga, Cd, Zn, Al, In, Fe[III], and Pb) b) Water (50 mℓ) (elution of Tl[I]) c) 3 M NH$_3$ solution (260 mℓ) (elution of Cd, Zn, and Al) d) 3 M Na$_2$CO$_3$ (70 mℓ) (elution of Ga) e) 2.5 M NH$_4$ carbonate (110 mℓ) (elution of In) f) 1.5 M HCl (340 mℓ) (elution of Fe) g) 1 M HNO$_3$ (100 mℓ) (elution of Pb)	22
Ga from Zn	Anionite EDE-10P (carbonate form) Column of 0.8 cm ID containing 8—10 g of the resin a) 2 M ammonium carbonate adjusted to pH 9—10 with NH$_3$ solution (adsorption of Ga as anionic carbonate complex; Zn passes into the effluent as cationic amine complex) b) 2.5 M HCl (elution of Ga) The method can be used to separate small amounts of Ga from large quantities of Zn when present in ratios of up to 1:10^3 For the separation of large amounts of Ga (10^3—10^4 parts) from small amounts of Zn (1 part) the solution, 2 M in ammonium carbonate, is passed through a column of the cationite KU-2 (NH$_4^+$ form) (adsorption of Zn as the amine complex; Ga passes into the effluent)	23

Table 9
DISTRIBUTION COEFFICIENTS OF Ga, In, AND Tl IN HYDROHALIC AND NITRIC ACID MEDIA ON STRONGLY ACID CATION EXCHANGE RESINS

Acid molarity (M)	Ga(III) HCl	Ga(III) HBr	Ga(III) HNO$_3$	In(III) HCl	In(III) HBr	In(III) HNO$_3$	Tl(III) HCl	Tl(III) HBr	Tl(III) HNO$_3$
0.1	>10^4	>10^3	>10^4	10	>10^2	>10^4	<1 (~4 × 10^2)	—	324 (173)
0.2	3 × 10^3	>10^3	4200	8	~10^2	>10^4	<1 (~130)	—	— (91)
0.5	260	>10^2	445	~5	~5	680	<1 (~40)	—	30 (41)
1	~42	~25	94	1	~1	118	1 (~11)	20	32 (~22)
2	~8	~3	20	<1	<1	23	~2	30	— (~10)
3	~3	~1.5	9	<1	<1	~10	~3	~10^2	— (~6)
4	~1	<1	~6	<1	<1	~6	~5	>10^2	~3 (~3)
5	~10	~1	—	<1	~1	—	~6	>10^2	—
6	~2 × 10^2	~50	—	<1	10	—	~8	>10^2	—
7	~4 × 10^2	~10^2	—	<1	~30	—	~10	>10^2	—
8	~6 × 10^2	~10^3	—	<1	~80	—	~10	>10^2	—
9	~6 × 10^2	—	—	<1	—	—	~10	—	—
10	~8 × 10^2	—	—	<1	—	—	~10	—	—

Note: Distribution coefficients for Tl(I) are listed in parentheses. It can be assumed that in the HBr media of comparable concentrations Tl(I) shows the same or very similar adsorbability as in the HCl systems.

Table 10
DETERMINATION OF Ga AND OTHER ELEMENTS IN INDUSTRIAL PRODUCTS AFTER CATION EXCHANGE SEPARATION[a]

Material	Ion exchange resin, separation conditions, and remarks	Ref.
High-purity Zn	I) Separation of short-lived radionuclides A) First column operation (Dowex® 50-X8, 100-200 mesh; H⁺-form) Column: 10 × 1.6 cm operated at a flow rate of 0.4 mℓ/cm² min a) 0.1 M HNO₃ sample solution containing a trace of HF (adsorption of Ga, Zn, Cu, Mn, Ni, rare earths, Na, and K; into the effluent pass Au, As, W, Sb, Mo, and Ge) b) 0.1 M HNO₃ (elution of residual nonadsorbed elements) c) 0.5 M HCl (as a rinse) d) 3.5 M HCl (elution of Ga and the other adsorbed metals) B) Second column operation (Dowex® 1-X8, 100—200 mesh; chloride form) Column: 10 × 1.6 cm containing 10—20 g of the resin and operated at a flow rate of 0.4 mℓ/cm² min a) Eluate (d) obtained by (A) adjusted to be 6 M in HCl by the addition of conc HCl (adsorption of Ga, Zn, and Cu; into the effluent pass Ni, Mn, Na, and K) b) 1 M HCl + H₂O₂ (elution of Ga together with Cu which is more readily eluted in the presence of H₂O₂ [see in the chapter on Copper, Volume III]) Subsequently, Ga is separated from Cu by precipitation of the latter as CuS; the elements contained in effluent (a) obtained by (A) are further fractionated by radiochemical separation steps (application of inorganic ion exchangers) which include adsorption of W on Dowex® 1 from 7 M HNO₃ to separate it from Sb, Mo, and Ge which pass into the effluent II) Separation of long-lived radionuclides A) First column operation (same experimental conditions as above under [I.A]) a) 0.1 M HNO₃ sample solution containing a trace of HF (adsorption of Zn, Fe, Co, Pa, Ag, Cr, Rb, Cs, and rare earths; into the effluent pass Se, Te, Hg, Zr, Hf, Ta, and Sc) b) 0.1 M HNO₃ (elution of residual nonadsorbed elements) c) 12 M HCl (elution of Zn and of the other adsorbed elements; Fe is only partially eluted) d) 3.5 M HCl (elution of residual Fe) B) Second column operation (same experimental conditions as above under [I.B]) a) Eluate (c) obtained by (A) (adsorption of Zn, Co, Pa, and Fe; Rb, Cs, Ag, and Cr pass into the effluent) b) 2 M HCl (elution of Co and Pa which is obtained by the neutron irradiation of Th) c) 0.5 M HCl (elution of Fe; this eluate is combined with eluate [d] obtained by [A]) The elements contained in effluent (a) obtained by (A) are further fractionated using radiochemical separations (use of inorganic ion exchangers) which also include a procedure based on the adsorption of Hg and Sb on a column of Dowex® 1 from 6 M HCl; Zr, Hf, and Sc pass into the effluent In the various fractions obtained, Ga and the other impurity elements are determined radiometrically Before the ion exchange separations, the sample (300—800 mg	7

Table 10 (continued)
DETERMINATION OF Ga AND OTHER ELEMENTS IN INDUSTRIAL PRODUCTS AFTER CATION EXCHANGE SEPARATION[a]

Material	Ion exchange resin, separation conditions, and remarks	Ref.
	containing parts per billion amounts of the impurity metals) is irradiated with neutrons and then dissolved in 7 M HNO$_3$ (4 mℓ) containing traces of HF ($\leq$0.3 M HF)	

[a] See also Table 13 in the chapter on Molybdenum, Volume IV; Tables 40, 49, 52, 54, and 55 in the chapter on Copper, Volume III; and Table 7 in the chapter on Bismuth.

Table 11
CATION EXCHANGE SEPARATION OF Ga IN MINERAL ACID-ACETONE MEDIA

Elements separated	Ion exchange resin, separation conditions, and remarks	Ref.
Ga from Al, In, and Tl	A) First column operation (Bio-Rad® AG 50W-X8, 200-400 mesh; H$^+$ form) Column of 2 cm ID containing the resin (60 mℓ) conditioned with 50% acetone - 0.1 M HCl a) 50% Acetone - 0.1 M HCl (25 mℓ) containing a small amount of Cl$_2$ (adsorption of Ga, Al, and In; Tl[III] passes into the effluent) b) Same eluent as (a) (200 mℓ) (elution of Tl[III]) (flow rate: 2.0 ± 0.3 mℓ/min) c) 50% Acetone - 0.5 M HCl (300 mℓ) (elution of In) (flow rate: 2.0 ± 0.3 mℓ/min) d) 70% Acetone - 2 M HCl (300 mℓ) (elution of Ga) (flow rate: 3.0 ± 0.3 mℓ/min) e) 3 M HCl (300 mℓ) (elution of Al at the same rate as Ga) B) Second column operation (isolation of Ga from eluate [d] obtained by [A]) (Bio-Rad® AG1-X8, 200-400 mesh, chloride form; column of 2 cm ID filled with 45 mℓ of the resin previously equilibrated with 40% acetone - 2 M HCl) a) 44% Acetone - 2 M HCl (480 mℓ, prepared by mixing eluate [d] [300 mℓ] with 2 M HCl [180 mℓ]) (adsorption of Ga) b) 0.1 M HCl (300 mℓ) (elution of Ga; an eluate virtually free from acetone is obtained so that no losses of Ga occur on evaporation; 1—2% of Ga is lost if HCl solutions containing high percentages of acetone are evaporated) The cation exchange separation procedure (A) can be used to separate millimole quantities of the elements and the recoveries are 100 ± 0.1% For large amounts of Tl(III), elution with 0.1 M HBr plus free Br$_2$ is recommended to prevent reduction to Tl(I) which is retained by the resin Elutions of In and Ga can also be effected with 40% acetone - 0.5 and 2.5 M HCl, respectively[39]	15
Ga from binary mixtures with Zn, Cu, In, Fe(III), Cd, Pb, Bi, Au(III), Pt(IV), Pd(II), Tl(III), and Sn(IV)	Bio-Rad® AG50W-X4 (200-400 mesh; H$^+$ form) A) Separations with 80% acetone - 0.5 M HBr Column of 2.0 cm ID containing 65 mℓ of the resin conditioned with 65% acetone - 0.3 M HNO$_3$ (100 mℓ) and operated at a flow rate of 3.0 ± 0.5 mℓ/min a) 65% Acetone - 0.3 M HNO$_3$ (25 mℓ) (adsorption of Ga and other elements) b) 80% Acetone - 0.5 M HBr (350 mℓ) (elution of Zn, Cu, Fe[III], In, Cd, Pb, and Bi) c) 0.3 M HNO$_3$ (100 mℓ) (removal of acetone) d) 3 M HCl (250 mℓ) (elution of Ga) Au(III), Pt(IV), Pd(II), Tl(III), and Sn(IV) are added to the column from 65% acetone - 0.5 M HBr (25 mℓ) and the effluent containing these elements is collected from the beginning of the sorption step B) Separations with 80% acetone - 0.2 M HBr Column: 5.5 × 1 cm containing 4.3 mℓ of the resin equilibrated with 80% acetone - 0.2 M HBr a) 80% Acetone - 0.2 M HBr (50 mℓ) (adsorption of Ga; into the effluent pass Zn, Cd, In, and Bi)	37

Table 11 (continued)
CATION EXCHANGE SEPARATION OF Ga IN MINERAL ACID-ACETONE MEDIA

Elements separated	Ion exchange resin, separation conditions, and remarks	Ref.
	b) 80% Acetone - 0.2 M HBr (80 mℓ) (elution of residual nonadsorbed elements) c) 0.2 M HCl (20 mℓ) (removal of acetone) d) 3 M HCl (40 mℓ) (elution of Ga) Ga is determined by atomic absorption spectrophotometry	
Ga from U(VI), Co, Al, Li, Na, Be, Mg, Ca, Ba, La, Ti(IV), Th, Mn(II), and Ni	Bio-Rad® AG50W-X4 (200-400 mesh; H⁺ form) Column of 2.15 cm ID containing 65 mℓ (15 g dry weight) of the resin conditioned with 70% acetone - 0.2 M HNO$_3$ (100 mℓ) and operated at a flow rate of 3.0 ± 0.5 mℓ/min a) 70% Acetone - 0.3 M HNO$_3$ (~25 mℓ) (adsorption of Ga and other elements) b) 84% Acetone - 0.2 M HCl (200 mℓ) (elution of Ga) c) 3 M HCl (200 mℓ) (elution of other elements) To the eluate (b) ~50 mℓ of water is added and the solution is passed through a column (7 × 2.0 cm) containing 22 mℓ (5 g) of the same resin, and acetone is washed from the resin bed with 0.1 M HCl and Ga is eluted with 3 M HCl; this separation of Ga from acetone is necessary for accurate work, because Ga is lost in small amounts (0.5—2%) when acetone-containing solutions of HCl (or HBr) are evaporated, even on a water bath Ga is determined by EDTA titration or atomic absorption spectrophotometry Recoveries are >99.9% The method is applicable to separations of microgram and millimolar amounts of Ga; when elements forming anionic bromide complexes, e.g., Zn and Cd, are eluted with 80% acetone - 0.5 M HBr prior to elution of Ga, the procedure separates Ga from almost all other elements	38

Table 12
CATION EXCHANGE SEPARATION OF Ga IN HYDROCHLORIC AND NITRIC ACID MEDIA

Elements separated	Ion exchange resin, separation conditions, and remarks	Ref.
Ga from Al, Cd, Cu, In, Mn, Ni, Pb, U(VI), and many other elements	Bio-Rad® AG50W-X4 (100-200 mesh; H$^+$ form) Column of 1.44 cm ID containing 13 mℓ (3 g) of the resin conditioned with 8 M HCl and operated at a flow rate of 4.0 ± 0.3 mℓ/min a) 8 M HCl sample solution (~50 mℓ) (adsorption of Ga; into the effluent pass essentially all other elements except Fe[III] and Sc which are coadsorbed with Ga) b) 8 M HCl (60 mℓ) (elution of residual nonadsorbed elements) c) 2.5 M HCl (4 × 10 mℓ) (elution of Ga) To separate Ga from Sc, 7 M HCl is used as the eluent for the latter element; when Fe(II) is separated from Ga, the eluent is 8 M HCl - 0.3% TiCl$_3$ solution (80 mℓ) Residual Ti is eluted with 8 M HCl (50 mℓ) before eluting the Ga Ga is determined by atomic absorption spectrophotometry The method can be used to separate trace amounts and up to 1.5 mg of Ga from up to gram quantities of the other elements	40
Ga from binary mixtures with In, Sb(III), Pb, Cu, Zn, and Fe(II)	Dowex® 50 (20-50 mesh; H$^+$ form) Column: 100 × 0.8 cm operated at a flow rate of 10 mℓ/min After adsorption of the elements the following eluents are used a) Water (100 mℓ) (as a rinse) b) <1 M HCl, e.g., 0.4 M (2 ℓ) (elution of In, Sb, Pb, Cu, Zn, and Fe) c) 1.5 M HCl (2 ℓ) (elution of Ga)	41
Ga from Al	Cationite KB-2x7P Column operated at a flow rate of 1 mℓ/cm^2 min a) Dilute HNO$_3$ sorption solution of pH 2.7 (adsorption of Ga and Al) b) 1 M NH$_4$NO$_3$ (elution of Al) c) 2 M HCl (elution of Ga)	42

Table 13
CATION EXCHANGE SEPARATION OF Ga IN THE PRESENCE OF ORGANIC ACIDS

Elements separated	Ion exchange resin, separation conditions, and remarks	Ref.
Ga from In	Amberlite® IR-120 (60-100 mesh; H$^+$ form) Column: 10 × 0.9 cm After adsorption of the elements the following eluents are used a) 0.05 M oxalic acid - 0.2 M HCl (50 mℓ) (elution of Ga) b) 0.05 M oxalic acid - 0.6 M HCl (50 mℓ) (elution of In)	45
Ga from Zn	Cationite SBS (grain size: 0.5—1 mm; NH$_4^+$ form) Column of 1 cm ID containing 10 g of the resin a) Tartrate solution of pH 3—10 (six- to tenfold excess of tartaric acid) (adsorption of Zn; Ga is eluted as anionic tartrate complex) b) 10% HCl solution (elution of Zn) The separation is just as satisfactory in the presence of oxalate, sulfosalicylate, or EDTA instead of tartaric acid	46
Ga from Pb and Cd	Cationite SBS (H$^+$ form) After adsorption of the elements on a column of this resin the following eluents are used a) Saturated ammonium acetate solution (elution of Pb) b) 0.2 M HCl (elution of Cd) c) 10% NaOH solution (elution of Ga) Any Cu and Fe is retained by the column	43

Table 14
CATION EXCHANGE SEPARATION OF Ga IN NEUTRAL AND ALKALINE MEDIA

Elements separated	Ion exchange resin, separation conditions, and remarks	Ref.
Ga from Zn, Cu, Co, and Ni	Cationite SBS (grain size: 0.25—0.5 mm; Na$^+$ form) Column: 25 × 1 cm operated at a flow rate of 5 mℓ/min a) Sample solution (100 mℓ) to which conc NH$_3$ solution (15 mℓ) and 2 M NaOH (10 mℓ) were added (adsorption of Ni, Co, Zn, and Cu as cationic amine complexes; Ga passes into the effluent as anionic hydroxide complex) b) Solution (100 mℓ) containing conc NH$_3$ (10 mℓ) and 2 M NaOH (5 mℓ) (elution of residual Ga)	47
Ga from Al	Cationite KU-2 (Na$^+$ form) Column of 1 cm ID containing 15 g of the resin and operated at a flow rate of 1 drop/4 sec a) 2 M NaCl (adsorption of Al; Ga passes into the effluent as anionic chloride complex) b) Water (elution of residual Ga) c) 2 M HCl (elution of Al) The method can be used for the separation of Ga (0.5 mg) from up to 531 mg of Al	44

Table 15
MICROCHEMICAL DETECTION OF Ga AND In BY RESIN SPOT TESTS

Ion exchange resin	Experimental conditions and remarks	Ref.
Dowex® 1-X1 (chloride form)	Beads of the resin (previously treated with 0.01% aq lumogallion, catechol violet, or stilbazol) are placed in the sample solution containing Ga or In made pH 5 and 6, respectively, with 1 M Na acetate and 1% acetic acid; lumogallion gives an orange fluorescence on irradiation with UV light; catechol violet and stilbazol give blue to olive-green and purple to dark-red colors The limits of detection and dilution for either Ga or In are 0.003 µg and 1 part in 10^7, with any one of these reagents	48
Dowex® 1-X1 (50—100 mesh; acetate form)	On a spot plate 1 drop each of the sample solution, saturated NH$_4$Cl solution, and 0.3 M NH$_3$ solution are mixed and one bead of the resin impregnated with alizarin red S (C.I. Mordant Red 3) is added The color changes from violet to red after 6 hr The limit of detection for Ga is 0.03 µg (limiting dilution 1 in 1×10^{-6}) Interfering ions are Al, In, Mo, W, Zr, Sn, V, Th, Bi, Cr, Fe, U, Ti, and Si	49

REFERENCES

1. **Evtimova, B.**, Analytical application of the reactions of gallium and indium with Chrome Azurol S and hexadecyltrimethylammonium bromide, *Dokl. Bolg. Akad. Nauk*, 36(7), 915, 1983.
2. **Zweidinger, R. A., Barnett, L., and Pitt, C. G.**, Fluorimetric determination of gallium in biological materials at nanogram levels, *Anal. Chem.*, 45, 1563, 1973.
3. **Gregory, G. R. E. C. and Jeffery, P. G.**, Determination of gallium, *Talanta*, 9(9), 800, 1962.
4. **Ad'yasevich, I. K. and Bel'skaya, E. P.**, Extractive chromatographic isolation and purification of isotopes of gallium after electromagnetic separation, *At. Energ.*, 28(1), 64, 1970.
5. **Nadezhina, L. S.**, Determination of small amounts of gallium, *Zh. Anal. Khim.*, 17(3), 383, 1962.
6. **DeLaeter, J. R.**, Isotopic composition and elemental abundance of gallium in meteorites and in terrestrial samples, *Geochim. Cosmochim. Acta*, 36(7), 735, 1972.
7. **Mousty, F. and Girardi, F.**, Determination of impurities in high-purity zinc by neutron activation, *J. Radioanal. Chem.*, 22, 29, 1974.
8. **Schaudy, R., Kiesl, W., and Hecht, F.**, Activation analysis for elements in meteorites, *Chem. Geol.*, 2(4), 279, 1967.
9. **Denisova, N. E. and Tsvetkova, E. V.**, Analysis of aluminum-antimony-gallium alloys, *Zavod. Lab.*, 27(6), 656, 1961.
10. **Stulzaft, O., Maziere, B., and Ly, S.**, Gallium determination in biological samples, *J. Radioanal. Chem.*, 55(2), 291, 1980.
11. **Korkisch, J., Steffan, I., Nonaka, J., and Arrhenius, G.**, Chemical analysis of manganese nodules. V. Determination of gallium after anion exchange separation, *Anal. Chim. Acta*, 109, 181, 1979.
12. **Miner, F. J. and DeGrazio, R. P.**, Ion-exchange separation and volumetric determination of gallium in plutonium-gallium alloys, *Anal. Chem.*, 37, 1071, 1965.
13. **Korkisch, J. and Hazan, I.**, Anion-exchange separations in hydrobromic acid-organic solvent media, *Anal. Chem.*, 37, 707, 1965.
14. **Korkisch, J. and Hazan, I.**, Anion-exchange separation of gallium, indium and aluminum, *Anal. Chem.*, 36, 2308, 1964.
15. **Strelow, F. W. E. and Victor, A. H.**, Quantitative separation of aluminum, gallium, indium, and thallium by cation-exchange chromatography in hydrochloric acid-acetone, *Talanta*, 19, 1019, 1972.
16. **Bagbanly, I. L., Luseinov, I. K., and Allakhverdieva, E. G.**, Use of lithium chloride in chromatographic analysis, *Azerb. Khim. Zh.*, No. 2, 166, 1969.
17. **Korkisch, J. and Hecht, F.**, Separation of gallium from tervalent iron by means of ion exchange, *Mikrochim. Acta*, No. 7-8, 1230, 1956.
18. **Kocheva, L. L. and Draganova, R.**, Separation of gallium from aluminum by ion exchange, *God. Sofii. Univ. Khim. Fak.*, 62, 129, 1967/1968.
19. **Shabana, R.**, Effect of resin cross-linking and temperature on anion exchange separation of gallium and indium, *J. Radioanal. Chem.*, 41(1-2), 53, 1977.
20. **Dybczyński, R., Polkowska-Motrenko, H., and Shabana, R. M.**, Influence of temperature and degree of resin cross-linking on the anion-exchange of phosphate complexes of gallium(III) and indium(III), *J. Chromatogr.*, 134(2), 285, 1977.
21. **Oguma, K., Maruyama, T., and Kuroda, R.**, Anion-exchange behavior of various metals in hydrazoic acid media, *Anal. Chim. Acta*, 74, 339, 1975.
22. **Eristavi, V. D.**, Separation of gallium from interfering elements on ion exchange resin in carbonate form, *Soobshch. Akad. Nauk Gruz. SSR*, 71(1), 105, 1973.
23. **Alimarin, I. P., Tsintsevich, E. P., and Gorokhova, A. N.**, Separation of gallium from zinc in ammonium carbonate solution by means of an ionite, *Zavod. Lab.*, 26(2), 144, 1960.
24. **Nazarenko, V. A., Vinkovetskaya, S. Ya., and Ravitskaya, R. V.**, Fluorimetric determination of trace amounts of gallium in semiconductor silicon and high-purity zinc, *Ukr. Khim. Zh.*, 28(6), 726, 1962.
25. **Tsintsevich, E. P. and Gorokhova, A. N.**, Separation of gallium from copper and nickel by ion exchange, *Izv. Vyssh. Uchebn. Zaved. Khim. Khim. Tekhnol.*, 3(2), 245, 1960.
26. **Brajter, K. and Olbrych-Sleszyńska, E.**, Application of xylenol orange to the separation of metal ions on Amberlyst A-26 macroreticular anion exchange resin, *Talanta*, 30, 355, 1983.
27. **Blasius, E. and Negwer, M.**, The separation of small quantities of gallium from iron by means of a strongly basic anion exchange resin, *Fresenius' Z. Anal. Chem.*, 143, 257, 1954.
28. **Suranova, Z. P., Morozov, A. A., and Grabchuk, O. Ya.**, Chromatographic separation of gallium from cadmium and zinc, *Tr. Nauchno Issled. Inst. Khromatograf. Voronezh. Univ.*, No. 2, 128, 1968.
29. **Kocheva, L. L.**, Anion-exchange separation of gallium from indium using iodide ions, *God. Sof. Univ. Khim. Fak.*, 62, 135, 1967/68.
30. **Kocheva, L. L.**, Separation of gallium and indium by ion-exchange chromatography, *C.R. Acad. Bulg. Sci.*, 22(4), 447, 1969.

31. **Nadezhina, L. S.,** Determination of small amounts of gallium, *Tr. Leningr. Politekh. Inst.,* No. 304, 145, 1970.
32. **Herber, R. H. and Irvine, J. W.,** Anion-exchange studies. I. Bromide complexes of Co(II), Cu(II), Zn(II), and Ga(III), *J. Am. Chem. Soc.,* 76, 987, 1954.
33. **Schindewolf, U. and Irvine, J. W.,** Preparation of carrier-free vanadium, scandium and arsenic activities from cyclotron targets by ion exchange, *Anal. Chem.,* 30, 906, 1958.
34. **Faris, J. P.,** Adsorption of the elements from hydrofluoric acid by anion exchange, *Anal. Chem.,* 32, 520, 1960.
35. **Faix, W. G., Caletka, R., and Krivan, V.,** Element distribution coefficients from hydrofluoric acid/nitric acid solutions and the anion exchange resin Dowex 1-X8, *Anal. Chem.,* 53, 1719, 1981.
36. **Caletka, R. and Krivan, V.,** Cation-exchange of 43 elements from hydrofluoric acid solution, *Talanta,* 30, 543, 1983.
37. **Strelow, F. W. E.,** Quantitative separation of gallium from zinc, copper, indium, iron(III), and other elements by cation-exchange chromatography in hydrobromic acid-acetone medium, *Talanta,* 27, 231, 1980.
38. **Strelow, F. W. E.,** Quantitative separation of gallium from uranium, cobalt, aluminum, and many other elements by cation-exchange chromatography in mixtures of hydrochloric or hydrobromic acid with acetone, *Anal. Chim. Acta,* 113, 323, 1980.
39. **Tsintsevich, E. P., Alimarin, I. P., Gorokhova, A. N., Murashko, M. S., and Kozyreva, G. V.,** Cation-exchange separation of gallium and indium in the presence of hydrochloric acid and organic solvents, *Vestn. Mosk. Gos. Univ. Ser. Khim.,* No. 1, 91, 1969.
40. **Van der Walt, T. N. and Strelow, F. W. E.,** Quantitative separation of gallium from other elements by cation-exchange chromatography, *Anal. Chem.,* 55, 212, 1983.
41. **Klement, R. and Sandmann, H.,** Separation of gallium, indium and germanium from other metals by ion exchange, *Fresenius' Z. Anal. Chem.,* 145, 325, 1955.
42. **Kudryavskii, Yu. P., Kazantsev, E. I., and Spiridonov, E. N.,** Sorption and separation of aluminum and gallium on cation-exchange resins, *Izv. Vyssh. Uchebn. Zaved. Tsvetn. Metall.,* 17(2), 62, 1974.
43. **Tsintsevich, E. P. and Nazarova, G. E.,** Separation of gallium from lead and cadmium by the method of ion-exchange, *Zavod. Lab.,* 23(9), 1068, 1957.
44. **Alimarin, I. P., Tsintsevich, E. P., and Usova, E. P.,** Investigation of the behavior of gallium halides in halogen acid solutions by ion exchange, *Vestn. Mosk. Univ. Ser. Khim.,* No. 2, 31, 1961.
45. **Nozaki, T., Takeuchi, K., Yamauchi, T., and Nomura, T.,** Cation-exchange separation of metals in mixed oxalic acid-hydrochloric acid media, *Jpn. Analyst,* 27, 454, 1978.
46. **Alimarin, I. P. and Tsintsevich, E. P.,** Use of the chromatographic method for separating gallium from other elements. Separation of gallium and zinc, *Zavod. Lab.,* 22(11), 1276, 1956.
47. **Zelyanskaya, A. I. and Bausova, N. V.,** Separation of gallium from zinc, copper, cobalt, nickel, and iron by ion-exchange, *Izv. Vost. Fil. Akad. Nauk SSSR,* No. 7, 51, 1957.
48. **Ichikawa, T., Kato, K., and Kakihana, H.,** Microchemical detection of gallium and indium with ion-exchange resin particles, *J. Chem. Soc. Jpn. Pure Chem. Sect.,* 89(11), 1071, 1968.
49. **Kato, K. and Kakihana, H.,** Micro-detection of gallium with alizarin red S in the presence of anion-exchange resin, *J. Chem. Soc. Jpn. Pure Chem. Sect.,* 80(3), 282, 1959.
50. **Marsh, S. F., Alarid, J., Hammond, C. F., McLeod, M. J., Roensch, F. R., and Rein, J. E.,** Anion-exchange of 58 elements in hydrobromic acid and hydroiodic acid media, Report LA-7084, Los Alamos Scientific Laboratory, Los Alamos, N.M., February 1978.
51. **Klakl, E. and Korkisch, J.,** Anion-exchange behavior of several elements in hydrobromic acid-organic solvent media, *Talanta,* 16, 1177, 1969.
52. **Kraus, K. A., Nelson, F., and Smith, G. W.,** Anion exchange studies. IX. Adsorbability of a number of metals in hydrochloric acid solutions, *J. Phys. Chem.,* 58, 11, 1954.
53. **Suzuki, T. M., Yokoyama, T., Matsunaga, H., and Kimura, T.,** Selective adsorption of gallium(III) and indium(III) using polystyrene resins with functional group having bis(carboxymethyl)amino-moiety, *Chem. Lett.,* No. 1, 41, 1985.

INDIUM

For the separation of In from accompanying elements anion and cation exchange resins of the strongly basic and acidic types are most frequently used. These resins are mainly employed in hydrochloric (HCl) and hydrobromic (HBr) acid media, but effective separations of In can also be performed in systems containing organic complex agents. Chelating resins have found only very limited application for separations involving In.

ANION EXCHANGE RESINS

General

The In(III) species predominating in 0.1 to 0.5, 0.5 to 2, 2 to 6, and 6 to 12 M HCl solutions are $InCl^{2+}$, $InCl_2^+$, $InCl_3$, and $InCl_4^-$, respectively. Among these, the anionic species is retained by strongly basic resins, e.g., Dowex® 1, to a moderate extent, reaching a maximum distribution coefficient of about 20 in 2 to 6 M HCl (see Table 1 in the chapter on Gallium). Consequently, adsorption of In from HCl media, as, for example, from 1 M HCl (see Table 1*),[1,2] 2 M HCl (see Table 2),[3] 3 M HCl,[4] 5 M HCl,[5] and 6 M HCl (see Table 2),[6-9] implies that when using somewhat larger volumes of these acid solutions the In will be eluted under the same conditions. This has been demonstrated by using 1,[10-13] 3,[14] and 6 M HCl[6] both as the sorption media and eluents. Coadsorbed with the In from 5 to 6 M HCl are Fe(III), $UO_2(II)$, Ga, Zn, Cd, Sn, and other elements forming stable anionic chloro complexes, so that in the presence of large amounts of these metals In may be displaced from the anion exchange resins. However, separations based on the adsorption of In from the 1 M acid make it possible to separate In from Fe, Ga, and U, but not from Zn, Cd, Sn, and Pb. Not adsorbed with the In from HCl media of any concentration are the alkali metals, alkaline earth elements, rare earths, Al, Ni, Th, and other elements.

On the weakly basic resin Amberlite® CG4B In was found to be strongly adsorbed from HCl solutions in which distribution coefficients for this element of 11, 1.5×10^2, and 2.1×10^2 have been measured in the 3.0, 6.0, and 9 M acid, respectively.[15] This makes it possible to separate In from Al which is not retained from, e.g., 6 M HCl. Coadsorbed Ga (K_d = 39) is then separated from the In by elution with 4.5 M HCl.

From the information presented in Table 1 (in the chapter on Gallium) it is seen that the adsorption of In on strongly basic resins from HBr media resembles that observed in HCl solutions.

From dilute HBr and also HCl media containing high concentrations of organic solvents such as aliphatic alcohols, tetrahydrofuran (THF), acetone, and acetic acid, In is retained by strongly basic resins, e.g., Dowex® 1.[16,17] This adsorption of In is very high in some 90% organic solvent HBr media. Thus, in 0.6 M HBr solutions containing 90% of methanol, ethanol, propanol, and acetic acid distribution coefficients for In of $>10^3$, $>10^3$, $>10^2$, and $>10^3$, respectively, have been measured.[17] Under the same conditions Al shows no or only negligible adsorption so that effective separations of these two elements are possible in these media. In the methanol-HBr system the separation factor α for these two elements is $>10^3$, while in pure aqueous solution of comparable acidity this factor is only ~>7 (see Table 1 in the chapter on Gallium). In the same methanol-HBr medium, In is also readily separated from the alkaline earth elements, rare earths, Th, V(IV), Mn(II), Co, and Ni. Coadsorbed are Fe(III), Bi, Pb, Zn, Cd, and Cu.[17]

From Table 1 (in the chapter on Gallium) it is seen that In is strongly adsorbed from 0.1 to 4 M hydroiodic acid media. Similarly, In is very strongly retained by anionic resins, e.g., Dowex® 1, from thiocyanate media both in the absence or presence of organic solvents such

---

* Tables for this chapter appear at the end of the text.

as acetone or THF.[18] This fact has been utilized for the separation of In from rare earth elements which are not adsorbed from, e.g., 0.1 M ammonium thiocyanate (K_{dIn} = >10^5). The adsorbed In is eluted with 0.05 M ethylenediaminetetraacetic acid (EDTA). Coadsorbed with the In are Hg, Ag, Co, Zn, Cd, Fe(III), UO$_2$(II), and other elements. Very strong adsorption of In has also been reported to occur from hydrazoic acid media, e.g., from the 0.5 M acid (K_{dIn} = >10^3).[19]

While In is slightly retained on strongly and also weakly basic resins from very dilute sulfuric acid media (K_{dIn} = <10; see Table 95 in the chapter on Actinides, Volume II), no adsorption of this element occurs from hydrofluoric (HF),[20,21] nitric, and perchloric acid media. In the presence of high concentrations of organic solvents, e.g., acetone, THF, and acetic acid, weak adsorption was observed from dilute nitric acid solutions.

Strong adsorption of In on basic resis takes place from solutions containing organic complexing agents as, for example, from tartaric and citric acid solutions of pH 2.5 to 3 (see Table 3),[10] malonic and ascorbic acid solutions of pH 4.5,[22] 0.01 to 0.4 M oxalic acid,[23] and EDTA solutions (see Table 3).[11,12,24] From most of these media a large number of elements is coadsorbed with the In, as is seen from the information presented in this table from which it is also evident that many of these elements can be separated from the In by use of suitable eluents. In the oxalic acid media distribution coefficients for In of >10^3 have been measured.[23]

After adsorption from the systems discussed in the preceding paragraphs, In was eluted with the following eluents: water,[2] 0.1 M HCl,[4,5] 0.15 M HCl,[3] 0.25 M HNO$_3$,[10] 0.4 M HCl,[1] 1 M HNO$_3$,[10] 1 M HCl,[10-13] and 3 M HCl (see Table 4).[14] Elution of In which had been adsorbed as a complex carbonate[25-27] can be effected with 4 M ammonium carbonate (see Table 4).[26] The adsorption of In as anionic carbonate complex on a carbonate form resin allows this element to be separated from Zn plus Cd, and Al plus Ga, which are eluted as the cationic amine and anionic hydroxide complexes, respectively, when using 3 M ammonia solution as eluent.

Applications

In Tables 1 and 2 separation procedures based on anion exchange are presented, which have been used in connection with the determination of In in geological materials (Table 1) and in metals and metallurgical dusts (Table 2; see also Table 5 in the chapter on Gallium). To separate In from synthetic mixtures with other elements the methods outlined in Tables 3 and 4 have been employed.

CATION EXCHANGE RESINS

General

From the distribution data presented in Table 9 (see in the chapter on Gallium) it is seen that In(III) is only very weakly retained by strongly acid cation exchange resins from 0.1 to 0.5 M HCl solutions and that the adsorbability rapidly decreases with increasing acid concentration. Consequently, for separations of In from accompanying elements the adsorption of In is never performed from pure aqueous HCl media which, however, are very effective eluents for In that has been adsorbed on cationic resins from other systems (see paragraphs below). Thus, elution of In has been performed with 0.4 M HCl (see Tables 5 and 8),[28-30] 0.5 M HCl,[30] 1 M HCl (see Table 7),[31] and 2 M HCl (see Table 5).[32] With the 0.4 to 0.5 M eluent any coadsorbed Fe, Al, Ga, rare earths, alkaline earths, Cu, Mn, Co, Ni, etc. are not eluted together with the In. Other chloride eluents that can be used for the elution of In include 30% acetone - 0.5 M HCl (see Tables 6 and 7),[33,34] 40% acetone - 0.5 M HCl,[35] and 0.033 M ethylenediamine dihydrochloride (see Table 8).[36] In the acetone system[33,34] the distribution coefficient of In has a value of 3.6 and with the latter eluent In is eluted before coadsorbed Zn. Also, negligible to no adsorption of In is observed in 75%

acetone - 0.2 M HBr (see Table 6)[37] as well as in EDTA solutions of pH 0.8 (see Table 5),[38] pH 2 to 3,[39] and in such containing dilute sulfuric acid.[40] Adsorbed from the acetone-HBr system are Fe(III), Ga, Al, and other elements (see Table 6). A similar adsorption behavior of In was observed in organic solvent systems with 0.6 M overall concentrations of HCl and in 90% organic solvent media of varied HCl molarity.[41] Thus, in systems of 0.15 to 1.2 M overall acidity containing 90% of the aliphatic alcohols, acetone, THF, and other solvents as well as in 0.6 M HCl systems containing from 0 to 90% of these solvents no measurable adsorption of In was observed on Dowex® 50. However, strong adsorption of In on this resin occurs from 50% acetone - 0.1 M HCl in which distribution coefficients for Al, Ga, In, and Tl(III) of $>10^4$, $>10^4$, 240, and 0.5, respectively, have been measured.[42] This fact was utilized for their sequential separation using 50% acetone - 0.5 M HCl, 70% acetone - 2 M HCl, and 3 M HCl for the elution of In, Ga, and Al, respectively. In the 50% acetone system In has an adsorption value of 2.0 so that ready separation from Ga (K_d = 780) and Al (K_d = 910) is achieved.

From Table 9 (see chapter on Gallium) it is evident that In is appreciably adsorbed on strongly acidic cation exchange resins from HBr media both at low and high acid concentrations. Coadsorbed with In from the very dilute acid are Fe(III) and Ga (see Table 5),[28,29] as well as Al, alkali metals, alkaline earth elements, rare earths, and many other elements. Therefore, selective separations of In cannot be achieved in this type of HBr systems and also not in dilute nitric acid media, e.g., 0.1[9] and 0.2 M HNO$_3$ (see Table 7)[31] from which In is also very strongly retained by, e.g., Dowex® 50 (see Table 9 in the chapter on Gallium). Similarly, adsorption of In from 30% acetone - 0.2 M nitric acid[33,34] or 50% acetone - 0.2 M HBr (see Tables 6 and 7)[31,33] is accompanied by the coadsorption of many elements which include Al, the alkaline earth elements, and the rare earths. If, on the other hand, In is adsorbed from 8 to 9 M HBr most of the elements mentioned above, except Fe(III), Ga, Tl(III), and Au(III), are not coadsorbed.[43] Subsequent elution of In with 9 M HCl separates this element from the coadsorbed metals, which, unlike In, are strongly retained by Dowex® 50-X4 from this acid.

In is not only adsorbed from hydrobromic and nitric acid media containing acetone (see above), but also in the presence of other organic solvents such as aliphatic alcohols, THF, and acetic acid.[44,45] Strong adsorption of In is also observed in dilute hydrofluoric and perchloric acid media, as well as in solutions containing complexing ligands such as 4 and 5% sulfosalicylic acid at pH 7 (see Table 8)[30] and pH 9 to 10 (see Table 5).[30] In the presence of this organic acid, Al, Fe, Bi, Sn, Pb, Cd, Cu, Zn, and other elements form stable anionic sulfosalicylate complexes which prevent the adsorption of these elements on cationic resins. Coadsorbed with the In are Ga, Co, Ni, and Cr(III).

Applications

Cation exchange separations of In from accompanying elements have been employed in connection with the determination of this element in rain water and industrial products (see prodceures in Tables 5 and 6) and to separate In from synthetic mixtures with numerous other elements (see procedures in Tables 7 and 8).

CHELATING RESINS AND ION EXCHANGERS USED FOR MICROCHEMICAL DETECTION

Preconcentration of In on chelating resins has been used in connection with the determination of this element in seawater,[4] Zn concentrates, and Pb cake.[46] For this purpose the procedures described in Table 9 have been employed. Dowex® A1 has also been used in an analytical scheme for the determination of In and Tl in high-purity Zn and Zn-base alloys (see Table 8 in the chapter on Thallium).

In Table 10 a method is presented for the microchemical detection of In by a resin spot test.

Table 1
DETERMINATION OF In IN GEOLOGICAL MATERIALS AFTER ANION EXCHANGE SEPARATION[a]

Material	Ion exchange resin, separation conditions, and remarks	Ref.
Sea- and riverwater	Bio-Rad® AG1-X8 (100-200 mesh; chloride form) Column: 20 × 0.8 cm conditioned with 1 M HCl and operated at a flow rate of 1 mℓ/min a) 1 M HCl sample solution (5 mℓ) (adsorption of In; Fe[III] passes into the effluent) b) 1 M HCl (25 mℓ) (elution of Fe) c) 0.4 M HCl (5 mℓ) (as a rinse) d) 0.4 M HCl (25 mℓ) (elution of In) In is determined by aniodic stripping voltammetry; before the ion exchange separation, the acidified and filtered sample (5 ℓ) is mixed with FeCl$_3$ solution (100 mg Fe) and In is coprecipitated with ferric hydroxide; the precipitate is dissolved in HCl to prepare solution (a) The detection limit for In in a 5-ℓ sample of seawater is 2.5 ng	1
Chondritic meteorites	Dowex® 1-X8 (100-200 mesh; chloride form) From 6 M HCl, Fe(III) + In are adsorbed on a column of this resin; then, with the same acid, In is eluted and determined radiometrically (following precipitation as a sulfide for yield determination) Before the ion exchange separation step, the sample (0.5—1 g) is irradiated with neutrons and then decomposed by fusion with Na$_2$O$_2$; the melt is dissolved in water (in the presence of 20 mg of In-carrier) and InBr$_3$ is extracted into ether from which it is back extracted with 6 M HCl	6

[a] See also Table 3 in the Chapter on Zinc, Volume IV.

Table 2
DETERMINATION OF In IN INDUSTRIAL PRODUCTS AFTER ANION EXCHANGE SEPARATION[a]

Material	Ion exchange resin, separation conditions, and remarks	Ref.
Sn	Dowex® 1-X8 (50-100 mesh; chloride form) Column of 1 cm ID containing 3 g of the resin (to a height of 7 to 8 cm) conditioned with eluent (b) and operated at a flow rate of 5—10 mℓ/min a) 0.5 M HCl sample solution (10—25 mℓ aliquot containing <120 mg of total metals and <5 mg of In) containing 1% of NH$_2$OH·HCl (adsorption of Sn; In passes into the effluent) b) 0.5 M HCl - 1% NH$_2$OH·HCl (~50—60 mℓ) (elution of residual In) In is determined spectrophotometrically Before the separation, the sample is dissolved in 12 M HCl	47
	Dowex® 1-X8 Column: 20 cm × 0.75 cm^2 operated at a flow rate of 0.1 mℓ/sec a) 10 M HF sample solution (10 mℓ) (adsorption of Sn and Sb as anionic fluoride complexes) b) 5 M HF (30 mℓ) (elution of In and Mn) ^{116m}In and ^{56}Mn are determined radiometrically	20, 21

Table 2 (continued)
DETERMINATION OF In IN INDUSTRIAL PRODUCTS AFTER ANION EXCHANGE SEPARATION[a]

Material	Ion exchange resin, separation conditions, and remarks	Ref.
Irradiated Cd	Before the separation, the sample (0.5 g) is irradiated with neutrons and then dissolved in 10 M HF (10 mℓ) containing 15% H_2O_2 Dowex® 1-X8 (200-400 mesh; chloride form) Column: 5 × 0.8 cm operated at a flow rate of 0.6 mℓ/min a) 2 M HCl sample solution containing 50 mg of Cd (adsorption of In and Cd) b) 0.15 M HCl (elution of In appearing in the eluate fractions from 2—12 mℓ) c) 1 M HNO$_3$ (elution of Cd in the 4—14-mℓ fraction) The separation takes <30 min, producing ^{115}In of 99.99% purity	3
High-purity U	Dowex® 1-X8 (Cl$^-$ form) a) 1 M HCl sample solution (adsorption of In; U[VI] passes into the effluent) b) Water (elution of In) In is determined spectrographically; amounts of In in the range of 0.02—0.2 µg were determined	2
Pb dusts and concentrates	Anionite EDE-10P (bromide form) Column containing 50 g of the resin a) 0.5 M HBr sample solution (50 mℓ) (adsorption of Pb and Cd as anionic bromide complexes; In together with Fe, Cu, Ga, Al, and Zn passes into the effluent) b) 0.1 M HBr (elution of residual In and of the other nonadsorbed elements) In is determined polarographically	48
Flue dust (from the processing of Zn-ores)	Before the separation, the sample (1—3 g, containing 0.007—0.128% of In) is dissolved in 5 M HBr The sample solution is adjusted to be 2—3 M in HCl, then applied to a column of Wofatit® SBW (chloride form) on which Zn, Cd, Sn, Bi, and some Sb are adsorbed; the effluent and washings (2.5 M HCl) are concentrated to be ≃6 M HCl, then passed through a column of Wofatit® L 150 (chloride form) to adsorb In, Pb, Cd, and Sb As passes into the effluent and In is eluted with water	7

[a] See also Table 9 in the chapter on Cadmium, Volume IV; Table 29 in the chapter on Zinc, Volume IV; Tables 26 and 27 in the chapter on Copper, Volume III; Table 2 in the chapter on Tungsten, Volume IV; Table 3 in the chapter on Tin; Table 5 in the chapter on Gallium; and Tables 9 and 10 in the chapter on Gold, Volume III.

Table 3
ANION EXCHANGE SEPARATION OF In IN MEDIA CONTAINING ORGANIC COMPLEXING AGENTS

Elements separated	Ion exchange resin, separation conditions, and remarks	Ref.
In from binary mixtures with other elements	Dowex® 21K (50-100 mesh; tartrate form) Column: 20 × 1.4 cm conditioned with 5% tartrate solution (200 mℓ) at pH 2.5—3.0 followed by water; flow rate: 1 mℓ/min a) Sample solution of pH 2.5—3.0 containing tartaric acid (~1.5 g) (adsorption of the anionic tartrate complexes of In, Zr, Cu[II], Fe[III], Al, V[IV], Th, U[VI]; Zn, Cd, Mn[II], Co, Ni, Mg, Sr, Ba, and Tl[I]; the alkali metals pass into the effluent) b) Water (200 mℓ) (elution of Zn, Cd, Mn, Co, Mg, Ni, Sr, Ba, and Tl) c) 0.25 M NaCl (200 mℓ) (elution of Cu) d) 0.25 M NaNO$_3$ (200 mℓ) (elution of Fe, U, and Al) e) 1 M NaCl (200 mℓ) (elution of V and Th) f) 0.25 or 1 M HNO$_3$ (200 mℓ) (elution of In) g) 2 M HNO$_3$ (200 mℓ) (elution of Zr) Separation from coadsorbed vanadate, molybdate, and chromate is effected by elution of the In with eluent (f); subsequently, 10% Na$_2$CO$_3$ solution and 1 M KCl can be employed for the elution of vanadate + molybdate and chromate, respectively	10
	Dowex® 21K (50-100 mesh; citrate form) Column: 20 × 1.4 cm conditioned with 5% citrate solution (200 mℓ) at pH 2.5—3.0 followed by water; flow rate: 1 mℓ/min a) Sample solution of pH 2.5—3.0 containing citric acid (2 g) (adsorption of In, Ti, Zn, Co, Ni, Al, Fe, Bi, Cu, and Cd as anionic citrate complexes; into the effluent pass alkali metals and the alkaline earths) b) Water (25 mℓ) (as a rinse) c) 0.25 M NaCl (200 mℓ) (elution of Zn, Co, and Ni) d) 1 M NaCl (200 mℓ) (elution of Fe[III] and Bi) e) 0.25 M NH$_4$NO$_3$ (200 mℓ) (elution of Al) f) 1 M ammonium sulfate (200 mℓ) (elution of Cu and Cd) g) 1 M HCl or 1 M HNO$_3$ (200 mℓ) (elution of In) h) 3 M HCl (200 mℓ) (elution of Ti)	10
In from binary mixtures with Al, Cd, and Zn	Wofatit® L-150 (grain size: 0.2—0.315 mm; chloride form) Column: 12 × 0.7 cm conditioned with 0.1 M HNO$_3$ (20 mℓ) containing EDTA (3 mg) a) 0.1 M HNO$_3$ sample solution (20 mℓ) containing EDTA (3 mg) (adsorption of In; Cd and Al pass into the effluent as anionic EDTA complexes) b) 0.1 M HNO$_3$ (40 mℓ) containing EDTA (6 mg) (elution of residual Cd and Al) c) 1 M HCl (80 mℓ) (elution of In at a flow rate of 0.5 mℓ/min) Separation of In from Zn is effected employing the same procedure, except that 0.1 M HCl is used in place of the HNO$_3$	11, 12

Table 4
ANION EXCHANGE SEPARATION OF In IN HCl AND AMMONIACAL MEDIA

Elements separated	Ion exchange resin, separation conditions, and remarks	Ref.
In, Sn, Sb, and Te	Amberlite® IRA-400 (chloride form) Column: 8—11 × 0.2 cm operated at a flow rate of 0.04 mℓ/min After adsorption of the elements, the following eluents are used a) 3 M HCl (elution of In and Sb) b) 1 M HCl (elution of Te) c) 2 M HClO$_4$ (elution of Sn) With this method carrier-free ^{125}Sb and ^{113m}In can be obtained and the column can be used repeatedly for the preparation of radiochemically pure ^{113m}In In place of the Amberlite® resin, the anionite ASD-2 may be employed[49]	14
In, Sn, Fe, Al, and Mn	Wofatit® L 150 (grain size: 0.08—0.15 mm; chloride form) Column: 65 × 0.75 cm a) 5 M HCl sample solution (minimum volume) (adsorption of In, Sn, and Fe; into the effluent pass Al and Mn) b) 4 M HCl (until all Fe is eluted) c) 0.1 M HCl (120—140 mℓ) (elution of In and Sn at a flow rate of 30—40 mℓ/hr)	5
In from Cd, Zn, Pb, Ga, Al, and Fe	Anionite AV-17 (carbonate form) After adsorption of the elements on a column of this resin, the following eluents are used a) 3 M NH$_3$ solution (elution of Cd, Zn, Al, and Ga) b) 3 M ammonium carbonate (elution of In) c) 1 M NaOH (elution of Pb) d) 1 M HCl (elution of Fe) In is determined polarographically or spectrophotometrically The method can be used to separate milligram amounts of In from 100-fold amounts of the other elements and to determine In with an error within ±10%	26

Table 5
DETERMINATION OF In IN PRECIPITATION AND IN INDUSTRIAL PRODUCTS[a] AFTER CATION EXCHANGE SEPARATION

Material	Ion exchange resin, separation conditions, and remarks	Ref.
Rainwater	Dowex® 50W-X8 (100-200 mesh; H⁺ form) Column: 3 × 1 cm conditioned with 0.4 M HCl a) Aqueous stripping solution (7 mℓ) (adsorption of In, Ga, and Fe[III]) b) 0.4 M HCl (20 mℓ) (elution of In but not of Fe and Ga) After coprecipitation with ferric hydroxide ^{116m}In is determined radiometrically Before the ion exchange separation, the In is preconcentrated from the sample (1 ℓ) by coprecipitation with ferric hydroxide which is then irradiated with neutrons; subsequently, As is removed as sulfide and InBr₃ is extracted into isopropyl ether from which it is back extracted with water to obtain solution (a)	28, 29
Pb and Zn dusts	Cationite SBS in a column operated at a flow rate of 4 or 5 mℓ/min a) Sample solution (50 mℓ) containing 5—7 g of sulfosalicylic acid and adjusted with NH₃ solution to an orange-yellow color (adsorption of In; Zn and other metals pass into the effluent as anionic sulfosalicylate complexes) b) 5% Sulfosalicylic acid solution adjusted to pH 9—10 with NH₃ solution (elution of residual Zn and Pb) c) Water (removal of sulfosalicylic acid) d) 2 M HCl (50—100 ml) (elution of In) In is determined by a fluorescent method Before the separation, the sample (0.5 g) is dissolved in aqua regia and H₂SO₄ Pb is removed as PbSO₄	32
Pb and Cd metals	Dowex® 50-X8 (100-200 mesh; Na⁺ form) Column: 15 cm × 1 cm² conditioned with acetate buffer (0.4 M Na acetate-HNO₃ buffer of pH 0.8) a) Sample solution (50 mℓ) of pH 0.8 containing acetate buffer (30 mℓ) and 0.05 M EDTA (5—10 ml) (adsorption of Pb and Cd; into the effluent passes In as anionic EDTA complex) b) Acetate buffer (150 mℓ) (elution of remaining In) After wet ashing of EDTA with HNO₃-HClO₄, In is determined polarographically Before the separation, the sample (0.1—0.5 g) is dissolved in HClO₄ The same separation procedure can be used for the determination of Pb and Cd in metallic In; after elution of In with acetate buffer from a small column of the resin (5 cm × 1 cm²) the adsorbed Pb and Cd are eluted with 4 M HCl (20—50 ml)	38
Ge dioxide	Dowex® 50 (50-100 mesh; Na⁺ form) Column: 10 × 0.7 cm The sample solution (2—3 mℓ of 0.03 M H₂SO₄ mixed with 2 mℓ of 0.001 M EDTA; substoichiometric amount) is passed through the resin bed to adsorb uncomplexed In(III); into the effluent in which In is present as anionic EDTA complex ^{114m}In is determined radiometrically Before the substoichiometric separation the sample is irradiated with neutrons and then dissolved in 8 M NaOH; subsequently, In is preconcentrated by dithizone-CCl₄ extraction in the presence of In-carrier	40

[a] See also Table 52 in the chapter on Copper, Volume III and Table 7 in the chapter on Bismuth.

Table 6
DETERMINATION OF In IN INDUSTRIAL PRODUCTS AFTER CATION EXCHANGE SEPARATION IN ACETONE-MINERAL ACID MEDIA[a]

Material	Ion exchange resin, separation conditions, and remarks	Ref.
Sulfide concentrates	Bio-Rad® AG50W-X8 (200-400 mesh; H⁺ form) Column: 12 × 2.1 cm conditioned with 30% acetone - 0.2 M HNO$_3$ (50 mℓ) and operated at a flow rate of 2.5 ± 0.5 mℓ/min a) 0.2 M HNO$_3$ sample solution (50 mℓ) which is 30% in acetone (adsorption of In, Cd, and numerous other elements) b) 30% Acetone - 0.1 M HNO$_3$ (50 mℓ) (as a rinse) c) 50% Acetone - 0.2 M HBr (350 mℓ) (elution of Cd, Bi, noble metals, Mo, and W) d) 30% Acetone - 0.5 M HCl (300 mℓ) (elution of In and Sn) e) 0.1 M HNO$_3$ (50 mℓ) (as a rinse) f) 3 M HCl (300 mℓ) followed by 3 M HNO$_3$ (300 mℓ) (elution of Mg, Ca, Be, Ti, Mn, Fe, Al, Na, Cr, Ni, and Co) The recovery of In after this separation is 100%; In is determined by atomic absorption spectrophotometry Before the separation, the sample is decomposed with Br$_2$-HCl, HNO$_3$, and H$_2$SO$_4$ and In is preconcentrated by coprecipitation with ferric hydroxide	33
Ag targets	Bio-Rad® AG50W-X8 (200-400 mesh; H⁺ form) Column of 1.1 cm ID containing 9 mℓ of the resin conditioned with eluent (b) (~50 mℓ) and operated at a flow rate of 2.5 ± 0.5 mℓ/min a) 0.2 M HBr sample solution (25 mℓ) which is 75% in acetone (adsorption of Fe[III], Ga, Al, La, Be, Mg, Ca, Co, and Mn[II]; In passes into the effluent) b) 75% Acetone - 0.2 M HBr (100 mℓ) (elution of residual In) ¹¹¹In is determined radiometrically Before the separation, the cyclotron-irradiated sample (~5 g) is dissolved in conc HNO$_3$ (~20 mℓ) and In is preconcentrated by coprecipitation with ferric hydroxide (~200 mg Fe) which is dissolved in HBr to obtain solution (a)	37

[a] See also Table 16 in the chapter on Lead and Table 37 in the chapter on Zinc, Volume IV.

Table 7
CATION EXCHANGE SEPARATION OF In IN ACETONE-MINERAL ACID MEDIA

Elements separated	Ion exchange resin, separation conditions, and remarks	Ref.
In from binary mixtures with Zn, Pb, Fe(III), and numerous other elements	Bio-Rad® AG50W-X8 (200-400 mesh; H$^+$ form) Column: 19 × 2.1 cm containing the resin (60 mℓ) conditioned with eluent (b) (50 mℓ) and operated at a flow rate of 2.5 ± 0.5 mℓ/min a) 0.2 M HNO$_3$ sample solution which is 30% in acetone (adsorption of In and accompanying elements) b) 30% Acetone - 0.2 M HNO$_3$ (as a rinse) c) 30% Acetone - 0.5 M HCl (300 mℓ) (elution of In) d) 0.1 M HNO$_3$ (~50 mℓ) (removal of acetone) e) 1 M HCl (300 mℓ) (elution of Li and Na) f) 3 M HCl (300 mℓ) (elution of Fe[III], Zn, Ca, Al, U, Cu, Ni, Be, Mg, and Ti) g) 3 M HNO$_3$ (300 mℓ) (elution of Pb and Ga) Sn accompanies In into the eluate In case of the In-Li pair 350 mℓ of 0.35 M HCl containing 45% of acetone is used for the elution of In Cd, Bi, Au, Pt, Pd, Rh, Mo, and W can be eluted with 50% acetone - 0.2 M HBr before elution of the In[31] (see also below and in Table 6) The method described above has been applied to the separation of milligram amounts of the elements	34
In from binary mixtures with Cd, Bi, Au(III), Pt(IV), Pd(II), Rh(III), Mo(VI), and W(VI)	Bio-Rad® AG50W-X8 (200-400 mesh; H$^+$ form) Column: 7.8 × 1.5 cm operated at a flow rate of 2.0 ± 0.3 mℓ/min a) 0.1 M HNO$_3$ sample solution (50 mℓ) (adsorption of In, Cd, and Bi) b) 0.1 M HNO$_3$ sample solution (50 mℓ) which contains 0.03% of H$_2$O$_2$ (adsorption of In; Mo and W pass into the effluent as anionic peroxy complexes) c) 0.2 M HNO$_3$ sample solution (100 mℓ) (adsorption of In and Bi) d) 50% Acetone - 0.2 M HBr (50 mℓ) (adsorption of In; Au, Pt, Pd, and Rh pass into the effluent) e) Same eluent as (a) (200 mℓ) (elution of Cd, Bi, and other elements) f) 1 M HCl (150 mℓ) (elution of In) The method is suitable for the separation of milligram amounts of the elements	31

Table 8
CATION EXCHANGE SEPARATION OF In IN MEDIA CONTAINING SULFOSALICYLIC ACID AND ETHYLENEDIAMINE

Elements separated	Ion exchange resin, separation conditions, and remarks	Ref.
In from Cu, Zn, Al, Fe, and other elements	Cationite KU-2 (NH_4^+ form) Column: 50 × 1.5 cm a) Test solution (100 mℓ) containing the sulfosalicylate complexes of the elements (adsorption of In, Ga, Co, Ni, and Cr[III]; into the effluent pass Cu, Zn, Al, Bi, Pb, Mo, Cd, Fe, Sn, As, and Sb) b) 4% Sulfosalicylic acid solution neutralized with NH_3 (elution of residual amounts of the elements forming sulfosalicylate complexes) c) Water (removal of sulfosalicylate) d) 0.4—0.5 M HCl (elution of In, Cr, Co, and Ni) The method is applicable to the determination of In in phosphate concentrates	30
In from Zn	Dowex® 50 (grain size: 0.05—0.1 mm) Column: 15 × 1.1 cm conditioned with 0.1 M ethylenediamine dihydrochloride to obtain the ethylenediammonium form of the resin a) Neutral sample solution (containing ≯10 mg of In and ≯13 mg of Zn) b) Water (as a rinse) c) 0.033 M ethylenediamine dihydrochloride (sequential elution of In and Zn at a flow rate of 2 mℓ/min; In appears in the first 100 mℓ of eluate and Zn in the next 120—125 mℓ)	36

Table 9
DETERMINATION OF In IN SEAWATER AND INDUSTRIAL PRODUCTS AFTER SEPARATION BY CHELATING RESINS[a]

Material	Ion exchange resin, separation conditions, and remarks	Ref.
Seawater	A) First column operation (Dowex® A1 chelating resin, 50-100 mesh, Na⁺ form)	4
	Column: 17 × 1.2 cm operated at a flow rate of 4—5 mℓ/min	
	a) Sample ($\not>$20 ℓ) adjusted to pH 9.2 with 1 M NH$_3$ solution (adsorption of In and other cationic constituents)	
	b) Water (100 mℓ) (as a rinse)	
	c) 3 M HCl (150 mℓ) (elution of In and other adsorbed elements rejecting the first 50 mℓ of eluate)	
	B) Second column operation (Dowex® AG 2-X8, 200-400 mesh; Cl⁻ form)	
	Column: 30 × 0.4 cm conditioned with 3 M HCl (50 mℓ)	
	a) Eluate (c) (100 mℓ) (obtained by [A]) (adsorption of In; into the effluent pass Na and other elements)	
	b) 3 M HCl (50 mℓ) (elution of remaining Na and other elements)	
	c) 0.1 M HCl (120 mℓ) (elution of In)	
	After evaporation of eluate (c) In is determined by neutron activation analysis (minor elements are removed by a series of postirradiation solvent extraction steps)	
	Down to 0.1 ng of In per liter of water was determined; the sensitivity was 6 pg/ℓ, and the coefficient of variation for replicate determination was ±5%	
	Before the first ion exchange separation step inshore waters are filtered through a 0.45-μm membrane filter	
Zn concentrates and Pb cake	Phosphonate resin SF-5 (grain size: 0.315—0.63 mm)	46
	The 0.1—0.2 M H$_2$SO$_4$ sample solution (50—60 mℓ) is treated with ascorbic acid (0.2 g) to reduce Fe(III) and agitated with the resin (3 mℓ) for 30 min on a vibrator to adsorb In; after filtration, the resin is washed with 0.1 M H$_2$SO$_4$ ($\not<$40 mℓ) and In is eluted with 2—3 M HCl (50 mℓ)	
	In is determined polarographically	
	Before the ion exchange separation, the sample (0.5—2 g; containing $\not>$2 mg of In) is dissolved in HCl-HNO$_3$-H$_2$SO$_4$ and Pb is removed by precipitation as PbSO$_4$	

[a] See also Table 78 in the chapter on Copper, Volume III and Table 8 in the chapter on Thallium.

Table 10
MICROCHEMICAL DETECTION OF In BY RESIN SPOT TEST[a]

Ion exchange resin	Experimental conditions and remarks	Ref.
Dowex® 1-X8 (50-100 mesh; acetate form)	One drop of the sample solution is mixed with 1 drop each of saturated ammonium acetate solution, 1 M ammonia solution, and 0.01% alizarin red S solution, and to the mixture are added 3—5 beads of the resin; the color becomes an intense purple with 10 hr	50
	The limit of detection is 0.03 μg of In; an increase of sensitivity is attained by the use of resin beads impregnated with the reagent solution; interference is caused by Zr, U, As, Mo, W, Al, Sn, and Ge; Al can be masked with fluoride	

[a] See also Table 15 in the chapter on Gallium.

REFERENCES

1. **Florence, T. M., Batley, G. E., and Farrar, Y.,** Determination of indium by anodic-stripping voltammetry: application to natural waters, *J. Electroanal. Chem.*, 56(2), 301, 1974.
2. **Schoenfeld, I.,** Determination of microgram quantities of indium in high-purity uranium, *Isr. J. Chem.*, 1(3), 136, 1963.
3. **Törkö, J.,** Separation of carrier-free indium-115m from cadmium-115m by ion-exchange chromatography, *Magy. Kem. Foly.*, 72(1), 17, 1966.
4. **Matthews, A. D. and Riley, J. P.,** Determination of indium in seawater, *Anal. Chim. Acta*, 51, 287, 1970.
5. **Jentzsch, D., Frotscher, I., Schwerdtfeger, G., and Sarfert, G.,** Quantitative analysis of indium, *Fresenius' Z. Anal. Chem.*, 144, 8, 1955.
6. **Schindewolf, U. and Wahlgren, M.,** The rhodium, silver and indium content of some chondritic meteorites, *Geochim. Cosmochim. Acta*, 18, 36, 1960.
7. **Scheffler, E. and Ziegenbalg, S.,** Use of ion-exchangers for the separation of indium from industrial solutions, *Freib. Forschungsh. B*, No. 83, 111, 1966.
8. **Sunderman, D. N., Ackermann, I. B., and Meinke, W. W.,** Radiochemical separations of indium, *Anal. Chem.*, 31, 40, 1959.
9. **Sunderman, D. N.,** The development and evaluation of radiochemical separation procedures for barium, calcium, strontium, silver, and indium, *Diss. Abstr.*, 17(6), 1207, 1957.
10. **Sitaram, R. and Khopkar, S. M.,** Anion-exchange behavior of indium(III) in citrate and tartrate solutions: separation from mixtures, *Anal. Chim. Acta*, 71, 472, 1974.
11. **Kocheva, L. L. and Koleva, E.,** Ion-exchange separation of indium from accompanying elements. Separation of small amounts of indium from cadmium and aluminum, *God. Sof. Univ.*, 60, 163, 1965/1966.
12. **Kocheva, L. L. and Tomova, T.,** Ion-exchange separation of indium from accompanying elements. Separation of indium from zinc, *Ann. Univ. Sofia Cl.D'Ochrida Fac. Chim.*, 57, 97, 1962/1963.
13. **Atrashkevich, V. V., Sidoruk, E. I., and Bilimovich, G. N.,** Determination of indium in indium phosphide by isotopic dilution, *Zavod. Lab.*, 44(8), 977, 1978.
14. **Stroński, I.,** Investigation of the anion-exchange method for the separation of indium, tin, antimony, and tellurium with radioactive indicators, *Rocz. Chem.*, 34(2), 709, 1960.
15. **Kuroda, R., Ishida, K., and Kiriyama, T.,** Adsorption behavior of a number metals in hydrochloric acid on a weakly basic anion exchange resin, *Anal. Chem.*, 40, 1502, 1968.
16. **Korkisch, J. and Hazan, I.,** Anion exchange behavior of uranium, thorium, the rare earths and various other elements in hydrochloric acid-organic solvent media, *Talanta*, 11, 1157, 1964.
17. **Klakl, E. and Korkisch, J.,** Anion-exchange behavior of several elements in hydrobromic acid-organic solvent media, *Talanta*, 16, 1177, 1969.
18. **Singh, D. and Tandon, S. N.,** Anion-exchange studies of metal thiocyanates in aqueous and mixed solvent systems, *Talanta*, 26, 163, 1979.
19. **Oguma, K., Maruyama, T., and Kuroda, R.,** Anion-exchange behavior of various metals in hydrazoic acid media, *Anal. Chim. Acta*, 74, 339, 1975.
20. **Maenhaut, W., Adams, F., and Hoste, J.,** Determination of trace impurities in tin by neutron-activation analysis. II. Determination of indium and manganese, *J. Radioanal. Chem.*, 9(1), 27, 1971.
21. **Maenhaut, W., Adams, F., and Hoste, J.,** Neutron-activation analysis of high-purity tin: chemical separations and nuclear interferences, *Anal. Chim. Acta*, 59, 209, 1972.
22. **Sitaram, R. and Khopkar, S. M.,** Anion-exchange studies of indium(III) in malonate and ascorbate solutions: separation from mixtures, *Chromatographia*, 6(4), 198, 1973.
23. **DeCorte, F., Van den Winkel, P., Speecke, A., and Hoste, J.,** Distribution coefficients for twelve elements in oxalic acid medium on a strong anion-exchange resin, *Anal. Chim. Acta*, 42, 67, 1968.
24. **Lebecka, J.,** Determination of indium content of saline water in hydrogeological tracer investigations by activation with a californium-252 neutron source, Report Inst. Fiz. Jad., INT-94/1, Chief Institute of Mining, Krakow, Poland, 1976, 55.
25. **Alimarin, I. P., Tsintsevich, E. P., and Burlaka, V. P.,** Study of the behavior of complex compounds of indium, zinc and cadmium in ammonium carbonate solution on ion-exchange resins, *Zavod. Lab.*, 25(11), 1287, 1959.
26. **Eristavi, D. I., Eristavi, V. D., and Kutateladze, G. Sh.,** Separation of indium from cadmium, zinc, lead, gallium, aluminum, and iron by means of AV-17 anion-exchange resin in its carbonate form, *Zh. Anal. Khim.*, 26(11), 2234, 1971.
27. **Mgaloblishvili, M. G., Kutateladze, G. Sh., and Eristavi, V. D.,** Separation of indium, gallium and thallium(I) on the carbonate form of anionite AV-17, *Tr. Gruz. Politekh. Inst.*, 4(177), 51, 1975.
28. **Bhatki, K. S. and Dingle, A. N.,** Tracer indium determination in rain samples by neutron activation and radiochemical analysis, *Radiochem. Radioanal. Lett.*, 3(1), 71, 1970.

29. **Bhatki, K. S. and Dingle, A. N.**, The measurement of tracer indium in rain samples, *J. Appl. Meteorol.*, 9(2), 276, 1970.
30. **Verdizade, A. A. and Mekhtiev, M. M.**, Chromatographic separation of indium from elements that form periodates of low solubility, and its subsequent determination, *Uch. Zap. Azerb. Gos. Univ. Ser. Khim. Nauk*, No. 3, 42, 1970.
31. **Strelow, F. W. E., Weinert, C. H. S. W., and Boshoff, M. D.**, Quantitative separation of indium from cadmium and other elements by cation-exchange chromatography in HBr-acetone, *J. S. Afr. Chem. Inst.*, 26(3), 118, 1973.
32. **Ginzburg, L. B. and Shkrobot, E. P.**, Use of ion-exchange methods for determining thallium and indium in products from the treatment of non-ferrous metal ores, *Zavod. Lab.*, 21(11), 1289, 1955.
33. **Jones, E. A. and Lee, A. F.**, Determination of thallium and indium in sulfide concentrates, Report NIM-2022, National Institute of Metallurgy, Randburg, South Africa, 1979.
34. **Strelow, F. W. E., Weinert, C. H. S. W., and Van der Walt, T. N.**, Selective separation of indium from zinc, lead, gallium, and many other elements by cation-exchange chromatography in hydrochloric acid-acetone medium, *Talanta*, 21, 1183, 1974.
35. **Fritz, J. and Rettig, T. A.**, Separation of metals by cation exchange in acetone-water-hydrochloric acid, *Anal. Chem.*, 34, 1562, 1962.
36. **Mosheva, P., Topalova, E., Zagorchev, B., and Kobarelova, S.**, Separation of indium and zinc by ion-exchange, *C.R. Acad. Bulg. Sci.*, 16(1), 73, 1963.
37. **Strelow, F. W. E. and Van der Walt, T. N.**, Separation of indium from iron(III) and other elements by cation exchange chromatography in hydrobromic acid-acetone; application to separation of indium-111 from cyclotron-irradiated silver targets, *S. Afr. J. Chem.*, 32(1), 13, 1979.
38. **Doležal, J., Povondra, P., Štulik, K., and Šulcek, Z.**, Rapid analytical method for investigation of metals and inorganic raw materials. XV. Determination of small amounts of indium in lead and cadmium metals and of impurities in metallic indium by ion-exchange, *Collect. Czech. Chem. Commun.*, 29(7), 1538, 1964.
39. **Ružička, J. and Starý, J.**, Isotopic-dilution analysis by ion-exchange. II. Sub-stoicheiometric determination of traces of indium, *Talanta*, 11, 691, 1964.
40. **Zeman, A., Starý, J., and Ružička, J.**, New principle of activation-analysis separations. V. Substoicheiometric determination of traces of indium, *Talanta*, 10, 981, 1963.
41. **Korkisch, J. and Ahluwalia, S. S.**, Cation-exchange behavior of several elements in hydrochloric acid-organic solvent media, *Talanta*, 14, 155, 1967.
42. **Strelow, F. W. E.**, Partly non-aqueous media for accurate chemical analysis by ion-exchange, *Ion Exch. Membr.*, 2(1), 37, 1974.
43. **Nelson, F. and Michelson, D. C.**, Ion-exchange procedures. IX. Cation-exchange in HBr-solutions, *J. Chromatogr.*, 25, 414, 1966.
44. **Korkisch, J. and Klakl, E.**, Cation-exchange behavior of several elements in hydrobromic acid-organic solvent media, *Talanta*, 16, 377, 1969.
45. **Korkisch, J., Feik, F., and Ahluwalia, S. S.**, Cation-exchange behavior of several elements in nitric acid-organic solvent media, *Talanta*, 14, 1069, 1967.
46. **Pavlov, Yu. I., Komarova, L. A., Zdanovich, I. D., and Pozhiganova, G. V.**, Adsorption-polarographic determination of indium, *Zavod. Lab.*, 39(11), 1317, 1973.
47. **Raby, B. A. and Banks, C. V.**, Direct spectrophotometric determination of indium in tin, *Anal. Chim. Acta*, 29, 532, 1963.
48. **Bausova, N. V.**, Separation of indium from lead and cadmium by ion-exchange chromatography, *Tr. Inst. Metall. Ural. Fil. Akad. Nauk SSSR*, No. 8, 107, 1963.
49. **Stronski, I. and Rybakow, W. N.**, Anion exchange of radio-isotopes of indium, tin and antimony and the preparation of carrier-free indium-113m and antimony-125, *Chem. Anal. (Warsaw)*, 4(5—6), 877, 1959.
50. **Katou, K. and Kakihana, H.**, Detection of a micro amount of indium with an anion exchanger and alizarin red S, *J. Chem. Soc. Jpn. Pure Chem. Sect.*, 79(6), 762, 1958.

THALLIUM

Highly selective separations of Tl from accompanying elements are achieved by adsorption of the trivalent element on anion exchange resins from hydrochloric (HCl) or hydrobromic (HBr) acid media. By means of this technique Tl can be isolated quantitatively from geological, biological, and industrial materials. Suitable separations of Tl from matrices of this type are also obtained by the use of cation exchangers or chelating resins.

ANION EXCHANGE RESINS

General

In the range of HCl concentrations of 0.1 to 4 and 4 to 12 M the predominating Tl(III) species are $TlCl_4^-$ (tetrachlorothallate) and $TlCl_5^{2-}$ (pentachlorothallate), respectively. The species retained by strongly basic resins, e.g., Dowex® 1, is $TlCl_4^-$ and, as is evident from the distribution data shown in Table 1 (in the chapter on Gallium), this adsorption is very high at all HCl concentrations. Under the same conditions Tl(I) is only slightly or negligibly adsorbed, so that essentially all anion exchange separations involving Tl are based on the adsorption of Tl(III) from media such as 0.1 M HCl (see Tables 1 and 2*),[1,2] 0.5 M HCl (see Table 1),[3] 1 M HCl (see Table 1),[4] 1.5 M HCl (see Table 3),[5] 2 M HCl (see Table 4),[6] 3 to 6 M HCl (see Table 3),[7] and 12 M HCl (see Table 1).[8] To ensure that Tl is completely present in the adsorbable trivalent oxidation state, the sorption solutions listed above have to contain a holding oxidant as, for example, Br (added a saturated Br water)[1-5,9] or Cl (added until saturation of the solution).[7,8] The addition of a large excess of Br should be avoided, because elemental Br is adsorbed on anion exchange resins (see in the chapter on Bromine) and may, if present in a very high concentration, occupy a large zone in the resin bed and thus prevent the complete retention of Tl by the anion exchanger. Furthermore, before and during the ion exchange separation light and dust must be excluded, e.g., by using amber bottles and/or by wrapping the ion exchange column etc. into Al foil. If this is not done, continual addition of Br, over the time interval needed for the separation, is required. It is also recommended to add the holding oxidant to all wash solutions used after adsorption of Tl to prevent any reduction of this element to the monovalent state in the resin bed (for examples see Tables 1 and 2).[2,4,9] Coadsorbed with Tl(III) from very dilute HCl media (0.1 to 2 M) is only a very limited number of metals which includes Zn, Cd, Hg, Pb, Bi, Au, Ag, and Pt metals so that highly selective separations of Tl can be achieved. Since the alkaline earth elements, alkali metals, rare earths, Al, Fe, Mn, Ti, and numerous other elements are not adsorbed together with the Tl, this separation principle is very well suited for the isolation of Tl from complex matrices such as geological and biological materials (see Tables 1 to 3).

From Table 1 (see in chapter on Gallium) it is seen that Tl(III) is also very strongly adsorbed from HBr media so that its adsorption (most probably as $TlBr_4^-$), e.g., from 0.15 M HBr, can also be used for highly selective separations of the type indicated in Table 2.[9] Coadsorbed with the Tl from HBr media are essentially the same elements as are retained from HCl systems of comparable acid concentrations. Strong adsorption of Tl(III) on basic resins is also observed in hydroiodic acid solutions at both very low and very high acid concentrations (see Table 1 in chapter on Gallium). The minimum in the adsorption at 3 to 6 M HI is probably caused by reduction of Tl(III) to nonadsorbed Tl(I) through the action of iodide.

After adsorption of Tl(III) from the dilute HCl or HBr media mentioned above, some

---

* Tables for this chapter appear at the end of the text.

coadsorbed elements can be removed using the following eluents: water,[5] 0.5 M HCl,[1] 0.5,[1] 1,[3] and 2 M HNO$_3$,[1,2] and 8[5] and 12 M HCl.[8] With water and the nitric acid eluents Zn, Cd, Pb, and Ag are eluted, while desorption of Pb and Ag (but not of Cd and Zn) is effected with the 8 and 12 M HCl solutions. In the 0.5 and 2 M nitric acid eluents the concentration of nitric acid is not sufficient to decompose the anionic chloro or bromo Tl(III) complex, so that this element is not eluted under these conditions, although the adsorption of Tl(III) nitrate on a resin in the nitrate form is very low ($K_d \simeq 1$ to 10) in these media (much higher adsorption of Tl[III] has been observed in 10 M nitric acid; $K_d \simeq 10^2$). Elution of Tl can, however, be effected in the presence of H$_2$O$_2$ which destroys the anionic chloro complex of Tl, so that this element is eluted (as trivalent Tl) with, e.g., 0.25 M nitric acid (see Table 1).[4]

For the elution of Tl(III) that has been adsorbed on anion exchange resins from the systems discussed above, usually aqueous solutions containing SO$_2$ are employed as the eluents (see Tables 1 to 3).[1-3,5,7-9] In these solutions the Tl is reduced on the resin to the nonadsorbed monovalent state so that complete and rapid elutions can be achieved.* Reductive elution of Tl(III) can also be effected in the presence of thiourea, e.g., by using 0.5 M HNO$_3$-0.1 M thiourea as the eluent.[10] In this medium Tl(I) forms a stable cationic complex with thiourea.

Negligible to no adsorption of mono- and trivalent Tl on strongly basic resins is observed in media containing thiocyanate, hydrofluoric, sulfuric, and perchloric acids. Tl(I) is also not retained on a resin in the carbonate form (see Table 4).[11] The nonadsorbability of Tl(I) from 1 and 2 M HCl has been used to separate this element from ^{212}Bi [12] and Zn, Cd, Hg, Pb, and Sb (see Table 4),[6] respectively.

Applications

In Tables 1 to 3 separation methods based on anion exchange are described which have been used in connection with the determination of Tl in rocks, minerals, sediments, meteorites, soils, and natural waters, as well as in biological and industrial materials. To separate Tl from synthetic mixtures with other elements the procedures outlined in Table 4 may be used.

CATION EXCHANGE RESINS

General

From the distribution coefficients presented in Table 9 (see in the chapter on Gallium) it is seen that Tl(III) is not retained on strongly acid resins to any appreciable extent from HCl media, while, however, its adsorption increases considerably with increasing acid concentrations in the range from 1 to 8 M HBr. From the same table it is evident that the adsorption of Tl(III) in nitric acid media is high at low concentrations of this acid and then decreases with increasing acid molarity. A similar decrease of adsorption is also shown by Tl(I) in HCl and nitric acid solutions, and it is very probable that this element shows similar adsorption characteristics in HBr systems of comparable acid concentrations. This is because Tl(I) is not known to form anionic complexes with any of these acids (and also not with organic complexing agents; see later), so that its adsorption behavior resembles that of other simple monovalent ions such as those of the alkali metals, as, for instance, Cs (see in the chapter on Rubidium, Cesium, and Francium).

When utilizing the nonadsorbability of Tl(III) from dilute HCl media, e.g., 0.1 M (see Table 5 and also Table 16 in the chapter on Lead),[13,14] it is necessary to carry out the separations in the presence of a holding oxidant, e.g., chlorine,[13,14] to prevent reduction of the Tl to the adsorbable monovalent state. Coeluted with the Tl are all anionic constituents of the solution, while essentially all cations are retained by the cation exchange resin. If,

* See also Table 14 in the chapter on Sulfur.

on the other hand, Tl is adsorbed as the monovalent cation, the separations are best performed in the presence of complexing agents to prevent the coadsorption of as many elements as possible. For this purpose Tl(I) has been adsorbed from the following media: 0.05 M HCl - 5% citric acid (see Table 5),[15] weakly acid tartaric or citric acid solutions containing Na pyrophosphate (see Table 6),[16,17] alkaline tartrate solution (see Table 6),[16,17] 0.2 M ethylenediaminetetraacetic acid (EDTA) solution (see Table 6),[18] EDTA solution of pH 4 (see Table 7),[19] and 0.5 to 1 M NaOH - 1 M glycerol (see Table 7).[20] These complexing agents form anionic complexes with Fe(III), Al, Sb(V), Hg, Cu, Zn, Cd, Pb, Bi, etc. so that rather selective separations are achieved which can be improved further by elution of coadsorbed elements using the eluents mentioned in Table 7.

For the elution of adsorbed Tl(I), solutions of mineral acids as, for example, 6[15-17] and 2 M HCl,[18,19] and 7% HNO$_3$ solution[21] have been employed.

Tl(III) is also not retained by strongly acid resins from dilute HCl or HBr solutions containing up to 95% of acetone[22] or ethanol.[23]

Strong adsorption on cationic resins, e.g., Dowex® 50W-X8, has been reported for Tl(I) from hydrofluoric acid (HF) solutions, with distribution coefficients decreasing from a value of ~10^3 in the <1 M acid to a value of ~20 in 20 M HF.[24] Appreciable adsorption of Tl(I and III) also occurs from sulfuric acid media. Thus, in the 0.1 N acid distribution coefficients for Tl(I) and Tl(III) of 452 and 6500, respectively, have been measured.

Applications

Cation exchange separation procedures have been used in connection with the determination of Tl in rocks, minerals, and soils (see Table 5), and in industrial products (see Table 6). To separate Tl from synthetic mixtures with other elements the methods outlines in Table 7 were employed.

CHELATING RESINS AND ION EXCHANGERS USED FOR MICROCHEMICAL DETECTION

An example for the application of chelating resins to the analysis of Zn matrices for Tl and In is presented in Table 8.

In Table 9 a procedure is outlined which can be used for the detection of traces of monovalent Tl by a resin spot test.

Table 1
DETERMINATION OF Tl IN GEOLOGICAL MATERIALS AFTER ANION EXCHANGE SEPARATION[a]

Material	Ion exchange resin, separation conditions, and remarks	Ref.
Rocks and sediments	Deacidite® FF (50-100 mesh; chloride form) Column: 7.5 × 0.6 cm operated at a flow rate of ≯3 mℓ/min a) 0.1 M HCl sample solution (500 mℓ, containing 7.7 mℓ of 6.5 M HCl to which saturated Br$_2$-water [5 mℓ] was added) (adsorption of Tl[III] as anionic chloro complex; matrix constituents pass into the effluent) b) Water (20 mℓ) (as a rinse) c) 0.5 M HNO$_3$ (350 mℓ) followed by 0.5 M HCl (250 mℓ) (elution of coadsorbed elements) d) Water (25 mℓ) (as a rinse) e) Saturated aqueous solution of SO$_2$ (35 mℓ) (reductive elution of Tl) Tl is determined fluorimetrically with rhodamine B Before the separation, the sample (0.05—1.5 µg of Tl) is decomposed with HF and HNO$_3$ and fluoride and nitrate are removed sequentially by repeated evaporation with HNO$_3$ and HCl, respectively The same ion exchange procedure can be used to isolate Tl from seawater (10 or 20 ℓ) filtered through 0.5-µm membrane filter and made 0.1 M with respect to HCl and containing Br$_2$ equivalent to 10 ppm; then the sample is passed through a 7.5 cm × 0.3 cm^2 column of the resin to adsorb Tl(III); in place of eluent (c), 2 M HNO$_3$ (55 mℓ) is used to remove coadsorbed elements In the eluate Tl is determined by neutron activation analysis	1
Rocks and minerals	Bio-Rad® AG1-X8 (100-200 mesh; chloride form) Column of 0.75 cm ID containing the resin to a height of 7 cm and operated at a flow rate of 1 mℓ/min a) 0.5 M HCl sample solution (100 mℓ) containing saturated Br$_2$-water (1 drop) (adsorption of Tl[III] as anionic chloro complex; matrix elements pass into the effluent) b) Water (20 mℓ) (as a rinse) c) 1 M HNO$_3$ (50 mℓ) (elution of coadsorbed elements) d) Water (20 mℓ) (removal of HNO$_3$) e) 1:1 Mixture (50 mℓ) of a saturated solution of SO$_2$ and water (reductive elution of Tl) Tl is determined by differential pulse anodic stripping voltammetry The procedure shows a coefficient of variation of ≃3% (for 700 ng/g of Tl) to ≃22% (for 50 ng/g of Tl); before the separation, the sample (containing ≮50—100 ng of Tl) is decomposed with HF and HClO$_4$	3
Silicate rocks	Dowex® 1-X8 (100-200 mesh; chloride form) Column containing the resin conditioned with 1 M HCl treated with Br$_2$ and operated at a flow rate of 50 mℓ/hr a) 1 M HCl sample solution (∼200—250 mℓ) containing Br$_2$-water (adsorption of Tl[III]; matrix elements pass into the effluent) b) 1 M HCl containing Br$_2$ (as a rinse and removal of residual nonadsorbed elements) c) 0.25 M HNO$_3$-1% H$_2$O$_2$ (600 mℓ) (elution of Tl) Tl is determined spectrographically; the detection limit corresponds to ≃3 parts of Tl per 10^9 in a 10-g sample Before the separation, the sample (1—10 g) is decomposed with HClO$_4$-HF (1:3)	4

Table 1 (continued)
DETERMINATION OF Tl IN GEOLOGICAL MATERIALS AFTER ANION EXCHANGE SEPARATION[a]

Material	Ion exchange resin, separation conditions, and remarks	Ref.
Meteorites and lunar soil	Dowex® 1-X8 (chloride form) Column: 8 × 0.4 cm a) 12 M HCl sample solution saturated with Cl_2 (adsorptin of Tl[III] and Pb) b) 12 M HCl (16 drops) saturated with SO_2 (as a rinse) c) 12 M HCl + SO_2 (75 drops) (elution of Pb) d) 0.1 M HCl + SO_2 (after the passage of 12 drops, Tl is collected in the next 35 drops) Tl and Pb are determined mass spectrometrically (stable isotope dilution analysis) Before the separation Tl, Pb, and other elements are extracted from the sample by heating in vacuo and then treated with $HClO_4$ and HNO_3; purification of Pb can also be effected on a column (4 × 0.4 cm) of the same resin using the adsorption of Pb from 1.8 M HCl; after rinsing with 1.8 M HCl (2.5 mℓ) Pb of suitable purity is collected by elution with water (2.5 mℓ)	8

[a] See also Table 1 in the chapter on Cadmium, Volume IV; Table 1 in the chapter on Molybdenum, Volume IV; and Table 86 in the chapter on Rare Earth Elements, Volume I.

Table 2
DETERMINATION OF Tl IN NATURAL WATERS AFTER ANION EXCHANGE SEPARATION

Material	Ion exchange resin, separation conditions, and remarks	Ref.
Mineral, hydrothermal, and oil field waters	Bio-Rad® AG1-X8 (100-200 mesh; chloride form) Column of 0.8 cm ID containing 4 g of the resin conditioned with eluent (b) (100 mℓ) and operated in the dark at a flow rate of 0.7 mℓ/min a) Water sample (adsorption of Tl[III] as anionic bromide complex, i.e., TlBr$_4^{3-}$; into the effluent pass all major and minor constituents of the water matrix) b) 0.15 M HBr - 1% saturated aqueous Br$_2$ (20 mℓ) (elution of residual nonadsorbed elements) c) 1:1 Mixture (100 mℓ) of a 5—6% solution of SO$_2$ and water (reductive elution of Tl) Tl is determined by atomic absorption spectrophotometry Before the ion exchange separation, the sample (1 ℓ) plus conc HBr (20 mℓ) is boiled to expel CO$_2$, cooled, and filtered; then 10 mℓ of saturated aqueous Br$_2$ is added to the filtrate, thus obtaining solution (a) The method can also be applied to the analysis for Tl of other waters such as tap and seawater	9
Sea- and riverwater	Bio-Rad® AG1-X8 (100-200 mesh; chloride form) Column: 8 × 0.5 cm operated in the dark at a flow rate of 1 mℓ/min a) Water sample (adsorption of Tl[III] as anionic chloride complex, i.e., TlCl$_4^{3-}$; into the effluent pass all major and minor constitutents of the water) b) Distilled water (25 mℓ) containing saturated aqueous Br$_2$ (0.05 mℓ) c) 2 M HNO$_3$ (100 mℓ) containing saturated aqueous Br$_2$ (0.2 mℓ) (elution of coadsorbed elements such as Zn and Cd) d) Distilled water (25 mℓ) (removal of HNO$_3$) e) 5—6% Sulfurous acid solution (35 mℓ) (reductive elution of Tl) Tl is determined by aniodic stripping voltammetry Before the separation, the sample (4 ℓ) is filtered through a 0.45-μm membrane filter and the filtrate is mixed with 10 M HCl (40 mℓ) and saturated aqueous Br$_2$ (4 mℓ) to prepare solution (a) With this method levels of ≃3 ng/ℓ of Tl may be determined	2

Table 3
DETERMINATION OF Tl AND Pb IN BIOLOGICAL MATERIALS AND INDUSTRIAL PRODUCTS AFTER ANION EXCHANGE SEPARATION[a]

Material	Ion exchange resin, separation conditions, and remarks	Ref.
Plants and milk powder	Bio-Rad® AG1-X10 (200-400 mesh; chloride form) Column: 19 × 1.3 cm conditioned with 1.5 M HCl a) 1.5 M HCl sample solution (2—3 mℓ) containing Br$_2$ solution (1—2 drops) (adsorption of Tl[III], Pb, Zn, and Cd as anionic chloro complexes) b) 1.5 M HCl (100 mℓ) (elution of alkalies, alkaline earth metals, Cu, and Mn) c) 8 M HCl (elution of Pb appearing in the 25—45-mℓ fraction) d) Water (120 mℓ) (elution of Zn and Cd) e) 5—6% SO$_2$ solution (30 ml) (reductive elution of Tl) Tl and Pb are determined by mass spectrometric isotope dilution analysis Before the ion exchange separation, the sample (containing parts per billion and parts per million amounts of Tl and Pb, respectively) is wet ashed with HNO$_3$ and H$_2$O$_2$	5
CsI scintillators	Dowex® 1 (chloride form) Column: 5 × 1 cm a) 3—6 M HCl sample solution containing Cl$_2$-water (≤5 mℓ) (adsorption of Tl[III]; Cs passes into the effluent) b) Dilute HCl (one column volume) (elution of residual Cs) c) 0.2% Sulfurous acid solution (4 column volumes) (reductive elution of Tl) Tl is determined polarographically; this separation is necessary when the Tl content of the sample is <0.01%	7

[a] See also Table 16 in the chapter on Copper, Volume III.

Table 4
ANION EXCHANGE SEPARATION OF Tl(I)

Elements separated	Ion exchange resin, separation conditions, and remarks	Ref.
Tl(I) from Zn, Cd, Pb, Sb, and Hg	Anionite EDE-10P (chloride form) Column of 1 cm ID containing 10 g of the resin conditioned with 2 M HCl and operated at a flow rate of 1—2 ml/min a) 2 M HCl test solution (50 mℓ) containing ≯10 mg of Tl and ≯30 mg each of the other elements) (adsorption of Zn, Cd, Pb, Hg, and Sb; Tl[I] passes into the effluent) b) Hot water (elution of Zn, Cd, and Pb) c) 5 N H$_2$SO$_4$ (elution of Sb) d) 2 M HNO$_3$ (elution of Hg) If Tl is present as Tl(III) it is coadsorbed with the other elements, and for the fractionation of the retained metals the following eluents are employed a) 0.65 M HCl (elution of Zn) b) 0.3 M HCl (elution of Cd) c) Hot water (elution of Pb) d) 2 M HNO$_3$ (elution of Tl)	6
Tl(I) from numerous elements	Anionite AV-17 (carbonate form) Column: 15 × 1.6 cm a) Sample solution (25 mℓ, containing 0.1—1 mg of Tl per milliliter, and <20 mg of the other elements) (retention of In, Be, U, Fe, Zn, Ga, Cd, Cu, Ni, Pb, Al, Cr, Ti, Mn, Mo, V, and W; Tl[I] passes into the effluent) b) Water (50 mℓ) (elution of residual Tl) Tl is determined spectrophotometrically with Rhodamine B	11

Table 5
DETERMINATION OF Tl IN GEOLOGICAL MATERIALS AFTER CATION EXCHANGE SEPARATION[a]

Material	Ion exchange resin, separation conditions, and remarks	Ref.
Rocks and minerals, e.g., pyrites, galena, chalcopyrite, and sphalerite	Wofatit® P (H⁺ form) Column: 16 × 1 cm conditioned with 0.05 M HCl containing 5% citric acid and operated at a flow rate of 120 mℓ/hr (elutions [a—c]) a) 0.05 M HCl sample solution (~20—25 mℓ) containing 5% citric acid (adsorption of Tl[I]; into the effluent pass Sb[V], Hg, etc.) b) 5% Citric acid (as a rinse) c) Water (as a rinse) d) 6 M HCl (80—100 mℓ) (elution of Tl at a flow rate of 40 mℓ/h) Tl is determined photometrically with crystal violet; before the separation, the sample (0.1—1 g) is decomposed with aqua regia (rock samples are treated with HF-HNO$_3$)	15
Rocks and soils	Bio-Rad® AG50W-X8 (200-400 mesh; H⁺ form) Column: 6.5 × 1 cm a) 0.1 M HCl sample solution (25 mℓ) containing a little Cl$_2$-water (adsorption of traces of metals which were coextracted with Tl; Tl[III] passes into the effluent) b) Water (15 mℓ) (elution of residual Tl) Tl is determined fluorimetrically; the estimated relative standard deviation was 9.2% (eight samples) Before the separation, the sample (1 g, containing 0.28—1.78 ppm of Tl) is decomposed with HClO$_4$ and HF, fluoride is masked with H$_3$BO$_3$, and after oxidation with Cl$_2$, the Tl(III) is extracted with hexone from 1 M HBr; the organic extract is evaporated and solution (a) is prepared	13

[a] See also Table 43 in the chapter on Copper, Volume III.

Table 6
DETERMINATION OF Tl IN INDUSTRIAL PRODUCTS AFTER CATION EXCHANGE SEPARATION[a]

Material	Ion exchange resin, separation conditions, and remarks	Ref.
Zn and Pb dusts	Cationite SBS in a column (e.g., 14 × 1.8 cm) operated at a flow rate of 4—5 mℓ/min A) Adsorption from alkaline medium a) Sample solution (50—60 mℓ) containing tartaric acid (5 g) and an excess of NaOH (adsorption of Tl[I]; into the effluent pass Fe, Cu, Zn, Cd, Pb, Al, and Sb as anionic tartrate complexes) b) 5% Tartaric acid solution (as a rinse) c) Water (removal of residual tartaric acid) d) 6 M HCl (50—100 mℓ) (elution of Tl) B) Adsorption from acid medium In this case the sorption solution is prepared by adding to the HCl sample solution (~25 mℓ) tartaric or citric acid (5—7 g each) and then Na-pyrophosphate ($Na_4P_2O_7$) (2—3 g); on percolating this solution through the column, Tl is adsorbed, while the other elements pass into the effluent as anionic complexes with these reagents Tl is determined spectrophotometrically Before the separation, the sample (0.5—1 g) is decomposed with HCl and HNO_3 and $PbCl_2$ is filtered from the final sample solution (20—25 mℓ of 10% HCl) Very similar procedures are used in connection with the determination of Tl in Zn electrolytes and in metallic Cd	16, 17
Refined Pb	Dowex® 50 or Amberlite® IR-120 a) 0.2 M EDTA sample solution (50 mℓ) (adsorption of Tl[I]; Pb passes into the effluent as anionic EDTA complex) b) Acetate buffer solution (elution of residual Pb) c) 2 M HCl (200 mℓ) (elution of Tl) Tl is determined polarographically; before the separation, the sample (1 g) is dissolved in 1:3 HNO_3 (30 mℓ) The error of the method is $\not> \pm 2.4\%$	18
TlCl or $Tl_2Cr_2O_7$	A mixture of cationite (H^+ form) and anionite (OH^- form) is added to the aqueous suspension of the Tl-salt and, after dissolution of the salt (at 50°C for $Tl_2Cr_2O_7$), the resins are transferred to CCl_4-dichloroethane of such density that the cationite separates by sedimentation; the cationite is dried at 80°C, then the Tl is desorbed with 7% HNO_3 solution for subsequent complexometric determination For 0.5 g of either salt, the relative error was $< \pm 0.5\%$	21

[a] See also Tables 52 and 54 in the chapter on Copper, Volume III.

Table 7
CATION EXCHANGE SEPARATION OF Tl(I) FROM SYNTHETIC MIXTURES

Elements separated	Ion exchange resin, separation conditions, and remarks	Ref.

Tl(I) from binary mixtures with other elements	Dowex® 50W-X8 (20-60 mesh; H⁺ form)
	Column: 19 × 1.4 cm
	After adsorption of the elements the following separations can be performed
	A) Separation of Tl from Cu, Cd, Al, Zn, Mn, Co, Ni, U, V, and Ag
	  a) 1 M H_2SO_4 (150 mℓ) (elution of all adsorbed metals except Tl)
	  b) 4 M H_2SO_4 (200 mℓ) (elution of Tl)
	In the case of V, however, 0.5 M H_2SO_4 is used for the elution of this element
	B) Separation of Tl from Sr, Sb, As, Mo, and Cr(VI)
	  a) 5% Citric acid (300 mℓ) of pH 2.2 or 5% tartaric acid (200 mℓ) of pH 1.0 (elution of Sb or Sr)
	  b) Water (elution of As, Mo, and Cr[VI])
	  c) 4 M H_2SO_4 (200 mℓ) (elution of Tl)
	C) Separation of Tl from Pb
	  a) 1 M ammonium acetate (200 mℓ) (elution of Pb)
	  b) 1 M ammonium acetate (300 mℓ) or 3 M ammonium acetate (200 mℓ) (elution of Tl)
	The procedures can be used for the separation of milligram amounts of the elements
		25

Tl(I) from binary mixtures with Hg(II), Ag, and Cu(II)	Amberlite® IR-120 (Na⁺ form) Column: 15 × 1 cm operated at a flow rate of 3 mℓ/cm² min After adsorption of the elements present in concentration ratios of 9:1—1:9 and rinsing of the resin with water, the following eluents are used a) 2% $NaNO_2$ solution (elution of Hg, Ag, and Cu as anionic nitrite complexes) b) 10% Na_2SO_4 solution (elution of Tl)	26
Tl(I) from Pb	Dowex® 50 (16-40 mesh; Na⁺ form) Column containing ~20 g of the resin conditioned with eluent (b) a) Test solution (~100 mℓ) which is 0.5—1 M in NaOH and 1 M in glycerol (adsorption of Tl; Pb passes into the effluent as anionic hydroxo complex) b) 0.5—1 M NaOH - 1 M glycerol (~150 mℓ) (elution of residual Pb) c) Distilled water (until effluent is neutral) d) 1 M $Ca(NO_3)_2$ (400—500 mℓ) (elution of Tl) Similarly, Tl can also be separated from Cu, Sn, Zn, Bi, As, Sb, Se, Te, Fe, and Al	20
Tl(I) from Pb, Hg, Bi, Cu, Fe, and Zn	Amberlite® IR-120 in a column containing 2 mℓ of this resin to a height of 5 cm a) Sample solution of pH 4 containing EDTA (20 mg/10 mℓ) (adsorption of Tl[I]; into the effluent pass the anionic EDTA complexes of the other metals) b) 2 M HCl (50 mℓ) (elution of Tl at a flow rate of 1 mℓ/min) Tl is determined by indirect colorimetry	19

Table 8
DETERMINATION OF Tl AND In IN INDUSTRIAL PRODUCTS AFTER SEPARATION ON CHELATING RESINS

Material	Ion exchange resin, separation conditions, and remarks	Ref.
High-purity Zn and Zn-base alloys	Zeo-Karb® 226 (carboxylic acid resin) (H⁺ form) and Dowex® Al chelating resin (100-200 mesh; H⁺ form) A) Separation on Zeo-Karb® 226 Column of 1 cm ID containing 2 g of the resin and operated at a flow rate of 1 mℓ/min a) Sample solution (100 mℓ) adjusted to pH 3.0 (adsorption of Tl[III], Fe[III], and In; Zn and the other elements pass into the effluent) b) Water of pH 3.0 (200 mℓ) (elution of residual Zn) c) 2 M HCl (50 mℓ) (elution of Tl, In, and Fe) B) Separation on Dowex® Al Column: 4 × 1.0 cm operated at a flow rate of 1 mℓ/min a) Sample solution (100 mℓ) adjusted to pH 2.0 (adsorption of Tl[III], In, Fe[III], and Cu; Zn and Al pass into the effluent) b) Water of pH 2.0 (elution of residual Zn and Al) c) 0.01 M o-phenanthroline of pH 2.0 (100 mℓ) (elution of all, but traces of Cu) d) 1 M HCl (50 mℓ) (elution of In and Fe) e) 2 M HCl (50 mℓ) containing H_2SO_3 (1.2%) (reductive elution of Tl) Tl and In are determined spectrophotometrically Before separation (A) or (B) the sample (~5 g) is dissolved in 6 M HCl, Tl is oxidized to Tl(III) with Br_2-water, and the pH is adjusted to 3 or 2, respectively	27

Table 9
MICROCHEMICAL DETECTION OF Tl BY A RESIN SPOT TEST

Ion exchange resin	Experimental conditions and remarks	Ref.
Dowex® 1-X1 (iodide form)	When resin beads are placed in an HNO_3 solution of pH 1 containing Tl(I) a yellow coloration appears after 1.5 hr The limit of detection is 0.01 µg and the concentration limit is 1 in 1.3 × 10⁶ Cu, Ag, Hg, Ti, Pb, Bi, U(VI), VO_3^-, Fe(III), and $S_2O_3^{2-}$ interfere	28

REFERENCES

1. **Matthews, A. D. and Riley, J. P.,** Determination of thallium in silicate rocks, marine sediments and seawater, *Anal. Chim. Acta,* 48, 25, 1969.
2. **Batley, G. E. and Florence, T. M.,** Determination of thallium in natural waters by anodic stripping voltammetry, *J. Electroanal. Chem.,* 61(2), 205, 1975.
3. **Calderoni, G. and Ferri, T.,** Determination of thallium at sub-trace level in rocks and minerals by coupling differential pulse anodic-stripping voltammetry with suitable enrichment methods, *Talanta,* 29, 371, 1982.
4. **de Albuquerque, C. A. R. and Muysson, J. R.,** Determination of parts-per-billion levels of thallium in silicate rocks by anion-exchange and spectrographic analysis, *Chem. Geol.,* 9(3), 167, 1972.
5. **Heumann, K. G., Kastenmayer, P., and Zeininger, H.,** Pb and Tl trace determination in the ppm and ppb range in biological samples by mass spectrometric isotope dilution analysis, *Fresenius Z. Anal. Chem.,* 306, 173, 1981.
6. **Efremov, G. V., Zvereva, M. N., and Tsedevsuren, Ts.,** Separation of thallium from accompanying elements on an anion-exchange column, *Zavod. Lab.,* 28(2), 159, 1962.
7. **Kubota, H.,** Determination of thallium in caesium iodide scinillators, *Anal. Chim. Acta,* 35, 534, 1966.
8. **Huey, J. M. and Kohman, T. P.,** Search for extinct natural radioactivity of lead-205 via thallium-isotope anomalies in chrondrites and lunar soil, *Earth Planet. Sci. Lett.,* 16, 401, 1972.
9. **Korkisch, J. and Steffan, I.,** Determination of thallium in natural waters, *Int. J. Environ. Anal. Chem.,* 6(2), 111, 1979.
10. **Strelow, F. W. E. and Toerien, S.,** Accurate determination of thallium by direct titration with EDTA using methylthymol blue as indicator, *Anal. Chim. Acta,* 36, 189, 1966.
11. **Eristavi, V. D. and Mgaloblishvili, M. G.,** The separation of thallium(I) on anion-exchangers in carbonate form, *Zh. Anal. Khim.,* 28(2), 375, 1973.
12. **Abrão, A.,** Radiochemical separation of thallium from thorium by anion exchange resin. A Tl-208 reservoir, *J. Chem. Educ.,* 41, 600, 1964.
13. **Boehmer, R. G. and Pille, P.,** Determination of thallium in rock and soil samples, *Talanta,* 24, 521, 1977.
14. **Jones, E. A. and Lee, A. F.,** The determination of thallium and indium in sulphide concentrates, Report NIM 2022, National Institute of Metallurgy, Randburg, South Africa, July 31, 1979.
15. **Panchev, B. N.,** Ion exchange separation of thallium with Wofatit P and its photometric determination with crystal violet, *Izv. Geol. Inst. Strashimir Dimitrov,* 12, 237, 1963.
16. **Gur'ev, S. D. and Shkrobot, E. P.,** New method for the determination of thallium in products from zinc production, *Sb. Nauchn. Tr. Gos. Nauchno Issled. Inst. Tsvet. Met.,* No. 12, 79, 1956.
17. **Ginzburg, L. B. and Shkrobot, E. P.,** Use of ion-exchange methods for determining thallium and indium in products from the treatment of non-ferrous metal ores, *Zavod. Lab.,* 21(11), 1289, 1955.
18. **Ziemba, S.,** Polarographic determination of thallium in refined lead, *Pr. Inst. Hutn.,* 16, 261, 1964.
19. **Nozaki, T.,** Indirect colorimetric determination of thallium, *J. Chem. Soc. Jpn. Pure Chem. Sect.,* 77(3), 493, 1956.
20. **Carobene, G. and Vicedomini, M.,** Ion-exchange separation of thallium(I) from lead(II), *J. Chromatogr.,* 33(3-4), 566, 1968.
21. **Bogatÿrev, V. L., Vulikh, A. I., and Sokolava, S. I.,** Determination of ions of precipitates by means of a mixture of ion-exchange resins, *Zh. Anal. Khim.,* 22(6), 837, 1967.
22. **Strelow, F. W. E.,** Partly non-aqueous media for accurate chemical analysis by ion exchange, *Ion Exch. Membr.,* 2(1), 37, 1974.
23. **Strelow, F. W. E., van Zyl, C. R., and Bothma, C. J. C.,** Distribution coefficients and the cation exchange behavior of elements in hydrochloric acid-ethanol mixtures, *Anal. Chim. Acta,* 45, 81, 1969.
24. **Caletka, R. and Krivan, V.,** Cation-exchange of 43 elements from hydrofluoric acid solution, *Talanta,* 30, 543, 1983.
25. **Rangnekar, A. V. and Khopkar, S. M.,** Cation-exchange studies of thallium(I) on Dowex 50W-X8; separation from mixtures, *Indian J. Chem.,* 4(7), 318, 1966.
26. **Bhatnagar, R. P. and Trivedi, R. G.,** Separation of thallium(I), mercury(II), silver(I) or copper(II) by ion-exchange chromatography, *J. Indian Chem. Soc.,* 42(1), 53, 1965.
27. **Iyer, S. G., Padmanabhan, P. K., Nair, L. D., and Venkateswarlu, C.,** Use of Zeo-Karb 226 and Dowex A-1 in the analysis of high purity zinc and zinc-based alloys for thallium and indium, *Talanta,* 23, 525, 1976.
28. **Kato, K. and Kakihana, H.,** Microchemical detection of thallium with the aid of an ion-exchange resin bead, *J. Chem. Soc. Jpn. Pure Chem. Sect.,* 81(3), 452, 1960.

SILICON

Anion and cation exchange resins have been used for analytical separations of Si from accompanying elements. Most of the anion exchange procedures are based on the adsorption of silicate or fluorosilicate, while the nonadsorption of these two anionic species on cationic resins serves to separate Si from essentially all cationic components of the solutions.

ANION EXCHANGE RESINS

Soluble silicate is appreciably retained by strongly basic anion exchange resins in the hydroxide forms from weakly alkaline media, e.g., at pH 8 to 10 (see Tables 1 to 3*).[1-4] Coadsorbed with the silicate are all elements which are present as anions as, for instance, phosphate, sulfate, nitrate, chloride, fluoride, and borate. Elution of the adsorbed silicate can be effected with 0.4,[1] 1.5,[5] and 2.5 M NaOH[2] or by contacting the resin with a solution of ammonium molybdate.[3] In the latter case yellow molybdosilicate is formed which may also be obtained by reactive ion exchange, using a resin in the molybdate form that has been equilibrated with the silicate containing sample solution at pH 3 to 9 (see Table 1).[6]

Strong adsorption of Si as the anionic fluorosilicate complex, i.e., SiF_6^{2-}, takes place on anionic resins from very dilute mineral acid media containing fluoride, as, for instance, from 0.1 M HF - 0.05 M HCl (see Table 4),[7-9] dilute HF-H$_2$SO$_4$ solution (see Table 2),[10] dilute H$_3$PO$_4$ - 1% NaF solution (see Table 2),[11] and very dilute hydrofluoric acid (HF) (see Table 1).[12] Coadsorbed from media of this type are all elements forming stable anionic fluoride complexes as, for example, Mo, W, Nb, Ta, Ti, Zr, and Hf. For the elution of Si that has been adsorbed from fluoride systems the following eluents can be used: 8 M HCl,[10] 8 M HCl - 1 M HF,[7-9] 0.05 M HF - 4% Na sulfate solution,[11] and a saturated solution of boric acid.[12] In the latter eluent the reaction $2H_2SiF_6 + 3H_3BO_3 \rightleftarrows 3HBF_4 + 2SiO_2 + 5H_2O$ is assumed to occur in the resin bed.

No adsorption of silicate occurs on strongly basic resins in the sulfate form, as, for instance, from oxalic acid solution of pH $\simeq 1$ (see Table 2)[13] or from very dilute sulfuric acid media.[5] From these systems Ti and W, respectively, are retained as anionic complexes, while silicate is not adsorbed. Also, no adsorption of silicate is observed on weakly basic resins in the chloride forms, e.g., on Bio-Rad® AG3 in ammonium tartrate solution (see Table 2)[14] and on Amberlite® IR-45 in very dilute hydrochloric acid (HCl) (see Table 3).[15,16] On these resins, even if used in the free-base forms, silicate and also other weak electrolytes such as boric acid are not adsorbed (see in the chapter on Boron), while, however, anions such as sulfate, phosphate, and nitrate are strongly retained, so that effective separations of these two groups of anions can be achieved.

The systems discussed in the preceding paragraphs have been used in anion exchange separation procedures which were employed in connection with the determination of Si in waters (see Table 1) and industrial products (see Table 2).

CATION EXCHANGE RESINS

Since Si is in solution usually present as an oxyanion species, i.e., silicic acid, this element is not adsorbed on cation exchange resins. Therefore, separations of Si on exchangers of this type are based on the adsorption of accompanying constituents in solutions from which strong adsorption of the metal cations occurs, e.g., at pH values of 1.0,[17,18] 2,[19] and 3 to 4.[20-23] These acidities are attained using sorption solutions containing dilute mineral acids

---

* Tables for this chapter appear at the end of the text.

as, for example, dilute HCl (e.g., 0.04 M HCl),[15,16,18,24] 0.2 M HNO$_3$,[25] and 0.02 M HClO$_4$.[26] From these media as well as from citric acid solution,[27] Si and other elements forming anions under these conditions (e.g., sulfate, phosphate, nitrate, etc.) are not adsorbed. Also, no retention of Si is observed on cationic resins from solutions containing fluoride, because in such media the stable anionic species SiF$_6^{2-}$ is formed. Systems of this type include dilute HCl containing Na fluoride,[28] acetone solutions which are 0.1 to 0.3 M in HCl and 0.1 M in HF,[7-9] and methanol solutions containing HF.[29]

From the cation exchange procedures illustrated in Tables 3 and 4 it is seen that separations based on the nonadsorbability of silicate and hexafluorosilicate have variously been employed in connection with the determination of Si in geological materials (Table 3)[15,16,20-22,24,27,29-31] and industrial products (Table 4).[7-9,17,25,26,32]

Table 1
DETERMINATION OF Si AND OTHER ELEMENTS IN WATERS AFTER ANION EXCHANGE SEPARATION

Material	Ion exchange resin, separation conditions, and remarks	Ref.
Tap- and riverwater	Dowex® 1-X8 (100-200 mesh; molybdate form which is prepared by immersion of the chloride form in ammonium molybdate solution) To the water sample (25 mℓ, pH 3—9) the dried resin (0.5 g) is added and the mixture is stirred for 20 min to effect adsorption of silicic acid as molybdosilicate;[a] subsequently, the resin is equilibrated for 20 min at 25°C with a reducing agent solution (containing 1% ascorbic acid, 0.01% of Sb-K-tartrate, and 0.1 M H_2SO_4); the solution is discarded, the resin (which is now blue) is washed with water, and Si is determined by ion exchanger colorimetry (700 nm) Up to 50 μg of Si can be determined; the coefficient of variation was 3.39% for 12 solutions each containing 1 ppm of Si	6
Tapwater	Strongly basic anion exchange resin (Hitachi® custom ion exchange resin no. 2630, OH^- form) Separation column: 8 × 0.9 cm operated at 40°C and at a flow rate of 1.5 mℓ/min After adsorption of the elements the following eluents are used a) 0.4 M NaOH - 0.02 M Na_2SO_4 (elution of Si followed by P) b) 0.4 M NaOH (elution of Si followed by Ge) Si, Ge, and P are determined by coulometric detection (via reaction with ammonium molybdate) Before the anion exchange separation, interfering cations should be removed using a cation exchange resin	1
Commercial deionized water	Permutit® ES (strongly basic anion exchange resin; chloride form) Column of 2 cm ID containing 10 mℓ of the resin and operated at a flow rate of 1 ℓ/20—25 min a) Sample containing 7.5 g of 1 M HF per liter (adsorption of Si as SiF_6^{2-}) b) Saturated aqueous solution (5 mℓ) of H_3BO_3 (this solution is passed through the column after 15 min of contact with the resin which is stirred intermittently) (elution of Si) Si is determined spectrophotometrically	12

[a] From 0.2 M HCl silicate is not retained by Dowex® 1-X8 (molybdate form), but phosphate is adsorbed (formation of AMP) so that separation of these two can be achieved.[33]

Table 2
DETERMINATION OF Si IN INDUSTRIAL PRODUCTS AFTER ANION EXCHANGE SEPARATION[a]

Material	Ion exchange resin, separation conditions, and remarks	Ref.
Ag-infiltrated W sample	Dowex® 1-X8 (100—200 mesh) Column of 1 in. ID filled with 20 g of the resin conditioned with eluent (b) (50 mℓ) and operated at a flow rate of 150—200 mℓ/hr a) Sample solution (35 mℓ) containing 5 mℓ of 30% H_2O_2 and 10 mℓ of strong H_2SO_4-HF solution (adsorption of Si and W) b) H_2SO_4-H_2O_2-HF solution (96 mℓ) (elution of Ag) c) Weak H_2SO_4-HF solution (30 mℓ) (removal of H_2O_2) d) 8 M HCl (15 mℓ) (as a rinse) e) 8 M HCl (40 mℓ) (elution of Si) Si is determined spectrophotometrically; the scope of the method is 0.02—0.2% with a relative standard deviation of ≯3%; the overall sensitivity is 3.6 ppm of Si Before the separation, the sample is dissolved in HNO_3-H_2O_2 The strong H_2SO_4-HF solution is prepared by adding conc HF (10 mℓ) and 1:1 H_2SO_4 (10 mℓ) to 380 mℓ of water; the H_2SO_4-H_2O_2-HF solution (eluent [b]) is obtained by mixing strong H_2SO_4-HF solution (12 mℓ) with 500 mℓ of H_2SO_4-H_2O_2 solution (30% H_2O_2 [35 mℓ] and 5 N H_2SO_4 [40 mℓ] diluted to 1 ℓ with water); on dilution of strong H_2SO_4-HF solution (12 mℓ) to 500 mℓ with 0.1 M H_2SO_4 the eluent (c) is made	10
Ti	Anionite EDE-10P (sulfate form) Column containing 12 g of the resin a) Sample solution adjusted to pH ≃ 1 with 3% oxalic acid (after neutralization of alkali using the same acid) (adsorption of Si and of Ti as anionic oxalate complex) b) 0.2 N H_2SO_4 (100 mℓ) (elution of Si) Si is determined spectrophotometrically as Mo-blue Before the separation, the sample (containing ≮0.001% of Si) is decomposed by fusion with KOH	13
Analytical reagent-grade orthophosphoric acid	Dowex® 1-X8 Column: 10 × 1.9 cm a) Sample solution (100 mℓ) containing 6 g of 86% H_3PO_4 and 1 g of NaF (adsorption of Si; H_3PO_4 passes into the effluent) b) 0.05 M HF - 4% Na_2SO_4 (elution of Si; the first 130 mℓ is discarded and Si is collected in the next 200 mℓ) Si is determined spectrophotometrically	11
Na alginate solutions	Amberlite® IRA-400 (OH$^-$ form) Column: ~20 × 1.3 cm operated at a flow rate of ≯4 mℓ/min a) Sample solution (25 mℓ) of pH 8—10 and containing 10—250 µg of SiO_2 (adsorption to silicate) b) Water (3 × 25 mℓ) (as a rinse) c) 2.5 M NaOH (10 mℓ) (elution of silicate) Si is determined spectrophotometrically using the silicomolybdate method	2
Solutions containing 0.02—0.1 µg Si per milliliter	Bio-Rad® AG2-X8 (20—50 mesh; OH$^-$ form) Microcolumn operated at a flow rate of 9—10 mℓ/min a) Alkaline sample solution (250 mℓ) (adsorption of silicate) b) Dilute NH_3 solution (10 mℓ) (as a rinse) c) The resin is taken from the column and mixed with NH_4-molybdate solution and other reagents and finally Si is determined spectrophotometrically	3

Table 2 (continued)
DETERMINATION OF Si IN INDUSTRIAL PRODUCTS AFTER ANION EXCHANGE SEPARATION[a]

Material	Ion exchange resin, separation conditions, and remarks	Ref.
Ga phosphide	Recovery was >90% for down to 20 ppb of Si; the coefficient of variation was 3% Weakly basic resin Bio-Rad® AG3-X4 (25—50 mesh; Cl⁻ form) Column: 17 × 2 cm a) Sample solution (~10 mℓ) containing 10% ammonium tartrate (1 mℓ) to prevent precipitation of Ga(OH)₃ (adsorption of phosphate; Si passes into the effluent) (flow rate: 2 mℓ/min) b) Water (~45 mℓ) (elution of remaining Si at a flow rate of ~5 mℓ/min) Si is determined spectrophotometrically Before the separation, the sample (0.1 g containing ≯40 μg of soluble Si) is dissolved in HCl-HNO₃	14

[a] See also Table 4 in the chapter on Aluminum.

Table 3
DETERMINATION OF Si IN GEOLOGICAL MATERIALS AFTER CATION EXCHANGE SEPARATION[a]

Material	Ion exchange resin, separation conditions, and remarks	Ref.
Silicates	A) Batch separation (Amberlite® IR-120) The sample (0.1 g) is fused with B_2O_3 (1 g) and Li_2CO_3 (0.1 g) at 1000°C for 30 min, the melt is dissolved in 0.04 M HCl (1 ℓ) containing 30% H_2O_2 (2 mℓ) and cationic constituents are adsorbed on 10 g of the resin with the aid of ultrasonic agitation; after filtration, the filtrate is subjected to the following separation (B) B) Column separation (composite column [37 × 1.4 cm] packed with Amberlite® IR-120 (H^+ form, upper resin bed of 10 cm length) and Amberlite® IR-45 (30-50 mesh; chloride form) (lower column) operated at a flow rate of 2 mℓ/min) A 50-mℓ aliquot of the filtrate obtained by A is passed through this set of columns and in the effluent Si is determined gravimetrically	15, 16
Silicate rocks	Amberlite® CG-120 (100-200 mesh; H^+ form) To the fusion product (600 mg), distilled water (200 mℓ), resin (5.5 g), and 3% of citric acid solution (10 mℓ) are added and the mixture is stirred magnetically for 3 hr; then the resin (on which cations are adsorbed) is filtered off and in the filtrate Si is determined by atomic absorption spectroscopy Before this batch separation, the sample (650 mg) is fused with Li_2CO_3 (650 mg) and H_3BO_3 (1.3 g)	27
Silicates	Cationite SDV-3 (H^+ form) Column contained in a polyethylene tube (50 × 2 cm) pretreated with 2 N H_2SO_4 followed by methanol (until neutral reaction to methyl orange) a) Solution of alkali metal silicate (= 2—7 mg of Si) in 35 mℓ of 26% methanol plus 1.5 mℓ of 40% HF (adsorption of cations; Si passes into the effluent) b) Methanol (140 mℓ) (elution of residual Si) Si is determined titrimetrically If the sample does not dissolve immediately in eluent (a) the supernatant solution is passed through the column and the residue is treated with methanol (3—5 mℓ) and 40% HF (1.5 mℓ); then the resulting solution is also passed through the column For fluorosilicates, the sample (0.1 g) is dissolved in water and the solution (25 mℓ) is passed through the column which is then washed with eluent (b)	29
Granite, labradorite, bauxite, and other minerals	Cationite KU-2 (H^+ form) in a column operated at a flow rate of 100 mℓ/hr a) Dilute HCl sample solution (250 mℓ) (adsorption of cationic constituents of the matrix, e.g., Al, Fe, Ti, Mg, and Ca; Si passes into the effluent) b) Titrimetric determination of Si in an aliquot of the effluent (after rejection of the first 50 mℓ) Before the separation, the sample (0.5 g) is decomposed by fusion with KOH + Na_2O_2	24

[a] See Also Table 4 in the chapter on Rare Earth Elements, Volume I.

Table 4
DETERMINATION OF Si IN INDUSTRIAL PRODUCTS AFTER CATION EXCHANGE SEPARATION[a]

Material	Ion exchange resin, separation conditions, and remarks	Ref.
Metals	A) Al (Dowex® 50-X8, H⁺ form) Column: 15 × 0.8 cm operated at a flow rate of 1.5 mℓ/min a) Sample solution (10 mℓ) which is 60% in acetone, 0.3 M in HCl, and 0.1 M in HF (adsorption of Al; Si passes into the effluent) b) 60% Acetone - 0.3 M HCl - 0.1 M HF (20 mℓ) (elution of Si) B) Fe (same resin and column as above under [A]) a) Sample solution (~10 mℓ) which is 50 or 60% in acetone, 0.2 or 0.3 M in HCl, and 0.1 M in HF (adsorption of Fe; Si passes into the effluent as SiF_6^{2-}) b) 50 or 60% Acetone - 0.1—0.3 M HCl - 0.1 M HF (20 mℓ) (elution of Si) C) Mo (Dowex® 1-X8, chloride form) Column: 15 × 0.8 cm operated at a flow rate of 2 mℓ/min a) Sample solution (15 mℓ) which is 0.1 M in HF and 0.05 M in HCl (adsorption of Mo and Si as anionic fluoride complexes) b) 0.1 M HF - 0.03 M HCl (25 mℓ) (as a rinse) c) 8 M HCl - 0.1 M HF (30 mℓ) (elution of Si) In each case the pH of the Si eluate is adjusted to 9 by adding NH₃, and after adsorption on a column of Al₂O₃, ³¹Si is determined radiometrically The limit of detection is 0.01 μg/g for a 100-mg sample Before the ion exchange separation, the sample (≥250 mg) is irradiated with neutrons and then dissolved in HCl (Al and Fe) or decomposed by fusion with NaOH-NaNO₃	7—9
Zircaloy and Al-U and Zr-U fuel solutions	Dowex® 50W-X8 (50-100 mesh; H⁺ form) A sample aliquot (≥25 mℓ, containing 5—100 μg of reactive Si) is mixed with the resin (~10 mℓ) to adsorb metal cations; then the resin is removed by filtration through a 0.45-μm membrane filter, and in the filtrate and water rinses Si is determined spectrophotometrically using the Mo-blue method In the presence of fluoride, the sample aliquot (≥5 mℓ, containing 5—100 μg of Si) is mixed with the same amount of resin; subsequently, saturated boric acid solution (40 mℓ) is added (to mask fluoride) and the mixture is let to stand for 30 min with intermittent mixing; afterwards the procedure is continued as above	32
Pu-silicides and Pu-Si carbides	Zeo-Karb® 225-X8 (100-200 mesh; H⁺ form) Column of 0.6 cm ID containing 2.5 mℓ of the resin conditioned with eluent (b) (20 mℓ) and operated at a flow rate of 0.5 mℓ/min a) Sample solution adjusted to pH 1.0 [60—70 mℓ; containing conc HCl [3 mℓ] and 20% NH₂OH·HCl [5 mℓ]) (adsorption of Pu[III]; Si passes into the effluent) b) 0.1 M HCl - 1% NH₂OH - HCl (20—25 mℓ) (elution of residual Si) With 60-mg amounts of Pu, effluents containing <30 μg of this element are easily obtained; Si is determined gravimetrically as quinoline molybdosilicate Before the separation, the sample is ignited at 850°C and decomposed by treatment with Na₂O₂ at 470—490°C	17
Pu	Dowex® 50W-X10 (200-400 mesh; H⁺ form) Column containing the resin to a height of 6 cm	25

Table 4 (continued)
DETERMINATION OF Si IN INDUSTRIAL PRODUCTS AFTER CATION EXCHANGE SEPARATION[a]

Material	Ion exchange resin, separation conditions, and remarks	Ref.
	From 0.2 M HNO$_3$ sample solution Pu(III) is adsorbed on the resin, and in the effluent Si is determined spectrographically; the lower limit of detection is 1 ppm of Si	
	The method is applicable to acid solutions containing Pu and to Pu metal or its sulfates and oxides	
Si in solutions containing cations (e.g., Fe) and anions	Dowex® 50-X8 (50-100 mesh; H$^+$ form)	26
	Column: 21 × 1 cm conditioned with 0.02 M HClO$_4$ and operated at a flow rate of 0.2 mℓ/min	
	a) Test solution (10 mℓ) which is 0.4—1.2 × 10^{-3} M in Si, 0.2 M in FeCl$_3$, and 0.02 M in HClO$_4$ (adsorption of Fe; Si passes into the effluent)	
	b) Water (5 × 10 mℓ) (elution of residual Si)	
	After this separation, the weakly basic anion exchange resin Amberlite® IR-4B (60-100 mesh) is used to remove Cl$^-$, NO$_3^-$, PO$_4^{3-}$, and AsO$_4^{3-}$	

[a] See also Table 9 in the chapter on Boron.

REFERENCES

1. **Takata, Y. and Muto, G.,** Application of controlled-potential coulometry to automatic recording in liquid chromatography. IX. Liquid chromatographic determination of silicate, phosphate and germanate ions with coulometric detection, *Jpn. Analyst,* 28(1), 15, 1979.
2. **Brown, E. G. and Hayes, T. J.,** Quantitative collection and recovery of silica by means of an ion-exchange column, *Mikrochim. Acta,* No. 5, 522, 1954.
3. **Bazzi, A. and Boltz, D. F.,** Pre-concentration, separation and spectrophotometric determination of traces of silicate, *Microchem. J.,* 20(4), 462, 1975.
4. **Fisher, S. and Kunin, R.,** Ion-exchange preparations of low-silica hydroxide solutions for colorimetric determinations of total silica, *Nature,* 177, 1125, 1956.
5. **Ma, C., Pau, C.-T., and Li, C.-C.,** Determination of micro amounts of silicon in tungsten, *Fen Hsi Hua Hsueh,* 8(2), 150, 1980.
6. **Tanaka, T., Hiiro, K., and Kawahara, A.,** Colorimetric determination of silicic acid using anion-exchange resin in molybdate form, *Jpn. Analyst,* 30(2), 131, 1981.
7. **Rouchaud, J. C., Fedoroff, M., and Revel, G.,** Determination of silicon in metals by thermal-neutron activation, *J. Radioanal. Chem.,* 38(1-2), 185, 1977.
8. **Rouchaud, J. C., Fedoroff, M., and Revel, G.,** Determination of silicon in iron by thermal-neutron activation, *J. Radioanal. Chem.,* 55(2), 283, 1980.
9. **Rouchaud, J. C., Debove, L., and Fedoroff, M.,** Recent developments in the determination of silicon by neutron activation, *J. Radioanal. Chem.,* 72, 537, 1982.
10. **Toy, C. H. and Van Santen, R. T.,** Ion exchange separation and determination of silicon in silver-infiltrated tungsten, *Anal. Chem.,* 36, 151, 1964.
11. **Takagi, T., Hashimoto, T., and Sasaki, M.,** Determination of traces of silicon in analytical reagent-grade orthophosphoric acid by ion exchange chromatography, *Jpn. Analyst,* 12(7), 618, 1963.
12. **Wickbold, R.,** The enrichment of very small amounts of silica by ion-exchange, *Fresenius' Z. Anal. Chem.,* 171, 81, 1959.
13. **Bausova, N. V. and Lebedeva, E. M.,** Determination of silicon in titanium by ion exchange, *Tr. Inst. Khim. Sverdlovsk,* No. 14, 129, 1967.
14. **Luke, C. L.,** Spectrophotometric determination of silicon in gallium phosphide, *Anal. Chem.,* 36, 2036, 1964.
15. **Ohlweiler, O. A., de Oliveira Meditsch, J., Santos, S., and Oderich, J. A.,** Determination of silica in silicates, containing phosphorus, titanium and zirconium by a modified procedure, *Anal. Chim. Acta,* 69, 224, 1974.
16. **Ohlweiler, O. A., de Oliveira Meditsch, J., Porto da Silveira, C. L., and Silva, S.,** Determination of silica in silicates containing phosphorus, titanium and zirconium, *Anal. Chim. Acta,* 61, 57, 1972.
17. **Milner, G. W. C., Jones, I. G., and Phillips, G.,** Analysis of plutonium silicides and plutonium silicon carbides, Report AERE-R5280, U.K. Atomic Energy Authority, 1966.
18. **Degtyarenko, Ya. A. and Oshchapovskii, V. V.,** Colorimetric determination of silicon in steel by ion exchange, *Nauchn. Zap. Lvov. Politekh. Inst.,* No. 22, 107, 1956.
19. **Doerffel, K. and Schlichting, P.,** Spectrochemical determination of silicon with a cascade-stabilised arc, *Fresenius' Z. Anal. Chem.,* 217, 107, 1966.
20. **Doerffel, K. and Yu Koe-Hue,** The use of a plasma jet in the spectrographic determination of silicon, *Talanta,* 13, 856, 1966.
21. **Lin, C. I. and Huber, C. O.,** Determination of phosphate, silicate, and sulfate in natural and waste water by atomic absorption inhibition titration, *Anal. Chem.,* 44, 2200, 1972.
22. **Looyenga, R. W. and Huber, C. O.,** Determination of silicate in waste water by atomic-absorption-inhibition titration, *Anal. Chem.,* 43, 498, 1971.
23. **Flaschka, H. and Amin, A. M.,** Rapid determination of silica in glass, *Chemist Analyst,* 43(1), 6, 1954.
24. **Khalizova, V. A., Alekseeva, A. Ya., and Smirnova, E. P.,** Rapid determination of silicic acid in mineral raw materials, *Zavod. Lab.,* 30(5), 530, 1964.
25. **Pietri, C. E. and Wenzel, A. W.,** Cation-exchange separation and spectrographic determination of soluble silicon in plutonium, *Anal. Chem.,* 35, 209, 1963.
26. **Andersson, L. H.,** Studies in the determination of silica. VI. Separation of some anions and cations from silicic acid by ion-exchange, *Ark. Kemi,* 19(3), 243, 1962.
27. **Govindaraju, K. and L'homel, N.,** Direct and indirect atomic absorption determination of silica in silicate rocks, *At. Absorpt. Newsl.,* 11(6), 115, 1972.
28. **Morimoto, Y. and Ashizawa, T.,** Rapid determination of silicon and phosphorus in metallic uranium, *Jpn. Analyst,* 10(5), 532, 1961.
29. **Kreshkov, A. P., Myshlyaeva, L. V., Sayushkina, E. N., Stolbov, V. F., Krasnoshchekov, V. V., and Sedova, I. V.,** Determination of silicon in silicates by ion-exchange, *Vestn. Tekh. Ekon. Inf. Nauchno Issled. Inst. Tekh. Ekon. Issled. Gos. Kom. Khim. Promsti. Gosplane SSSR,* No. 9, 35, 1964.

30. **Kuroda, R., Ida, I., and Kimura, H.,** Spectrophotometric determination of silicon in silicates by flow-injection analysis, *Talanta,* 32, 353, 1985.
31. **Yoshimura, K., Motomura, M., Tarutani, T., and Shimono, T.,** Micro-determination of silicic acid in water by gel-phase colorimetry with molybdenum blue, *Anal. Chem.,* 56, 2342, 1984.
32. **Duce, F. A. and Yamamura, S. S.,** Versatile spectrophotometric determination of silicon, *Talanta,* 17, 143, 1970.
33. **Ogata, K., Soma, S., Koshiishi, I., Tanabe, S., and Imanari, T.,** Determination of silicate by flow injection analysis coupled with a suppression column and solvent extraction system, *Jpn. Analyst,* 33, E535, 1984.

GERMANIUM

Ion exchange separations of Ge from accompanying elements can be achieved by use of anionic and cationic resins as well as chelating polymers. On anion exchange resins of the strongly basic type anionic Ge species are retained, while on nonchelating cation exchangers Ge is not adsorbed. These two facts have been utilized in connection with the determination of Ge in geological materials and industrial products.

ANION EXCHANGE RESINS

General

On strongly basic anion exchange resins, e.g., Dowex® 1, Ge(IV) is appreciably adsorbed as an anionic complex (probably as $GeCl_4^{2-}$ or $Ge[OH]_4Cl_2^{2-}$) from media containing high concentrations of hydrochloric acid (HCl). This adsorption is negligible from the 0.1 to 2 M acid, but at higher acid concentrations distribution coefficients for Ge of 1.5, ~10, ~60, ~180, and 200 have been measured in the 4, 6, 8, 10, and 12 M acid, respectively.[1] Coadsorbed with the Ge from 10 to 12 M HCl media are Fe(III), Ga, UO_2(II), Bi, Zn, Cd, Cu, and all other elements forming anionic chloro complexes. However, under these conditions Ge can be separated from a large number of other metals such as the alkalies, alkaline earth elements, rare earths, Al, and also from essentially all elements present as simple anions or oxyanions, as, for example, from As, phosphate, sulfate, and nitrate. Because Ge is not appreciably retained at lower HCl concentrations, e.g., from 4 M HCl (see above), its separation from more strongly adsorbed elements such as Fe(III), Ga, Au(III), U(VI), etc. is also possible, and, in addition, the more dilute acid solutions can be employed for the elution of Ge that has been adsorbed, e.g., from 10 to 12 M HCl (see Table 1*).[2] Separations based on the nonadsorbability of Ge from dilute HCl media can also be achieved in the presence of organic solvents such as acetic acid or methyl glycol-tributyl phosphate (see Table 2).[3,4] From these systems, elements forming anionic chloro complexes are strongly adsorbed on Dowex® 1, while Ge is not retained.

Strong adsorption of Ge on basic resins has been observed in hydrofluoric acid solutions (like Si, Ge is probably adsorbed as GeF_6^{2-}).[5] Thus, in the 0.5, 1, 2.5, 5, 10, and 15 M acid, distribution coefficients for Ge of 5×10^4, 3×10^4, 3×10^3, 6×10^2, 20, and 12, respectively, have been measured.

From alkaline solutions Ge is adsorbed, e.g., on Amberlite® XE-243 as a germanate, i.e., GeO_3^{2-} or $GeO(OH)_3^-$, and this fact can be utilized to separate Ge from accompanying silicate (see Table 1).[6] Similarly, adsorption of germanate also takes place on anionic resins in the hydroxide or chloride form from solutions adjusted to pH 4 to 6 (see Table 2).[7,8] The adsorption of germanates on Amberlite® IRA-400 (chloride form) over the pH range from 6 to 12, in the presence of ethylene glycol, glycerol, and mannitol, has also been studied,[9] but no analytical applications of these investigations have been reported.

Applications

Separation methods utilizing anion exchange resins have been employed in connection with the determination of Ge in industrial products (see Table 1) and to separate this element from synthetic mixtures with numerous other elements (see Table 2).

CATION EXCHANGE RESINS

General

Like Si, Ge is not adsorbed on cation exchange resins of the sulfonic acid type at any

---

* Tables for this chapter appear at the end of the text.

concentration of mineral acid such as hydrochloric or nitric acids. Consequently, it is possible to separate Ge from virtually all cationic constituents of a sample provided that the adsorption of the metal ions is performed from dilute acids, e.g., 0.1 M HCl[10] or 0.5 M nitric acid,[11] or from solutions adjusted to pH 2[12] or pH $\simeq 1.4$.[13] Examples for separations of this type are presented in Tables 3 and 4. Coeluted with the Ge are oxyanions and anionic metal complexes as well as simple anions. However, most of these anionic impurities can be separated from the Ge by using a mixture of a strongly acid cation exchange resin with a strongly or weakly basic anion exchanger (see Table 3).[12,14] Under these conditions the anionic constituents are removed simultaneously by adsorption on the basic resins which do not retain the Ge at these acidities. However, germanate is adsorbed from solutions of higher pH values (see preceding section on anion exchange resins).

Applications

In Tables 3 and 4 methods are described which are based on the separation principles mentioned above, and that have been used in connection with the determination of Ge in geological materials and industrial products.

CHELATING RESINS

Selective adsorption of Ge can be achieved[15,16] on a chelating resin based on a phenyl-fluorone derivative such as a resin obtained by the alkali-catalyzed condensation of resorcylfluorone with formaldehyde. On this resin, Ge is strongly retained from azeotropic HCl, while, for example, Fe and As are not adsorbed. Water may be used for the elution of Ge.

For the separation of traces of Ge from Zn sulfate solutions at pH 3 to 5, polyhydric phenol (-formaldehyde) resins have been employed.[17] The adsorbed Ge is eluted with dilute sulfuric acid (1:4) that is 1 M in ammonium fluoride.

Table 1
DETERMINATION OF Ge IN INDUSTRIAL PRODUCTS AFTER ANION EXCHANGE SEPARATION[a]

Material	Ion exchange resin, separation conditions, and remarks	Ref.
High-purity Fe and Al	Dowex® 1-X8 (100-200 mesh; chloride form) Column: 10 × 1 cm a) 10—12 M HCl sample solution (adsorption of Ge; other elements pass into the effluent) b) 12 M HCl (elution of residual accompanying elements) c) 5.5 M HCl (elution of Ge) After precipitation as GeS_2 and toluene extraction of $GeCl_4$, the ^{71}Ge is determined radiometrically; the limit of determination is ≃1 ng of Ge Before the ion exchange separation step, the sample (0.5 g) is irradiated with neutrons and after dissolution, the Ge is preconcentrated by distillation of $GeCl_4$ from a solution containing HCl, H_2SO_4, and H_3PO_4 (to hold back Sn); the distillate (30 mℓ) is made 10—12 M in HCl to obtain solution (a)	2
Si-Ge thermoelectric alloys	Amberlite® XE-243 (basic form) Column: 30 × 1 cm conditioned in this order with 3 M HCl (50 mℓ), water (until neutral), and 3 M NH_3 solution and water (until neutral) and operated at a flow rate of 1—2 mℓ/min a) NaOH sample solution (~50 mℓ) (adsorption of Ge; silicate passes into the effluent) b) 0.1 M NaOH (50 mℓ) (elution of remaining silicate) c) Water (until effluent is neutral) d) 1 + 9 HNO_3 (100 mℓ) followed by water (50 mℓ) (elution of Ge) Ge is determined titrimetrically using the mannitol method Before the separation, the sample (30—60 mg) is fused with NaOH (2 g) and the melt is dissolved in water to prepare solution (a) B which accompanies the Ge through the entire procedure is determined spectrographically	6

[a] See also Table 10 in the chapter on Gallium.

Table 2
ANION EXCHANGE SEPARATION OF Ge FROM SYNTHETIC MIXTURES WITH OTHER ELEMENTS

Elements separated	Ion exchange resin, separation conditions, and remarks	Ref.
Ge from As(III and V) and other elements	Dowex® 1-X8 (100-200 mesh; chloride form) Column of 1 cm ID containing 10 g of the resin conditioned with eluent (b) (50 ml) and operated at a flow rate of 1 ml/4.5 min a) Sample solution (10 ml) which is 90% in acetic acid and 10% in 9 M HCl (adsorption of As and many other elements such as U, Th, Fe[III], Co, Cu, Mn, Au, Zn, Cd, Hg, In, Sn, Pb, Sb, Bi, and Mo; Ge passes into the effluent) b) 90% Acetic acid - 10% 9 M HCl (35 ml) (elution of residual Ge) Ge is determined spectrophotometrically with phenylfluorone Only Ca, Mg, Ni, and some Cr accompany Ge into the eluate	3
Ge from U(VI) and other elements	Dowex® 1-X8 (100-200 mesh; chloride form) Column of 0.8 cm ID containing 1 g of the resin conditioned with eluent (b) a) Sample solution (50 ml) consisting of 30% tributyl phosphate (TBP), 60% methylglycol, and 10% 12 M HCl (adsorption of U and other coextracted elements, e.g., Cu and Co; Ge passes into the effluent) b) 30% TBP - 60% methyl glycol - 10% 12 M HCl (elution of residual Ge) Ge is determined spectrophotometrically Before the separation, the Ge is extracted into 30% TBP in kerosine from 6 M HCl, and after evaporation of the kerosine, solution (a) is prepared	4
Ge from B	Anionite AV-17 (OH$^-$ form) a) Test solution of pH 5.3—6 (adsorption of Ge and B at a flow rate of 1 ml/min) b) Water (as a rinse) c) 0.2 M acetic acid (elution of Ge at a flow rate of 3 ml/min) d) 5% NaOH solution (elution of B)	7
Ge from As(III)	Anionite EDE-10P (chloride form) a) Sorption solution of pH 4 (adsorption of Ge; As passes into the effluent) b) 9 M HCl (elution of Ge)	8

Table 3
DETERMINATION OF Ge IN GEOLOGICAL MATERIALS AFTER CATION AND ANION EXCHANGE SEPARATION[a]

Material	Ion exchange resin, separation conditions, and remarks	Ref.
Coal	Amberlite® IR-120 (H⁺ form) and anionite AN-2F (chloride form) Column: 5 × 0.8 cm consisting of a 7:3 mixture of the cation and anion exchange resins, respectively; the column is operated at a flow rate of 1 mℓ/min a) Sample solution (10 mℓ) acidified with conc HCl (2 or 3 drops) (adsorption of Fe[III] on the cation exchanger and of Sn, Sb, As, and Mo on the anionite; Ge passes into the effluent) b) Water (elution of residual Ge) Ge is determined photometrically with phenylfluorone Before the separation, the sample (2 g) is decomposed by fusion with Na$_2$SO$_4$	14
	Nalcite® HCR-X8 (50-100 mesh; H⁺ form, a strongly acid cation exchange resin) and Amberlite® IR-45 (weakly basic anion exchange resin) Column of 1.4 cm ID containing a 7:3 mixture of the cation exchanger (7 g) and the basic resin (3 g) and operated at a flow rate of 1 mℓ/min a) Sample solution (50 mℓ) adjusted to pH 2.0 (Fe[III], Ni, Zn, Ca, Mg, and other elements are adsorbed on the cation exchange resin, while anions such as molybdate are retained by the anionic resin; Ge passes into the effluent) b) Water (150 mℓ) (elution of residual Ge) Ge is determined spectrophotometrically with phenylfluorone Before the separation, the sample (0.3 g) is dry ashed (in the presence of 0.15 g of Na$_2$CO$_3$) and the ash is dissolved in 6 M HCl	12
Mineral raw materials	Cationite KU-2 (H⁺ form) The sample solution is adjusted to 0.1 M HCl and is passed through a column of the resin to adsorb Fe(III), Ti, and V; in the effluent Ge is determined colorimetrically using phenylfluorone	10

[a] See also Table 1 in the chapter on Silicon.

Table 4
CATION EXCHANGE SEPARATION OF Ge FROM SYNTHETIC MIXTURES WITH OTHER ELEMENTS

Material	Ion exchange resin, separation conditions, and remarks	Ref.
Zn	Dowex® 50W (100-200 mesh; Na$^+$ form) Column: 25 × 2 cm containing 50 g of the resin and operated at a flow rate of 0.5 mℓ/cm^2 min a) Sample solution (~8 mℓ) adjusted to pH 1.42 with 10 M NaOH (adsorption of Zn and other metals, e.g., Co, Ni, Mn, and Al) b) Dilute HCl solution (175 mℓ) of pH 1.42 (elution of Ge) After precipitation as GeS$_2$, ^{71}Ge is determined radiometrically Before the separation, the sample (1—2 g, containing ≯500 ppm of Ge) is irradiated with neutrons and then dissolved in 1.4 M HNO$_3$ (5 mℓ), 4 M HF (0.5 mℓ), H$_2$PtCl$_6$ solution (2 drops), and water (3 mℓ) (in the presence of 10—15 mg Ge-carrier)	13
Low melting glasses	Cation exchange resin (H$^+$ form) a) Sample solution (~100 mℓ) containing 1:1 HNO$_3$ (3.5 mℓ) (adsorption of matrix constituents; Ge passes into the effluent) b) Water (200 mℓ) (elution of residual Ge) Ge is determined by potentiometric titration Before the separation, the sample (0.1 g) is decomposed with HF and HNO$_3$	18
Mixtures containing Ge, Sb, As, Se, and Fe(III)	Bio-Rad® AG 50W-X8 (H$^+$ form) a) 0.5 M HNO$_3$ sample solution (adsorption of Fe[III]; into the effluent pass the anions Ge(OH)$_3$O$^-$, SbO$_4^{3-}$, AsO$_4^{3-}$, and SeO$_3^{2-}$) b) 0.5 M HNO$_3$ (elution of remaining anions) Ge and the other elements are determined by atomic absorption spectroscopy Before the separation, Ge and the other elements are collected on ferric hydroxide using adsorbing colloid flotation	11
Mixtures of Ge and Fe(III)	Amberlite® IR-120 From 1 M HCl (3 mℓ), Fe is adsorbed on a column of the resin and in an aliquot of the effluent, Ge is determined spectrophotometrically using the quercetin method Before the separation Ge (<120 μg) is distilled from 6—7 M HCl and coprecipitated with Fe(OH)$_3$ (≃40 mg)	19

REFERENCES

1. **Nelson, F. and Kraus, K. A.**, Anion-exchange studies. XVIII. Germanium and arsenic in HCl solutions, *J. Am. Chem. Soc.*, 77, 4508, 1955.
2. **May, S. and Samadi, A. A.**, Determination of germanium in high-purity iron and aluminum by neutron activation, *Bull. Soc. Chim. Fr.*, No. 4, 1628, 1970.
3. **Korkisch, J. and Feik, F.**, Anion-exchange separation of germanium from arsenic (III and V) and other elements in hydrochloric acid-acetic acid medium, *Sep. Sci.*, 2(1), 1, 1967.
4. **Koch, W. and Korkisch, J.**, Anion-exchange separation of elements that are extractable by tributyl phosphate. II. Separation and spectrophotometric determination of germanium, *Mikrochim. Acta*, No. 1, 101, 1973.
5. **Schindewolf, U. and Irvine, J. W.**, Preparation of carrier-free vanadium, scandium and arsenic activities from cyclotron targets by ion exchange, *Anal. Chem.*, 30, 906, 1958.
6. **Conrad, F. J., Dosch, R. G., Merrill, R. M., and Wanner, D. E.**, Chemical characterization of silicon-germanium thermoelectric alloys, *Anal. Chim. Acta*, 61, 475, 1972.
7. **Dranitskaya, R. M., Tsȳbul'kova, L. P., and Gavril'chenko, A. I.**, Separation of germanium from boron by ion-exchange chromatography, *Zh. Anal. Khim.*, 22(3), 448, 1967.
8. **Dranitskaya, R. M. and Liu, C. C.**, Separation of germanium from arsenic(III) by ion-exchange chromatography, *Zh. Anal. Khim.*, 19(6), 769, 1964.
9. **Everest, D. A. and Harrison, J. C.**, The chemistry of quadrivalent germanium. V. Ion exchange studies of germanate solutions containing polyhydric alcohols, *J. Chem. Soc.*, p. 4319, 1957.
10. **Shishkov, D. A. and Peteva-Iordanova, S.**, Ion exchange separation of germanium(IV) from vanadium(V) and titanium(IV), *C.R. Acad. Bulg. Sci.*, 17(11), 1027, 1964.
11. **DeCarlo, E. H., Zeitlin, H., and Fernando, Q.**, Simultaneous separation of trace levels of germanium, antimony, arsenic, and selenium from an acid matrix by adsorbing colloid flotation, *Anal. Chem.*, 53, 1104, 1981.
12. **Cabbell, T. R., Orr, A. A., and Hayes, J. R.**, Ion-exchange separation of germanium, *Anal. Chem.*, 32, 1602, 1960.
13. **Machiroux, R. and Mousty, F.**, Determination of impurities in zinc by neutron-activation. II. Determination of germanium, *Anal. Chim. Acta*, 48, 219, 1969.
14. **Salikova, G. E. and Sevryukov, N. N.**, Determination of germanium in coal, *Zavod. Lab.*, 36(1), 25, 1970.
15. **Seidl, J., Štamberg, J., and Hrbková, E.**, Selective ion-exchanger based on a phenylfluorone derivative, *J. Appl. Chem.*, 12(11), 500, 1962.
16. **Seidl, J. and Štamberg, J.**, A new type of selective ion-exchanger, *Chem. Ind.*, No. 38, 1190, 1960.
17. **Kraft, G., Dosch, H., and Gabbert, K.**, Separation of traces of germanium from zinc-sulfate solutions by means of polyhydric phenol ion-exchange resins, *Fresenius' Z. Anal. Chem.*, 267, 106, 1973.
18. **Nemirovskaya, E. M. and Kolosova, G. M.**, Potentiometric determination of germanium dioxide in low-melting glasses, *Zavod. Lab.*, 45(4), 311, 1979.
19. **Oka, Y. and Matsuo, S.**, Studies on spectrophotometry. X. Determination of microgram quantities of germanium with quercetin, *J. Chem. Soc. Jpn. Pure Chem. Sect.*, 76(6), 610, 1955.

TIN

Ion exchange separations of Sn from complex materials such as geological and industrial samples are in most cases performed by the use of strongly basic anion exchange resins on which Sn(IV) is adsorbed as an anionic complex with chloride or fluoride. Occasionally, separations can also be effected by adsorbing tin chelates on anionic resins or by the application of cation exchange resins.

ANION EXCHANGE RESINS

General

The Sn(IV) species predominating in 0.2 to 2, 2 to 6, and 6 to 12 M hydrochloric acid (HCl) solutions are $SnCl_4$, $SnCl_5^-$, and $SnCl_6^{2-}$, respectively. The anionic species are appreciably retained by strongly basic anion exchange resins of the quaternary ammonium type.[1] Thus, on Dowex® 1-X8 distribution coefficients for Sn(IV) of ~20, ~100, ~500, ~2 × 10³, ~5 × 10³, ~8 × 10³, ~7 × 10³, ~5 × 10³, and ~2 × 10³ have been measured in 0.25, 0.5, 1, 2, 4, 6, 8, 10, and 12 M HCl, respectively. At the same acidities Sn(II) is less strongly retained and with increasing HCl concentration its adsorption decreases linearly from a distribution value of ~500 in 1 M HCl to a value of ~10 in the 12 M acid. At very low acidities, e.g., in 0.5 M HCl, the adsorption of Sn(II) is somewhat higher than that of Sn(IV) which hydrolyzes at <0.25 M HCl. Consequently, when adsorbing Sn from very dilute HCl solutions, as, for example, from 0.4,[2] 0.5 to 1,[3] and 2 M HCl,[2,4-6] the oxidation state of this element does not have much influence on its adsorbability (in 2 M HCl Sn[II] is still strongly adsorbed with a distribution coefficient of ~300). Coadsorbed with the Sn at these low acid concentrations are Sb(III), Bi, Pb, Te(IV), Tl(III), Zn, Cd, Hg(II), Cu(I), Ag, Au(III), and Pt metals, while virtually all other metal ions are not retained so that rather selective separations of Sn can be obtained (see Table 1*). At acid concentrations exceeding 2 M other elements, as, for example, Fe(III), Ga, Mo, UO_2(II), Co, Ti, Zr, and Hf, will be coadsorbed with Sn(IV) so that less selective separations are achieved at high HCl molarities. Nevertheless, for separations of Sn from complex matrices the following systems have been employed frequently: 3 M HCl (see Tables 2 and 6),[7,8] 5 M HCl (see Tables 2 and 6),[9,10] 6 M HCl (see Table 1),[11-13] 6.5 M HCl (see Table 3),[14] 7 M HCl (see Tables 3 and 6),[15,16] and 8 M HCl (see Table 2).[17,18] From most of these media, Pb, Cu(I), and Ag are not coadsorbed with the Sn so that more selective separations from these metals are achieved at high than at low acid concentrations. Before adsorbing Sn from solutions of high acid molarities any Sn(II) (which is only weakly adsorbed from such systems) can be oxidized by exposure of the solution to air or additional treatment with Br,[19] K chlorate or perchlorate,[20,21] or by using H_2O_2 as a holding oxidant.[16] This oxidation of Sn to the tetravalent state is also necessary whenever this element is to be retained on basic resins from solutions in which Sn(IV) forms adsorbable anionic chelates with organic complexing agents, as, for example, in 0.05[22] or 0.1 M[23] oxalic acid (see Table 7), malonic acid solutions of pH 4.8 (see Table 4),[20,21] and 0.02 M tartaric acid-0.1 M sulfuric acid (see Table 7).[24] Sn(II) is not complexed by these organic reagents.

From hydrofluoric acid (HF) solutions Sn(IV) is strongly adsorbed on anion exchange resins such as Dowex® 1, especially at relatively low acid molarities.[25,26] This adsorption of Sn (most probably as SnF_6^{2-}) decreases with increasing acid concentration. Thus, in the 2, 6, and 17 M acid, distribution coefficients for Sn(IV) of ~500, ~100, and ~10, respectively, have been measured. Coadsorbed with Sn from dilute HF solutions are Be, Sc,

---

* Tables for this chapter appear at the end of the text.

Ti, Zr, Hf, Nb, Ta, Mo, W, U, Sb, Hg, and a few other elements also forming adsorbable anionic fluoride complexes. On the other hand, Sn is readily separated from common elements, e.g., alkaline earth elements and the alkalies, so that this adsorption from HF media can be used for the isolation of Sn from complex mixtures (see Tables 1, 3, and 5).[11,12,14,27]

Adsorption of Sn as anionic thiostannate can be effected from an alkaline solution of Na polysulfide (see Table 2).[28]

The elution of Sn that has been adsorbed from the systems discussed above can be effected by the use of acids or solutions of alkali hydroxides. The most popular acid eluents are dilute solutions of sulfuric acid[4,8,11,20-24] and nitric acid[2,6,17-19] such as the 1 or 2 M acids. Also, 2 M perchloric acid[29] and 0.2 M oxalic acid - 2 M ammonium nitrate[14] have been used as eluents. In these acid systems Sn(IV) does not form adsorbable anionic complexes so that effective elutions can be achieved. Alkaline eluents which elute Sn as the anionic species $Sn(OH)_6^{2-}$ include 0.5 M KOH,[28] 0.5 M NaOH - 0.5 M NaCl,[5] and 1,[3] 2,[15] and 6 M NaOH.[16]

Applications

Adsorption of Sn on anion exchange resins has variously been employed in connection with the determination of this element in geological materials (see Table 1) and industrial products (see Tables 2 to 5) and to separate Sn from synthetic mixtures with other elements (see Tables 6 and 7). For the determination of Sn in biological materials the procedure illustrated in Table 5 (chapter on Mercury, Volume IV), can be used.

CATION EXCHANGE RESINS

General

Sn is adsorbable on strongly acid resins from very dilute HCl solutions, but the adsorbability rapidly falls off at higher acid concentrations, since nonadsorbable chloride complexes are formed (see preceding section). Thus, in 0.2, 0.5, 1, 2, and 3 M HCl solutions distribution coefficients for Sn(IV) of ~100, ~6, ~2, <1, and <0.5, respectively, have been measured. In hydrobromic acid (HBr) systems, the adsorbability of Sn(IV) decreases from a distribution value of 16 in the 0.5 M acid to a value of 1.3 near 4 M HBr. A very similar behavior of Sn is shown in HF solutions of comparable concentrations, which is not surprising since Sn(IV) forms anionic fluoride complexes in these media (see preceding section). Consequently, adsorption of Sn on cation exchange resins as a means for separating this element from accompanying elements can only be effected at very low acid concentrations or by the use of solutions containing noncomplexing acids, as, for example, very dilute sulfuric acid (see Table 8)[30] or at relatively high pH values, e.g., pH 4 (see Table 9),[31] or from 2% tartaric acid solution (see Table 9).[32] For the elution of the adsorbed Sn 2[31] or 5 M HCl[30] or alkaline solutions such as 1 M NaOH or ammonia solution[32] can be employed. By using 0.5 M HCl as the eluent, Sn can be separated from a number of coadsorbed elements which include Mn, Fe, Co, Ni, Cu, Zn, Mg, and Ca (see Table 9).[33]

Ionic species of organotin compounds can be adsorbed and fractionated on cation exchange resins using special procedures.[34,35]

Applications

From Table 8 it is seen that cation exchange separations of Sn from accompanying elements have been used in connection with the determination of this element in industrial products. Furthermore, Sn can be separated from synthetic mixtures with other elements by use of the procedures illustrated in Table 9.

To determine Sn in waters and plants, the procedures described in Tables 20 and 38 (see in the chapters on Cobalt [Volume V] and Zinc [Volume IV]) and 45 (see in the chapter on Copper [Volume III]), respectively, can be used.

Table 1
DETERMINATION OF Sn IN GEOLOGICAL MATERIALS AFTER ANION EXCHANGE SEPARATION[a]

Material	Ion exchange resin, separation conditions, and remarks	Ref.
Rocks, sediments, and soils	Amberlite® IRA-400 (100-200 mesh; Cl⁻ form) Column: 10 × 1 cm conditioned with 6 M HCl (20 mℓ) and operated at a flow rate of 1 mℓ/min a) ≃6 M HCl sample solution (25 mℓ) containing 5 mℓ of conc HF (adsorption of Sn; major constituents pass into the effluent) b) 6 M HCl (45 mℓ) (elution of remaining nonadsorbed elements) c) 1 M citric acid (25 mℓ) (as a rinse) d) Water (25 mℓ) (as a rinse) e) 1 M H$_2$SO$_4$ (45 mℓ) (elution of Sn) After further purification by toluene extraction of SnI$_4$, the Sn is determined spectrophotometrically with phenylfluorone; the limit of detection is 0.1 μg of Sn and in the determination of 1.8 ppm of Sn in basalt, the coefficient of variation was 1.8% (six determinations) Before the separation, the sample (∼0.1 g containing 0.74—4.4 ppm of Sn) is decomposed with HCl-HF; if organic matter is present H$_2$O$_2$ is added to the acids Limestone and Mn nodules are decomposed by shaking with 6 M HCl overnight	11
Silicate rocks	Dowex® 2 (200-400 mesh) a) 2 M HF sample solution (adsorption of ^{113}Sn and Sb; into the effluent pass Fe, In, and the alkalies) b) 2 M HF (elution of ^{113m}In which is formed on the resin by the reaction: ^{113}Sn -β⁻ → ^{113m}In) Afterwards, the eluted In is adsorbed on a column of cellulose phosphate and determined radiometrically Before the ion exchange separation step the sample (0.1—0.2 g) is irradiated with neutrons, then decomposed by fusion with Na$_2$O$_2$, and after dissolution of the melt in HNO$_3$, ^{113}Sn (formed by the reaction ^{112}Sn [n,γ] ^{113}Sn) is purified by repeated coprecipitation with ferric hydroxide; interfering ^{46}Sc is removed by precipitation as ScF$_3$	27
	Dowex® 1-X8 (50-100 mesh; oxalate form) a) Sample solution which is 0.05 M in oxalic acid (adsorption of Sn) b) Water (as a rinse) c) 1 M H$_2$SO$_4$ (elution of Sn) Sn is determined fluorometrically Before the separation, the sample is decomposed with mineral acids	22
Chondritic meteorites	Dowex® 1-X8 (200-400 mesh; thiocyanate form) Column: 7 × 1 cm containing 3 g of the resin a) Sample solution (20 drops) which is 0.5 M in NH$_4$SCN and 2 M in HCl (adsorption of Sn, Sb[III], and As) b) 0.5 M NH$_4$SCN - 0.5 M HCl (15 mℓ) (elution of As) c) 0.005 M NH$_4$SCN - 0.5 M HCl (10 mℓ) (as a rinse) d) 1 N H$_2$SO$_4$ (60 mℓ) (elution of >90% of Sb) e) 0.5 M NaCl - 0.5 M NaOH (30 mℓ) (elution of Sn) After further radiochemical purification steps, Sn, As, and Sb are determined radiometrically Before the ion exchange separation, the sample (0.2—0.5 g) is	5

Table 1 (continued)
DETERMINATION OF Sn IN GEOLOGICAL MATERIALS AFTER ANION EXCHANGE SEPARATION[a]

Material	Ion exchange resin, separation conditions, and remarks	Ref.
	irradiated with neutrons, decomposed by fusion with Na_2O_2 (6 g), and Sn, As, and Sb are preconcentrated by sulfide precipitation in the presence of carriers (10 mg each)	
	After dissolution of the sulfides in aqua regia, Sb is reduced to Sb(III) with $NH_2OH\text{-}HCl$	
Seawater	A) First column operation (Dowex® 1-X8, 100-200 mesh, chloride form)	2
	Column: 6 × 1.1 cm conditioned with 2 M HCl	
	a) Acidified sample (750 mℓ) (adsorption of Sn, Fe [partly], and Mo; into the effluent pass the components of the water matrix)	
	b) 2 M HNO_3 (100 mℓ) (elution of Sn; coeluted are Fe and Mo)	
	The Sn eluate is evaporated in the presence of 1 M H_2SO_4 (1.5 mℓ) and Sn is further purified using the following separation	
	B) Second column operation (same resin in a column: 6 × 1 cm conditioned with 0.4 M HCl)	
	a) 0.4 M HCl (20 mℓ) (adsorption of Sn)	
	b) 0.4 M HCl (200 mℓ) (elution of Fe and Mo)	
	c) 2 M HNO_3 (50 mℓ) (elution of Sn)	
	Sn is determined spectrophotometrically	
	Before the first column separation, the sample (500 mℓ) is made 2 M in HCl by the addition of 6 M HCl (250 mℓ)	

[a] See also Table 1 in the chapter on Silver, Volume III and Table 86 in the chapter on Rare Earth Elements, Volume I.

Table 2
DETERMINATION OF Sn IN METALS AND ALLOYS AFTER ANION EXCHANGE SEPARATION[a]

Material	Ion exchange resin, separation conditions, and remarks	Ref.
Pb	Dowex® 1-X8 (Cl⁻ form) a) 8 M HCl sample solution (adsorption of Sn; Pb and Mn pass into the effluent) b) 8 M HCl (elution of residual Pb) c) 1 M HNO$_3$ (40 mℓ) (elution of Sn) Sn is determined colorimetrically Before the ion exchange separation, the sample (10—15 g) is dissolved in HNO$_3$ (1 + 1) and Sn is preconcentrated by co-precipitation with MnO$_2$ (a small amount of Pb is also precipitated) which is dissolved in HCl to prepare the sample solution (a)	17, 18
High-purity Cr	Dowex® 1-X8 (100-200 mesh; chloride form) Column: 6 × 1 cm conditioned with 8 M HCl (30 mℓ) saturated with Br$_2$ a) 8 M HCl sample solution (adsorption of Sn[IV]; into the effluent pass Cr and Pb) b) 8 M HCl (50 mℓ) (elution of residual nonadsorbed elements) c) 1 M HNO$_3$ (35 mℓ) (elution of Sn) Sn is determined polarographically Before the separation, the sample (<5 g) is dissolved in 6 M HCl (<50 mℓ) and air is bubbled through the solution to oxidize Cr(II) to Cr(III); then conc HCl (saturated with Br$_2$) is added to prepare solution (a)	19
Heat-resistant Cu-based alloys	Anionite (Cl⁻ form) Column: 15—20 × 0.9—1.2 cm conditioned with 3 M HCl a) 3 M HCl sample solution (15 mℓ) (adsorption of Sn; Al, Cu, Ni, and Cr pass into the effluent) (flow rate: 0.5 mℓ/min) b) 3 M HCl (60 mℓ) (elution of remaining Cu and other nonadsorbed metals) (flow rate: 0.8—1 mℓ/min) c) 0.5 M HCl (150 mℓ) (elution of Sn at a rate of 1 mℓ/min) Sn is determined spectrophotometrically with phenylfluorone Before the separation, the sample (1—2 g) is dissolved in HCl + HNO$_3$	7
W products	Anionite EDE-10P (Cl⁻ form) Column containing 8—9 g of the resin conditioned with 5 M HCl and operated at a flow rate of 3 mℓ/min a) ~5 M HCl sample solution (~90 mℓ) (adsorption of Sn) b) 5 M HCl (as a rinse) c) 0.5 M HCl (150—300 mℓ) (elution of Sn) Sn is determined titrimetrically Before the separation, the sample (0.5—1 g containing ≯80 mg of Sn) is fused with NaOH	9
Alloys	Dowex® 2 (20-50 mesh; OH⁻ form) Column: 12—13 × 1.7 cm a) Alkaline sample solution (100—150 mℓ) containing 3% Na polysulfide solution (30—50 mℓ) (adsorption of Sn and Sb as anionic thio-complexes, e.g., thiostannate) b) 2% Na-polysulfide solution (as a rinse) c) Water (removal of residual polysulfide solution) d) 0.5 M KOH (900 mℓ) (elution of Sn) e) 3.5 M KOH (1200 mℓ) (elution of Sb) Sn is determined gravimetrically as SnO$_2$ and Sb by using a titrimetric method	28

Table 2 (continued)
DETERMINATION OF Sn IN METALS AND ALLOYS AFTER ANION EXCHANGE SEPARATION[a]

Material	Ion exchange resin, separation conditions, and remarks	Ref.
	Before the separation, the sample (0.2—0.3 g) is decomposed by fusion with S and K_2CO_3 to obtain the polysulfides of the elements	

[a] See also Table 11 in the chapter on Cadmium, Volume IV; Table 23 in the chapter on Zinc, Volume IV; Tables 16 and 26 in the chapter on Copper, Volume III; Table 9 in the chapter on Gold, Volume III; and Table 96 in the chapter on Rare Earth Elements, Volume I.

Table 3
DETERMINATION OF Sn AND OTHER ELEMENTS IN Fe AND STEEL AFTER ANION EXCHANGE SEPARATION[a]

Material	Ion exchange resin, separation conditions, and remarks	Ref.

High-purity Fe	Dowex® 1 (F^- and Cl^- forms)
	 A) First column operation (on F^--form resin)
	  a) 6 M HF sample solution (adsorption of Sn, As, and Te; into the effluent pass Co, Cu, In, and Zn)
	  b) 17 M HF (elution of Sn, As, and Te)
	 B) Second column operation (on Cl^--form resin) (fractionation of Sn, As, and Te)
	  a) 6.5 M HCl (adsorption of Sn, As, and Te)
	  b) 11 M HCl (elution of As)
	  c) 0.1 M oxalic acid (elution of Te)
	  d) 0.2 M oxalic acid - 2 M NH_4NO_3 (elution of Sn)
	 Sn and the other elements are determined radiometrically
	 Before the first column operation, the sample is irradiated with neutrons and Fe(III) is removed by extraction into 2-chloroethyl ether from 11.4 M HCl
		14

Steel	Bio-Rad® AG1-X8 (Cl^- form)
	  a) ≃7 M HCl sample solution (adsorption of Sn, Sb, Fe[III], and other elements)
	  b) 7 M HCl (as a rinse)
	  c) 1 M HCl (elution of Fe[III] and other coadsorbed elements)
	  d) 2 M NaOH (elution of Sn; Sb is not coeluted)
	 Sn is determined polarographically
		15

	Anionite Anex L
	 Column operated at a flow rate of 1 mℓ/min
	  a) ~2 M HCl sample solution (25 mℓ, containing 1 g of the sample) (adsorption of Sn; Fe passes into the effluent)
	  b) 2 M HCl (50 mℓ) (elution of residual Fe)
	  c) 0.8 M HNO_3 (98 mℓ) (elution of Sn)
	 Before the separation, the sample is dissolved in 6 M HCl + H_2O_2
		6

[a] See also Table 2 in the chapter on Tungsten, Volume IV and Table 3 in the chapter on Bismuth.

Table 4
DETERMINATION OF Sn AND OTHER ELEMENTS IN ALLOYS AFTER ANION EXCHANGE SEPARATION

Material	Ion exchange resin, separation conditions, and remarks	Ref.
Alloys (e.g., white metal)	Amberlite® IRA-400 or -410 (50-72 mesh BSS; malonate form) Column: 30 cm × 1 cm^2 a) Sample solution (10-mℓ aliquot equivalent to 10 mg of the sample) adjusted to pH 4.8 with 9 M NH$_3$ solution and containing malonic acid (2—3 g) (adsorption of Sn[IV], Sb[V], and Pb as anionic malonate complexes) (at a flow rate such that the whole sorption takes 20 min) b) 3% Malonic acid solution (3 × 1 mℓ) (as a rinse) c) 3% Malonic acid solution of pH 4.8 (120 mℓ) (elution of Sb and Pb at a flow rate of 0.5 mℓ/cm^2 min) d) 9 N H$_2$SO$_4$ (100 mℓ) (elution of Sn at a flow rate of 1 mℓ/cm^2 min) Sn, Sb, and Pb are determined colorimetrically Before the separation, the sample (0.1 g) is dissolved in aqua regia and KClO$_3$ or KClO$_4$ (1—2 g) is added to the cold solution to oxidize Sn and Sb to the higher valency states	20, 21

[a] See also Tables 18, 21, and 22 in the chapter on Copper, Volume III and Table 20 in the chapter on Zinc, Volume IV.

Table 5
DETERMINATION OF Sn IN SALT AND GLASS AFTER ANION EXCHANGE SEPARATION

Material	Ion exchange resin, separation conditions, and remarks	Ref.
Common salt	Dowex® 1-X8 (50-100 mesh; chloride form) Column: 7 × 1 cm operated at a flow rate of 1—2 mℓ/min a) 0.5—1 M HCl sample solution (adsorption of Sn) b) 0.5 M HCl (50 mℓ) (elution of Fe, Ge, Ti, Zr, Sb, and Se) c) 1 M NaOH (30 mℓ) (elution of Sn[IV] plus Bi and Hg if present) Sn is determined spectrophotometrically using the phenylfluorone method Before the ion exchange separation, Sn(IV) is coprecipitated with Fe(OH)$_3$ ($\simeq$1 mg of Fe) at pH 6—8, thus separating it from Mo The method has also been applied to seawater and bittern and allows the determination of as little as 1 μg Sn per kilogram	3
Glass[a]	Rohm and Haas® SB-2 anion exchange paper (chloride form) The 1% HF sample solution (10 mℓ, containing $\not>$250 μg of Sn and obtained by the etching of Sn-containing glass) is diluted with conc HCl (11 mℓ) and water (4 mℓ) to a mixture which is 6 M HCl; into this solution a circle (diameter: 1 5/16 in.) of the resin paper is immersed for 18 h to adsorb the Sn; subsequently, Sn is determined directly on the paper by X-ray fluorescence spectrography; 5 μg of Sn can be detected with a precision of ±10%; other elements present in glass do not interfere	12

[a] See also Table 10 in the chapter on Gold, Volume III.

Table 6
ANION EXCHANGE SEPARATION OF Sn IN HCL MEDIA[a]

Elements separated	Ion exchange resin, separation conditions, and remarks	Ref.
Sn from Ni, Cu, Zn, and Cr(III)	XAD-4 anion exchange resin (low-capacity strong-base resin, chloride form) Column: 2.5 × 0.2 cm (in a liquid chromatograph) operated at a forced flow rate of 1.5 mℓ/min a) Injection of ~5 M HCl sample solution (adsorption of Sn) b) 5 M HCl (elution of Ni and other metals) c) 1 M HCl (elution of Sn[IV])	10
Sn from phosphate and fluoride	Strong-base anion exchange resin (acetate form) Column: 5.5 × 0.8 cm operated at a flow rate of 1.4 mℓ/min a) Sample solution (aliquot containing 4—40 µg of Sn consisting of 7 M HCl [4 mℓ] and 3% H_2O_2 [1 mℓ]) (adsorption of Sn[IV]) b) 7 M HCl (18 mℓ) (elution of phosphate and fluoride) c) 6 M NaOH (12 mℓ) (elution of Sn[IV])	16
Sn, Sb, and Te	Anionite ASD-2 (or Dowex® 1-X8) (grain size: 30 µm; Cl⁻ form) Column: 10 × 0.2 cm containing 100—150 mg of the resin After adsorption of the elements the following eluents are used a) 3 M HCl (2 mℓ) (elution of Sb) b) 1 M HCl (2.2 mℓ) (elution of Te) c) 2 M $HClO_4$ (1 mℓ) (elution of Sn)	29

[a] See also Table 16 in the chapter on Copper, Volume III.

Table 7
ANION EXCHANGE SEPARATION OF Sn IN THE PRESENCE OF ORGANIC COMPLEXING AGENTS[a]

Elements separated	Ion exchange resin, separation conditions, and remarks	Ref.
Sn(IV) from As(III) and Sb(III)	Dowex® 1-X8 (100-200 mesh; tartrate form) Column: 5.5 cm × 0.79 cm² After adsorption of the elements the following eluents are used a) 0.02 M tartaric acid - 0.1 M H_2SO_4 (~100 mℓ) (elution of As[III]) b) 0.4 M tartaric acid - 0.1 M H_2SO_4 (100 mℓ) (elution of Sb[III]) c) 2 M H_2SO_4 (elution of Sn[IV])	24
Sn(IV) from Sb(V) and Te(IV)	Dowex® 1 (200-300 mesh; oxalate form) Column: 10.5 cm × 0.0384 cm² operated at a flow rate of 0.011—0.017 mℓ/min a) 0.1 M oxalic acid (2 mℓ) (adsorption of Sn and Sb; Te passes into the effluent) b) 0.1 M oxalic acid adjusted to pH 4.8 (10 mℓ) (elution of Sb) c) 1 M H_2SO_4 (10 mℓ) (elution of Sn) The method was used to separate tracer quantities of ^{113}Sn, ^{124}Sb, and ^{125}Te	23

[a] See also Reference 38.

Table 8
DETERMINATION OF Sn IN INDUSTRIAL PRODUCTS AFTER CATION EXCHANGE SEPARATION[a]

Material	Ion exchange resin, separation conditions, and remarks	Ref.
Conc H_2O_2	Zeo-Karb® 225 (50-100 mesh; H^+ form) Column: 15 × 0.75 cm a) 0.02 M H_2SO_4 sample solution (50 mℓ, equivalent to 20 mℓ of the sample) (adsorption of Sn) b) 0.02 M H_2SO_4 (20 mℓ) (as a rinse) c) 5 M HCl (15 mℓ) (elution of Sn; the eluate is recycled twice more and then the column is washed with 20 mℓ of water) Sn is determined polarographically	30
Cu alloys	Zerolite® 225 (H^+ form) Column: 12 × 1.2 cm operated at a flow rate of 5 mℓ/min a) Sample solution containing Sn as SnF_6^{2-} (adsorption of Cu and other elements; Sn passes into the effluent) b) Distilled water (elution of residual Sn) Sn is determined titrimetrically or gravimetrically Before the separation, the sample (1 g) is dissolved in HNO_3 and HF	36

[a] See also Tables 51 and 56 in the chapter on Copper, Volume III; Table 49 in the chapter on Rare Earth Elements, Volume I; and Table 7 in the chapter on Bismuth.

Table 9
CATION EXCHANGE SEPARATION OF Sn FROM SYNTHETIC MIXTURES WITH OTHER ELEMENTS

Elements separated	Ion exchange resin, separation conditions, and remarks	Ref.
Sn(IV), Sb(III), Pb, Cu(II), and Fe(III)	Merck I cation exchange resin (or Dowex® 50-X8 or Vionit® CS-3) After adsorption of the elements on a column of the resin the following eluents are used a) 2% Tartaric acid solution (elution of Sb, Pb, Cu, and Fe) b) 1 M NaOH (elution of Sn) Alternatively, if Sb is absent, the column is washed with water and the Sn can then be eluted with eluent (b), leaving the other elements on the resin; in place of 1 M NaOH the elution of Sn as stannate can also be effected with 1 M NH_3 solution[37] To determine Sn in bronze, the sample solution is passed through a column of Vionit® CS-3 (H^+ form), the column is washed with water, and the Sn is eluted with 1 M NH_3 solution and determined spectrophotometrically	32
Sn from binary mixtures with Mn, Fe, Co, Ni, Cu, Zn, Mg, and Ca	Zerolit® 225 (H^+ form) Column: 15 cm × 4.9 cm^2 After adsorption of the elements (3—300-mg amounts) the following eluents are used a) 0.5 M HCl (≯220 mℓ) (elution of Sn) b) 2 M HCl (≯700 mℓ) (elution of all other elements except Ca) c) 4 M HCl (200 mℓ) (elution of Ca)	33
Sn, Pb, and Cd from phosphate	Dowex® 50-X8 (20-50 mesh; H^+ form) Column: 20 × 1 cm a) Sample solution (20 mℓ) adjusted to pH 4 (adsorption of Sn, Pb, and Cd) b) Water (until a neutral effluent is obtained) (elution of phosphate) c) 2 M HCl (20 mℓ) followed by water (until neutral) (elution of Sn, Pb, and Cd) Sn, Pb, and Cd are determined polarographically	31

REFERENCES

1. **Everest, D. A. and Harrison, J. H.**, Anion exchange studies of solutions of stannic chloride in hydrochloric acid, *J. Chem. Soc.*, p. 1439, 1957.
2. **Kodama, Y. and Tsubota, H.**, Determination of tin in seawater by anion-exchange separation and spectrophotometry with catechol violet, *Jpn. Analyst*, 20(12), 1554, 1971.
3. **Shimizu, K. and Ogata, N.**, Determination of traces of tin in common salt with phenylfluorone, *Jpn. Analyst*, 12(6), 526, 1963.
4. **Khalizova, V. A., Alekseeva, A. Ya., Smirnova, E. P., and Krasyukova, N. G.**, Ion-exchange separation of tin in analysis of mineral raw materials. Photometric determination of tin, *Zh. Anal. Khim.*, 25(8), 1525, 1970.
5. **Hamaguchi, H., Onuma, N., Hirano, Y., Yokoyama, H., Bando, S., and Furukawa, M.**, The abundance of arsenic, tin and antimony in chondritic meteorites, *Geochim. Cosmochim. Acta*, 33, 507, 1969.
6. **Janoušek, I. and Študlar, K.**, Photometric determination of tin in steel, *Hutn. Listy*, 15(11), 889, 1960.
7. **Kurbatova, V. I. and Stepin, V. V.**, Ion-exchange chromatography in the determination of tin in heat-resistant and copper-based alloys, *Tr. Vses. Nauchno Issled. Inst. Stand. Obraztsov Spekr. Etalonov*, 1, 14, 1964.
8. **Marley, J. L. and Articolo, O. J.**, Determination of tin in zircaloy and uranium-zircaloy: colorimetric procedure, Report KAPL-M-JLM-2, U.S. Atomic Energy Commission, 1957.
9. **Zelenina, T. P., Gladÿsheva, K. F., and Zinov'eva, L. D.**, Determination of tin in tungsten products, *Sb. Nauchn. Tr. Vses. Nauchno Issled. Gornometall. Inst. Tsvetn. Met.*, No. 9, 124, 1965.
10. **Gjerde, D. T. and Fritz, J. S.**, Chromatographic separation of metal ions on macroreticular anion-exchange resins of low capacity, *J. Chromatogr.*, 188, 391, 1980.
11. **Smith, J. D.**, Spectrophotometric determination of traces of tin in rocks, sediments and soils, *Anal. Chim. Acta*, 57, 371, 1971.
12. **Chamberlain, B. R. and Leech, R. J.**, Determination of microgram quantities of tin(IV) by a combined ion-exchange X-ray fluorescence technique, *Talanta*, 14(5), 597, 1967.
13. **Gibalo, I. M., Kamenev, A. I., Agasyan, E. P., Mymrik, N. A., and Chernysheva, A. Yu.**, Analysis of lead tin telluride single crystals, *Zh. Anal. Khim.*, 28(10), 2056, 1973.
14. **Artyukhin, P. I., Gil'bert, E. N., and Pronin, V. A.**, Determination of impurities in iron by neutron activation, *Zh. Anal. Khim.*, 22(1), 111, 1967.
15. **Nens, C. and Molina, R.**, Determination of tin in steel by pulse polarography under reducing or oxydizing conditions, *Analusis*, 4(1), 1, 1976.
16. **Speirs, R. L.**, Photometric determination of small amounts of tin with phenylfluorone, *J. Dent. Res.*, 41, 909, 1962.
17. **Kojima, M.**, Colorimetric determination of tin in non-ferrous metals by the use of oxidised haematoxylin. I. Tin in metallic zinc, *Jpn. Analyst*, 6(3), 139, 1957.
18. **Kojima, M.**, Colorimetric determination of tin in nonferrous metals by the use of oxidised haematoxylin. II. Tin in lead, *Jpn. Analyst*, 6(3), 142, 1957.
19. **Kawase, A. and Ogawa, H.**, Chemical analysis of high purity chromium. X. Determination of tin, *Jpn. Analyst*, 11(11), 1155, 1962.
20. **Dawson, J. and Magee, R. J.**, The determination of antimony, tin and lead, *Mikrochim. Acta*, No. 3, 330, 1958.
21. **Dawson, J. and Magee, R. J.**, The anion-exchange separation of tin and antimony, *Mikrochim. Acta*, No. 3, 325, 1958.
22. **Huffman, C. and Bartel, A. J.**, Ion-exchange separation of tin from silicate rocks, U.S. Geological Survey, Prof. Pap., No. 501-D, 131, 1964.
23. **Smith, G. W. and Reynolds, S. A.**, Anion-exchange separation of tin, antimony and tellurium, *Anal. Chim. Acta*, 12, 151, 1955.
24. **Casal, A. R. and Zalba, M. S.**, Study of anion exchange processes in tartaric acid-sulfuric acid media, *Afinidad*, 38, 405, 1981.
25. **Faris, J. P.**, Adsorption of the elements from hydrofluoric acid by anion-exchange, *Anal. Chem.*, 32, 520, 1960.
26. **Faix, W. G., Caletka, R., and Krivan, V.**, Element distribution coefficients for hydrofluoric acid/nitric acid solutions and the anion exchange resin Dowex 1-X8, *Anal. Chem.*, 53, 1719, 1981.
27. **Das, H. A., van Raaphorst, J. G., and Umans, H. J. L. M.**, Routine determinations of Al, K, Cr, and Sn in geochemistry by neutron activation analysis, Report RCN-103, Reactor Centrum Nederland, February 1969.
28. **Klement, R. and Kühn, A.**, Anion-exchange separation of arsenic, antimony and tin, *Fresenius' Z. Anal. Chem.*, 152, 146, 1956.
29. **Rybakov, V. N. and Stronskii, I. I.**, Separation of tin, antimony and tellurium on anionites, *Zh. Neorg. Khim.*, 4(11), 2449, 1959.

30. **Reynolds, G. F.,** The polarographic determination of tin in hydrogen peroxide after use of a cation-exchange resin, *Analyst (London),* 82, 46, 1957.
31. **Bourbon, P., Esclassan, J., and Vandaele, J.,** Simultaneous analysis for tin, lead and cadmium by pulse polarography, *Analusis,* 10(2), 104, 1982.
32. **Vladescu, L. and Voicu, D.,** Utilization of sodium hydroxide in separation of tin with ion-exchange resins. II. Cation-exchange separations, *Rev. Roum. Chim.,* 25(8), 1269, 1980.
33. **Burriel-Marti, F. and Alvarez-Herrero, C.,** Cation-exchange separation of tin from manganese, from iron, from cobalt, from nickel, from copper, from zinc, from magnesium or from calcium. X, *Chim. Anal.,* 48(11), 602, 1966.
34. **Neubert, G. and Andreas, H.,** Analysis of organotin compounds, *Fresenius' Z. Anal. Chem.,* 280, 31, 1976.
35. **Jewett, K. L. and Brinckman, F. E.,** Speciation of trace di- and triorganotins in water by ion-exchange HPLC-GFAA, *J. Chromatogr. Sci.,* 19, 583, 1981.
36. **Burriel-Marti, F., Gárate, M. E., and Gárate, M. T.,** Determination of tin in alloys of high copper content, *An. Soc. Esp. Fis. Quim. B,* 58(3) 269, 1962.
37. **Vladescu, L. and Pop, C. D. M.,** Separation of tin from elements that interfere in its quantitative determination, *Rev. Chim. (Bucharest),* 30(12), 1245, 1979.
38. **Casal, A. R. and Zalba, M. S.,** Study of the anion exchange process in citric-sulfuric acid medium. Separation of arsenic(III), antimony(III) and tin(IV), *An. Quim. Ser. B,* 81(2), 203, 1985.

LEAD

Most methods so far reported for the ion exchange separation of Pb from solutions of complex composition, such as those from natural materials, are based on its adsorption on strongly basic resins. On exchangers of this type Pb is effectively retained from hydrochloric (HCl) and hydrobromic (HBr) acid media as well as from nitric acid systems containing organic solvents. By means of these techniques many more selective separations of Pb from accompanying elements can be achieved than by the application of cation exchange resins or chelating polymers. The latter have been employed rather frequently to isolate Pb together with transition elements.

ANION EXCHANGE RESINS

General

The anionic Pb(II) species $PbCl_3^-$ which is retained by strongly basic resins is predominant in 2 to 4 M HCl, while the species $PbCl_4^{2-}$ is mainly present in the 4 to 12 M acid solutions. The Pb species predominating in 0.1 to 0.5 and 0.5 to 2 M HCl media are $PbCl^+$ and $PbCl_2$, respectively. From dilute HCl solutions Pb is only moderately adsorbed on, e.g., Dowex® 1, and the adsorption decreases linearly when increasing the acid concentration from 2 to 12 M. Thus, in 0.5, 1, 1.5, 2, 3, 4, 6, 8, and 12 M HCl, distribution coefficients for Pb of <10, >10, ~30, >10, ~10, ~7, ~1, <1, and <1, respectively, have been measured. Consequently, the adsorption of Pb from HCl systems is usually performed from 1 M HCl (see Tables 1, 7, 9, and 12*),[1-8] 1.5 M HCl (see Tables 2, 7, and 9),[9-13] and 2 M HCl (see Tables 3, 6, 10, and 15).[8,14-23] Occasionally, Pb has also been adsorbed from 0.5 M HCl (see Table 8).[22,24] Coadsorbed with Pb from all these dilute HCl media are Zn, Cd, Hg(II), Cu(I), Ag, Au(III), Pt metals, Sn(IV and II), In, Tl(III), Sb(III), Bi, Te(IV), Re(VII), and Tc(VII), while virtually all other elements are not retained under these conditions. Therefore, Pb is readily separated from all major and minor constituents of natural and most industrial samples, which explains the high popularity of this separation method (see examples in Tables 1 to 3 and 6 to 10). A disadvantage of this procedure is, however, that only limited volumes of sorption and washing solutions of the above HCl concentrations can be passed through a column of anionic resin before Pb appears in the effluent. This is due to the low distribution coefficients of Pb in these systems and in the presence of larger amounts of salts, e.g., chloride and perchlorate; this adsorption is further decreased causing early breakthrough of the Pb. If the adsorption of Pb is performed from 2 M HCl solution a small portion of accompanying Fe(III) can only be washed off with difficulty. Therefore, it has been suggested to either perform the adsorption of Pb in the presence of ascorbic acid as a holding reductant (see Table 3)[14] or to use 2 M HCl containing this reducing agent as a wash solution.[17]

In 0.1 to 4 M HBr solutions the adsorbability of Pb on anionic resins of the quaternary ammonium type, e.g., Dowex® 1, is higher by orders of magnitude than that observed in HCl media of comparable acidity. Consequently, the cited difficulties when employing the HCl systems are largely eliminated by the use of HBr media. From these, the same elements are coadsorbed with Pb as from the HCl solutions. In 0.15, 1, 2, and 3 M HBr, distribution coefficients for Pb of 3.4 × 10³, 353, ~110, and 44, respectively, have been measured.[25,26] To effect the analytical separations described in Tables 4, 6 to 8, and 16 the Pb is adsorbed from 0.1[27] and 0.15 M HBr,[26,28] 0.2 M HBr - 0.1 M HNO₃,[29] and 0.3,[30] 1,[31,32] and 2 M HBr.[25,33,34]

---

* Tables for this chapter appear at the end of the text.

From nitric acid solutions Pb is only very weakly retained by strongly basic resins showing a maximum distribution coefficient of ~5 in the 5 M acid. This has been utilized for the separation outlined in Table 11[35] (see also Table 55 in the chapter on Actinides, Volume II). In the presence of high concentrations of organic solvents such as methanol[36-38] or tetrahydrofuran (THF),[39] the adsorbability of Pb from dilute nitric acid solutions is considerably higher than from the pure aqueous media. Thus, in 90% THF - 10% 5 M nitric acid a distribution coefficient of 250 has been measured for Pb.[39] This system and methanol-nitric acid media of similar acidities have been used in connection with the determination of Pb in geological materials of the type indicated in Table 5. Coadsorbed with Pb from these organic solvent-nitric acid systems are all elements forming stable anionic nitrate complexes such as Th, Bi, and rare earth elements.

Pb was also found to be adsorbed on an anion exchange resin from a malonic acid solution of pH 4.8 (see Table 11).[40]

After adsorption of Pb from the systems discussed above, remaining non- or only slightly adsorbed elements are removed by rinsing the resin bed with a wash solution of the same composition as the sorption solution, and then the Pb is eluted by means of one of the following eluents: water,[9,12,14,16,19,20,41-43] 0.005,[8,44] 0.01,[3,11,13,15] 0.02,[18,45-47] or 0.1 M HCl,[4] 0.1 M HCl - 1 M HF,[7] 0.3 M HNO$_3$ - 0.025 M HBr,[27,34] 1 M HCl,[36] 1[18,30] or 2 M HNO$_3$,[17,48] 2 M HNO$_3$ - 0.03 M HBr,[29] 3[49-51] or 6 M HCl,[25,26,31-33] 6 M HBr,[28] and 8[1,2,6,10,21,24,35,52] or 12 M HCl.[5] Although water has been employed by many investigators for the elution of Pb, its use, especially when applied at elevated temperatures,[20,41-43] entails the coelution of Zn, Cd, and other coadsorbed elements which are not eluted together with the Pb when using 8 M HCl or other HCl eluents. With the nitric acid solutions as eluents, Pb is eluted together with Zn, Cd, Cu, Ag, In, Sn, and other elements which were coadsorbed from HCl or HBr media, so that eluent systems of this type are of relatively low selectivity. Other eluents that have been employed for the elution of Pb include 80% THF - 20% 5 M nitric acid,[39] 1 M ammonium acetate,[40] tartaric acid solution,[53] and 3% NaOH solution.[53] The latter eluent has been used in connection with the separation of Pb by reactive ion exchange on a sulfate-form resin (see Table 13).[53,54] A similar precipitation of Pb can be effected on a resin in the bichromate form.[54]

Applications

Anion exchange separation procedures, which are based on the adsorption principles discussed above, have variously been employed in connection with the determination of Pb in geological materials (see Tables 1 to 6 and Tables 2 and 4 in the chapter on Zinc [Volume IV], Table 4 in the chapter on Actinides [Volume II], and Table 3 in the chapter on Copper [Volume III]), biological samples (see Table 7), and industrial products (see Tables 8 to 11 and Tables 3, 6, and 7 in the chapter on Nickel [Volume V], Table 16 in the chapter on Iron [Volume V], Tables 19 to 21 in the chapter on Cobalt [Volume V], and Table 23 in the chapter on Copper [Volume III]). To separate Pb from Bi and Po, and from synthetic mixtures with other elements, the methods outlined in Tables 12 and 13, respectively, have been used.

CATION EXCHANGE RESINS

General

From dilute mineral acid solutions Pb is strongly adsorbed on cation exchange resins of the sulfonic acid type. This adsorption decreases rapidly with increasing HCl concentration, from a distribution value of 2590 in the 0.1 M acid to values of ~80, ~50, ~6, ~1, and <1 in the 0.5, 0.6, 1, 2, and 3 M acid, respectively. In HBr media this decrease of adsorbability is even more pronounced, because Pb forms more stable anionic complexes with bromide than with chloride (see preceding section). Thus, in 0.5, 0.6, and 1 M HBr (or at higher concentrations) distribution coefficients for Pb of ~40, ~15, and <1, respectively, have been measured. Stronger adsorption of Pb than from these two acids takes place from nitric and hydrofluoric acid (HF) media. In nitric acid solutions distribution coefficients for Pb of >10^4, ~1400, ~180, ~36, ~10, ~7, and ~4 have been determined in the 0.1, 0.2, 0.5, 1, 2, 3, and 4 M acids, respectively. In HF media the distribution coefficient of Pb decreases with increasing acid concentration from a value of >10^4 in the 1 M acid to a value of ~10^2 in 15 M HF.

Since essentially all other divalent metals and also many other elements are coadsorbed with Pb from these dilute acid media, separations of Pb on cation exchange resins are of low selectivity. Therefore, methods based on this technique have been used for special separations only (see Tables 14 to 18). For this purpose Pb has been adsorbed from water (see Table 14),[55] wine (see Table 14),[56] sample solutions of pH 2[57] and 7[21] (see Table 15), 0.1 M HCl (see Table 16),[29,58] 0.5 M HCl (see Table 5),[37] 1 M HF (see Table 15),[59] and 20% dioxane-0.01 M HCl (see Table 16).[60] Adsorption of Pb from HF medium allows this element to be separated from all elements forming stable anionic fluoride complexes, which include Fe(III), Ti, Zr, Hf, and Nb. Similarly, an increase of the selectivity of separation of Pb from accompanying elements can be achieved by its adsorption from citric or tartaric acid solutions (see Table 17).[61,62] Under these conditions Pb is separated from Sn(IV), Sb(III), Ce(III), Zr, Mo, W, Nb, and Ta which form anionic chelates with the organic acids.

In most cases Pb, which has been adsorbed from the systems mentioned above, is eluted by means of 2 M HCl.[21,55,59,60,63] Other eluents that have been used for the elution of Pb include: 0.1 M HCl containing 82% of acetone,[29] 0.6 M HBr,[64] 1.5[37] and 3 M HCl,[58] 3 M HNO$_3$,[62] 4 M HCl,[56] 5% NaNO$_2$ solution,[65] 0.1 M Na thiosulfate,[66] 1 M ammonium acetate,[61] and 1 M glycine at pH 8.3 to 8.5.[67]

No elution is required when determning Pb directly on the resin, e.g., by X-ray fluorescence (see Table 15).[57]

From ethylenediaminetetraacetic acid (EDTA) solutions of pH < 1 Pb is not adsorbed on a cationic resin, a fact which has been employed for its separation from interferences (see Table 14).[68] Also, no adsorption of Pb on cation exchangers takes place from ammoniacal tartrate solutions from which, however, Cu and Cd are retained as cationic amine complexes.[69,70] Coeluted with the Pb are Bi and Sb.

Applications

In Tables 14 to 18 methods are presented that have been employed for the cation exchange separation of Pb from biological materials, natural water, industrial products, and synthetic mixtures with other elements. Cation exchange separation procedures that have been used in connection with the determination of Pb in rocks and sediments and in airborne particulate matter are described in the chapters on Zinc, Volume IV (see Table 37); Copper, Volume III (see Table 40); and Rare Earth Elements, Volume I (Table 86).

CHELATING RESINS

The chelating polymers most frequently employed for the isolation of Pb from complex matrices such as geological, biological, and industrial materials are resins containing the iminodiacetate grouping (e.g., Chelex® 100 [Dowex® A-1] and Wofatit® MC 50).[71-78] Examples of practical applications of this type of resin for the separation of Pb are shown in Table 19 as well as in the chapter on Copper, Volume III. Coadsorbed with the Pb are numerous transition metals which include Cu, Zn, Cd, Mn, Ni, and Co, so that all these elements are not only isolated as a group, but also eluted together with the Pb, e.g., with 2 M nitric or hydrochloric acid. Applications of other chelating resins that can be used for the separation of Pb are also outlined in Table 19.

RESIN SPOT TESTS

For the microchemical detection of Pb, the resin spot tests described in the chapter on Copper, Volume III (Table 80) can be used.

Table 1
DETERMINATION OF Pb IN GEOLOGICAL MATERIALS AFTER SEPARATION BY ANION EXCHANGE IN 1 M HCL MEDIA[a]

Material	Ion exchange resin, separation conditions, and remarks	Ref.
Pegmatitic feldspars	Dowex® 1-X8 (100-200 mesh; chloride form) Column: 28 × 0.9 cm conditioned with 1 M HCl and operated at a flow rate of 0.5 mℓ/min a) 1 M HCl sample solution (50 mℓ) (adsorption of Pb; Al and other elements pass into the effluent) b) 1 M HCl (70 mℓ) (elution of residual nonadsorbed elements) c) 8 M HCl (35 mℓ) (elution of Pb) For further purification of the eluted Pb, the separation is repeated on a small column (15 × 0.4 cm) of the resin employing 6-, 10-, and 7-mℓ volumes of eluents (a), (b), and (c), respectively (at a flow rate of 0.3 mℓ/min) After precipitation as PbS, isotopic abundance measurements are made on 10—12 µg of the sulfide Before the separation, the sample (0.5—4 g) is decomposed with HF + HClO$_4$ and Pb is coprecipitated with Al-hydroxide which is dissolved in HCl to prepare solution (a)	1
Sedimentary pyrite	Dowex® AG1-X8 (100-200 mesh; chloride form) a) 1 M HCl sample solution (adsorption of Pb; into the effluent pass Fe, U, and most other elements) b) 1 M HCl (elution of residual Fe and other nonadsorbed elements) c) 8 M HCl (elution of Pb) Pb is determined by mass spectrometry To isolate U, effluent (a) is made 8 M HCl and Fe(III) extracted into isopropyl ether; the aqueous phase is passed through a column of the same resin to adsorb U(VI), which, after elution with 1 M acid, can be determined mass spectrometrically Before the separation, the sample is decomposed with HNO$_3$ and HCl	2

[a] See also Table 13 in the chapter on Actinides, Volume II and Table 3 in the chapter on Zinc, Volume IV.

Table 2
DETERMINATION OF Pb IN GEOLOGICAL MATERIALS AFTER SEPARATION BY ANION EXCHANGE IN 1.5 M HCL MEDIA[a]

Material	Ion exchange resin, separation conditions, and remarks	Ref.
Marine sediments	Bio-Rad® AG1-X8 (100-200 mesh; chloride form) Column: 30 × 1 cm containing the resin (~10 g) pretreated with 1.5 M HCl (100 ml) and operated at a flow rate of 0.5 ml/min a) 1.5 M HCl sample solution (200 ml) (adsorption of Pb; matrix elements pass into the effluent) b) 1.5 M HCl (100 ml) (elution of residual matrix components) c) Deionized water (100 ml) (elution of Pb) Subsequently, Pb is precipitated as $PbSO_4$ and the ^{210}Pb is determined after 1 month by radiometric measurement of the ^{210}Bi daughter product Before the separation, the sample (2 g) is dried at 110°C and ashed at 400°C, and finally decomposed with aqua regia (in the presence of 50 mg of Pb-carrier)	9
Lake sediments	Dowex® 1-X8 (100-200 mesh; Cl^- form) a) 1.5 M HCl sample solution (adsorption of Pb; Ba and Ra pass into the effluent) b) 1.5 M HCl (3 column volumes) (elution of residual Ba and Ra) c) 8 M HCl (3 volumes) (elution of Pb) After precipitation as the chromates ^{210}Pb and ^{226}Ra are determined radiometrically; the recoveries are 80—85 and 35—40% for Ba-Ra and Pb, respectively Before the ion exchange separation, the sample (1—3 g) is dry ashed at <450°C and the ash is decomposed with $HF-HNO_3$ (in the presence of carriers; Ba is used as a carrier for Ra) and Fe(III) is removed by extraction into ethyl ether	10
Ores	Anionite AN-2F (grain size: 0.1—0.25 mm; Cl^- form) Column containing 1 g of the resin conditioned with 1.5 M HCl (40—50 ml) and operated at a flow rate of 2.5—3 ml/min a) 1.5 M HCl sample solution (50 ml) containing 2 g of NaCl (adsorption of Pb, Zn, and Cd) b) 1.5 M HCl (150 ml) (elution of nonadsorbed elements) c) 0.01 M HCl (100 ml) (elution of Pb) Pb is determined by EDTA titration Before the separation, the sample (0.1—0.25 g) is dissolved in conc HCl + H_2O_2	11

[a] See also Tables 10 and 55 in the chapter on Actinides, Volume II; Table 1 in the chapter on Silver, Volume III; and Table 5 in the chapter on Radium, Volume V.

Table 3
DETERMINATION OF Pb IN GEOLOGICAL MATERIALS AFTER SEPARATION BY ANION EXCHANGE IN 2 M HCL MEDIA[a]

Material	Ion exchange resin, separation conditions, and remarks	Ref.
Zr minerals	Strongly basic anion exchange resin (Cl⁻ form) a) 2 M HCl sample solution containing ascorbic acid (adsorption of Pb; Fe[II] passes into the effluent) b) 2 M HCl (as a rinse) c) Water (elution of Pb) Pb is determined mass spectrometrically Before the separation, the sample is decomposed by fusion with borax and the melt is dissolved in 2 M HCl; ^{207}Pb is used for isotope dilution	14
Chromite ores	Anionite EDE-10P (Cl⁻ form) conditioned with 2 M HCl Column operated at a flow rate of 3—4 mℓ/min a) 2 M HCl sample solution (100 mℓ) (adsorption of Pb; Fe[III] and Cr pass into the effluent) b) 2 M HCl (elution of residual Fe and Cr) c) 0.01 M HCl (160—200 mℓ) (elution of Pb) Pb is determined polarographically Before the separation, the sample (1 g) is fused with Na$_2$O$_2$ (5 g) at 500—600°C, the melt is extracted with water, and Pb is preconcentrated by coprecipitation with ferric hydroxide; the precipitate is dissolved in HCl to prepare solution (a)	15

[a] See also Table 86 in the chapter on Rare Earth Elements, Volume I and Table 1 in the chapter on Thallium.

Table 4
DETERMINATION OF Pb IN GEOLOGICAL MATERIALS AFTER SEPARATION BY ANION EXCHANGE IN HBr MEDIA

Material	Ion exchange resin, separation conditions, and remarks	Ref.
Silicate rocks and U oxides	Dowex® 1-X8 (100-200 mesh; bromide form) Column of 0.8 cm ID containing the resin (4 g) conditioned with 2 M HBr (50 mℓ) and operated at a flow rate of 1 mℓ/min a) 2 M HBr sample solution (50 mℓ) (adsorption of Pb, Bi, Zn, Cd, In, Au[III], Pt[IV], and Pd[II]; essentially all other elements including U pass into the effluent) b) 2 M HBr (30 mℓ) (elution of residual nonadsorbed elements) c) 6 M HCl (50 mℓ) (elution of Pb; only In is coeluted) Pb is determined by atomic absorption spectrophotometry or spectrophotometrically with dithizone Before the separation, the rock sample (1 g) is decomposed by repeated evaporation with HF and HClO$_4$ and finally the residue is dissolved in 2 M HBr; perchlorate is removed by precipitation as KClO$_4$; U oxide samples (U$_3$O$_8$ and yellow cake [Na-diuranate]) (1 g) are dissolved by treatment with HClO$_4$ and HBr	25, 33
Minerals	Bio-Rad® AG1-X8 (100-200 mesh; bromide form) Column: 7 × 2 cm containing the resin (10 g) conditioned with 0.1 M HBr and operated at a flow rate of 3—4 mℓ/min a) Sample solution which is between 0.1 and 4 M in bromide and/or HBr (adsorption of Pb, Bi, Tl[III], Cd, Hg[II], Au[III], Pt[IV], and Pd[II]; virtually all other elements pass into the effluent) b) 0.1 M HBr (250 mℓ) (elution of remaining nonadsorbed elements) c) 0.3 M HNO$_3$ - 0.025 M HBr (400 mℓ) (selective elution of Pb; the coadsorbed metals are not eluted) Pb is determined either by EDTA titration or by mass spectrometric isotope dilution There is some interference by Ag, Ta, Ir(IV), Rh(III), ClO$_4^-$, and SO$_4^{2-}$, but otherwise the method is highly accurate	27
Tantaloniobate	Bio-Rad® AG1-X8 (100-200 mesh; bromide form) Column of 2 cm ID containing 10 g (23 mℓ) of the resin conditioned with 0.1 M HBr and operated at a flow rate of 3.0 ± 0.2 mℓ/min a) 2 M HBr sample solution (50 mℓ) (adsorption of Pb) b) 0.1 M HBr (250 mℓ) (elution of rare earths, U, Th, and other elements) c) 0.3 M HNO$_3$ - 0.025 M HBr (400 mℓ) (elution of Pb) Pb is determined by EDTA titration Before the separation, the sample (50 g) is dissolved in 2 M HCl - 12 M HF and the insoluble fluorides of Pb, Th, and U are converted into bromides From the eluate (b), Th and U are isolated by cation and anion exchange procedures, respectively For the cation exchange separation of Th a column containing 10 g (23 mℓ) of Bio-Rad® AG50W-X12 resin (200-400 mesh, H$^+$ form) is employed; through this column the following solutions are passed a) Eluate (b) obtained by the anion exchange separation described above (adsorption of Th, U, rare earths, and other elements)	34

Table 4 (continued)
DETERMINATION OF Pb IN GEOLOGICAL MATERIALS AFTER SEPARATION BY ANION EXCHANGE IN HBr MEDIA

Material	Ion exchange resin, separation conditions, and remarks	Ref.
	b) 0.5 M HCl (200 mℓ) (elution of remaining traces of Nb and Ta) c) 2 M HCl (250 mℓ) (elution of U, Cu, and traces of some other elements) d) 4 M HCl (400 mℓ) (elution of rare earths, Zr, and Hf) After ashing of the resin Th is determined by EDTA titration From eluate (c) U is isolated by using the procedure outlined in Table 96 in the chapter on Actinides, Volume II	

Table 5
DETERMINATION OF Pb AND OTHER ELEMENTS IN GEOLOGICAL MATERIALS AFTER SEPARATION BY ANION EXCHANGE IN ORGANIC SOLVENT-NITRIC ACID MEDIA

Material	Ion exchange resin, separation conditions, and remarks	Ref.
Marine sediments	Dowex® 1-X8 (100-200 mesh; nitrate form) Column: 12 × 0.6 cm conditioned with eluent(b) (50 mℓ) and operated at a flow rate of 0.5 mℓ/min a) Sample solution (20 mℓ) which is 95% in methanol and 5% in 5 M HNO$_3$ (adsorption of Pb, Bi, Th, and rare earth elements; essentially all other elements pass into the effluent) b) 95% Methanol - 5% 5 M HNO$_3$ (100 mℓ) (elution of weakly and nonadsorbed elements) c) 90% Methanol - 10% 6 M HCl (100 mℓ) (elution of Th and rare earth elements) d) 1 M HCl (100 mℓ) (elution of Pb) e) 1 M HNO$_3$ (50 mℓ) (elution of Bi) Pb and Bi are determined spectrophotometrically or by EDTA titration Before the separation, the sample is decomposed with HNO$_3$ and HF The same separation technique can be used for the determination of Pb and Bi in steel	36
	Dowex® 1-X8 (100-200 mesh; nitrate form) Column: 10 × 0.6 cm conditioned with eluent (b) (30—50 mℓ) and operated at a flow rate of 0.25—0.3 mℓ/min a) Sample solution (20 mℓ, equivlaent to 250 mg of the sample) which is 90% in THF and 10% in 5 M HNO$_3$ (adsorption of Pb and rare earth elements; into the effluent pass all matrix elements) b) 90% THF - 10% 5 M HNO$_3$ (160 mℓ) (elution of Mg, Ca, Tl, Bi, Th, U, etc.) c) 80% THF - 20% 2.5 M HNO$_3$ (50 mℓ) (elution of Pb and rare earth elements, e.g., La) Pb is determined by standard methods The technique is also applicable to the analysis of a variety of other materials such as optical glass, canned food, wine, urine and other body fluids, gasoline, and high-purity metals or alloys	39
Lunar material	A) First column operation (Bio-Rad® AGl-X8, 100-200 mesh; NO$_3^-$ form) Column: 4 × 0.2 cm conditioned with 94% methanol - 6% conc HNO$_3$ (~500 μℓ) a) Sample solution (200 μℓ) which is 94% in methanol and 6% in conc HNO$_3$ (adsorption of Pb, Th, Ba, and other elements) b) 70% Methanol - 30% 3.3 M HNO$_3$ (800 μℓ) (elution of U, Cu, Co, and major rock-forming elements including the bulk of Fe) c) 0.5 M HNO$_3$ (1 mℓ) (elution of Pb together with Th and other elements) After evaporation of eluate (c) the Pb is further purified by using the following separation procedure (B) B) Second column operation (Bio-Rad® AG50W-X8, 100-200 mesh; H$^+$ form)	37

Table 5 (continued)
DETERMINATION OF Pb AND OTHER ELEMENTS IN GEOLOGICAL MATERIALS AFTER SEPARATION BY ANION EXCHANGE IN ORGANIC SOLVENT-NITRIC ACID MEDIA

Material	Ion exchange resin, separation conditions, and remarks	Ref.
	Column: 2 × 0.2 cm conditioned with 1.5 M HCl (300 µℓ) a) 0.5 M HCl solution (25 µℓ) of the Pb-containing evaporation residue obtained from (c) (A) (adsorption of Pb, Th, Fe, Ca, Ba, and Ti) b) 1.5 M HCl (200 µℓ) (elution of Pb) c) 4 M HCl (500 µℓ) (elution of Fe, Ca, Ba, and Ti) d) 4 N H$_2$SO$_4$ (500 µℓ) (elution of Th) Pb and U are determined by mass spectrometry; before the determination of U this element is isolated from eluate (b) obtained by (A) using the following procedure (C) C) Third column operation (Bio-Rad® AG1-X8, 100-200 mesh; Cl$^-$ form) Column: 6 × 0.2 cm conditioned with 4 M HCl (1 mℓ) containing ascorbic acid (1 g in 20 mℓ) of the acid a) 4 M HCl solution (500 µℓ, containing ascorbic acid) of the evaporation residue of eluate (b) obtained by (A) (adsorption of U[VI]; Fe[II] passes into the effluent together with Co and Cu) b) 4 M HCl ascorbic acid solution (1 mℓ) (elution of remaining Fe, Co, and Cu) c) 4 M HCl (2 mℓ) (removal of ascorbic acid) d) 1 M HCl (1 mℓ) (elution of U) In the case where only procedure (B) is used, U is eluted from the column by 400 µℓ of eluent (b) after Pb was eluted; then the U is further purified by application of procedure (C) Before the first column operation, the sample (1 mg) is decomposed with 1:1 HNO$_3$-HF (100 µℓ)	

Table 6
DETERMINATION OF Pb IN NATURAL WATERS AFTER ANION EXCHANGE SEPARATION[a]

Material	Ion exchange resin, separation conditions, and remarks	Ref.
Tap- and riverwater	Dowex® 1-X8 (100-200 mesh; bromide form) Column of 0.8 cm ID containing the resin (4 g) conditioned with 0.15 M HBr (50 mℓ) and operated at a flow rate of ~1.5 mℓ/min a) 0.15 M HBr sample solution (200 mℓ, containing 1 g of KBr) (adsorption of Pb; into the effluent pass the matrix elements, i.e., Mg, Ca, and alkalies) b) 0.15 M HBr (50 mℓ) (elution of residual nonadsorbed elements) c) 6 M HCl (50 mℓ) (elution of Pb) Pb is determined spectrophotometrically with dithizone or by means of atomic absorption spectroscopy Before the separation, the sample (1 ℓ) is acidified with conc HBr (18.5 mℓ) and after filtration, a 200-mℓ aliquot is evaporated to dryness; subsequently, organic material is destroyed by fuming with $HClO_4$ and the perchlorates are converted into bromides by evaporation with conc HBr (5 mℓ)	26
Natural waters	Dowex® 1-X8 (100-200 mesh; Br^- form) Column: 8.5 × 1.2 cm operated at a flow rate of 1.5 mℓ/min a) Water sample (100 mℓ) acidified with conc HBr (2 mℓ) (adsorption of Pb; Ca, Mg, Fe, and other elements pass into the effluent) b) 0.15 M HBr (50 mℓ) (elution of residual nonadsorbed elements) c) 6 M HBr (50 mℓ) (elution of Pb) Pb is determined spectrophotometrically; recovery ranges from 95—105% The same procedure can be used for the analysis of sediments and Al alloys	28
Rainwater and atmospheric particulates	Amberlite® IRA-400 (chloride form) Column: 15 × 1 cm a) 2 M HCl sample solution (adsorption of ^{210}Pb + carrier; into the effluent pass P, Be, Sr, and other elements) b) 2 M HCl (as a rinse) c) Water (elution of Pb) ^{210}Pb and the other radionuclides are determined radiometrically Before the ion exchange separation, the air filter sample is decomposed with HNO_3 + Br_2 and ^{210}Po is removed by deposition on Ag; in the case of water samples the ^{210}Pb and the other radionuclides, i.e., ^{7}Be, ^{32}P, ^{90}Sr, and ^{210}Po, are first preconcentrated by coprecipitation with ferric hydroxide Then Fe is removed by ether extraction from 8 M HCl	16

[a] See also Table 4 in the chapter on Cobalt, Volume V, Tables 7 and 8 in the chapter on Copper, Volume III, and Tables 19 and 36 in the chapter on Actinides, Volume II.

Table 7
DETERMINATION OF Pb IN BIOLOGICAL MATERIALS AFTER ANION EXCHANGE SEPARATION[a]

Material	Ion exchange resin, separation conditions, and remarks	Ref.
Food	Amberlite® IRA-400 (60-100 mesh BSS; chloride form) Column: ~8 × 0.8 cm conditioned with 1 M HCl (10 mℓ) a) 1 M HCl sample solution (5 mℓ, containing ⊁40 µg of Pb) (adsorption of Pb; matrix elements pass into the effluent) (flow rate: 1 mℓ/min) b) 1 M HCl (25 mℓ) (elution of residual nonadsorbed elements at a rate of 2 mℓ/min) c) 0.01 M HCl (elution of Pb at a rate of 1 mℓ/min rejecting the first 2.5 mℓ and collecting the next 10 mℓ) Pb is determined spectrophotometrically with dithizone Before the separation, the sample (suitable amount containing ⊁5 g of dry matter) is dry ashed at ⊁500°C and the ash is dissolved in HCl	3
Animal tissue, blood, and excreta	Dowex® 1-X8 (100-200 mesh; chloride form) Column: 8 × 0.4 cm conditioned with 1 M HCl (20 mℓ) a) 1 M HCl sample solution (2 mℓ) (adsorption of ^{210}Pb and its daughter nuclides ^{210}Bi and ^{210}Po; matrix elements pass into the effluent) b) 0.1 M HCl (10 mℓ) (elution of ^{210}Pb) c) 2 N H$_2$SO$_4$ (12 mℓ) (elution of ^{210}Bi and ^{210}Po) ^{210}Pb is determined radiometrically Before the separation, the sample is wet ashed at 50—117°C with HNO$_3$ + HClO$_4$	4
Blood	Dowex® 1-X8 (200-400 mesh; Br$^-$ form) Column: 7 × 1 cm a) 1 M HBr sample solution (40 mℓ) (adsorption of Pb, Cd, and Bi; main constituents of the matrix including Fe pass into the effluent) b) 0.5 M HBr (15 mℓ) (elution of remaining nonadsorbed elements) c) 1.2 M HCl (15 mℓ) (as a rinse) d) 6 M HCl (elution of Pb) Pb is determined by mass spectrometry; recovery of Pb is between 80 and 90% Before the separation, the sample is wet ashed with HNO$_3$ and the organic residue is charred at 300°C; this method can also be used for the determination of Pb in granulite facies[32]	31
Biological materials	Bio-Rad® AGl-X8 (50-100 mesh; Cl$^-$ form) The sample is adjusted with HCl to give a Cl$^-$ concentration of ≃1.5 M and is passed through a column of the resin to adsorb Pb; then Pb is eluted with water and determined by atomic absorption spectrophotometry	12

[a] See also Table 15 in the chapter on Zinc, Volume IV and Table 3 in the chapter on Thallium.

Table 8
DETERMINATION OF Pb IN INDUSTRIAL PRODUCTS AFTER ANION EXCHANGE SEPARATION IN <1 M HCl AND HBr MEDIA[a]

Material	Ion exchange resin, separation, conditions, and remarks	Ref.
Alloys and glazed pottery	Dowex® 1-X8 (250-325 mesh; Cl⁻ form) Column: 7.25 × 0.635 cm operated at a forced flow rate of 3 mℓ/min a) 0.5 M HCl sample solution containing 10—30 µg Pb per milliliter (adsorption of Pb; matrix elements pass into the effluent) b) 8 M HCl (elution of Pb) The elution is monitored at 270 nm, the peak height being directly proportional to Pb concentration; the lower limit of determination is ≃0.4 µg/mℓ; the upper limit is 40 µg/mℓ	24
High-purity Co	Dowex® AG1-X8 (100-200 mesh) Column: 10 × 1.2 cm conditioned with 0.3 M HBr and operated at a flow rate of 1 mℓ/min a) 0.3 M HBr sample solution (80 mℓ) (adsorption of Pb; Co passes into the affluent) b) 0.3 M HBr (50 mℓ) (elution of residual Co) c) 1 M HNO$_3$ (100 mℓ) (elution of Pb) Pb is determined by atomic absorption spectrophotometry; for 0.35 ppm of Pb the precision is ≃30% Before the separation the sample (3 g) is dissolved in conc HNO$_3$	30

[a] See also Table 5 in the chapter on Bismuth.

Table 9
DETERMINATION OF Pb IN INDUSTRIAL PRODUCTS AFTER ANION EXCHANGE SEPARATION IN 1 AND 1.5 M HCl MEDIA[a]

Material	Ion exchange resin, separation conditions, and remarks	Ref.
Steel	Dowex® 1-X8 (100-200 mesh; chloride form) Column: 13 × 1.3 cm conditioned with 1 M HCl (50 mℓ) a) 1 M HCl sample solution (~100 mℓ) (adsorption of Pb; Fe[III] and other elements pass into the effluent) b) 1 M HCl (75 mℓ) (elution of residual Fe and other nonadsorbed elements) c) 12 M HCl (150 mℓ) (elution of Pb; the first 75 mℓ of this eluate is returned to the column to remove traces of Fe eluted from the resin bed) For the elution of Pb water (e.g., 140 mℓ) can also be used[79,80] Pb is determined by atomic absorption spectrophotometry Before the separation, the sample (5 g) is dissolved in a 4:1 mixture of HCl and HNO$_3$; the recovery of Pb is >95%	5
Co	Amberlite® IRA-400 (20-50 mesh; chloride form) Column: 15 × 1 cm conditioned with 1 M HCl a) 1 M HCl sample solution (adsorption of Pb; Co passes into the effluent) b) 1 M HCl (30 mℓ) (elution of residual Co) c) 8 M HCl (50 mℓ) (elution of Pb) Pb is determined spectrophotometrically with dithizone The method gives good results for 0.3—4-g samples of Co that contain 20—1 μg of Pb, respectively	6
"Pure" Fe and alloys	Anionite AN-2F (grain size: 0.1—0.25 mm; Cl⁻ form) Column containing 1 g of the resin conditioned with 1.5 M HCl (40—50 mℓ) and operated at a flow rate of 1—1.5 mℓ/min a) ≃1.5 M HCl sample solution (90—140 mℓ) (adsorption of Pb; Fe passes into the effluent) b) 1.5 M HCl (70 mℓ) (elution of remaining Fe) c) 0.01 M HCl (100 mℓ) (elution of Pb) Pb is determined polarographically Before the separation, the sample (3—5 g) is dissolved in HCl and HNO$_3$	13

[a] See also Table 3 in the chapter on Silver, Volume III; Table 9 in the chapter on Cadmium, Volume IV, and Table 36 in the chapter on Actinides, Volume II.

Table 10
DETERMINATION OF Pb AND Bi IN INDUSTRIAL PRODUCTS AFTER ANION EXCHANGE SEPARATION IN 2 M HCl MEDIA[a]

Material	Ion exchange resin, separation conditions, and remarks	Ref.
Steel	Anionite EDE-10P (Cl$^-$ form) Column: 12 × 0.9 cm operated at a flow rate of 2 mℓ/min a) ≃ 2 M HCl sample solution (80 mℓ) (adsorption of Pb; the bulk of Fe together with Cr, Ni, and Cu passes into the effluent) b) 2 M HCl (60 mℓ) (elution of more Fe) c) 2 M HCl - 1% ascorbic acid solution (20 mℓ) (reduction of adsorbed Fe[III] to Fe[II]) d) 2 M HCl (20 mℓ) (elution of Fe[II]) e) 2 M HNO$_3$ (250 mℓ) (elution of Pb and Bi) Pb and Bi are determined by atomic absorption spectrometry Before the separation, the sample (5 g containing 1—30 ppm of Pb or Bi) is dissolved in HCl and HNO$_3$	17
Chromic oxide	Anionite AN-31 (Cl$^-$ form) a) 2 M HCl sample solution (60 mℓ) (adsorption of Pb and Bi; Fe passes into the affluent) b) 2 M HCl (as a rinse) c) 0.02 M HCl (elution of Pb) d) 1 M HNO$_3$ (250 mℓ) (elution of Bi) Pb is determined polarographically and Bi by extractive titration with dithizone solution Before the separation, the sample is repeatedly fumed with HClO$_4$ and Pb + Bi are preconcentrated by coprecipitation with ferric hydroxide; the precipitate is dissolved in 4 M HCl (30 mℓ)	18
Th	Deacidite® FF (−100 to +200 mesh; chloride form) Column: 10 × 1.5 cm a) 2 M HCl sample solution (50 mℓ, containing 10 g of Th) (adsorption of Pb; Th passes into the effluent) b) 2 M HCl (25 mℓ) (elution of residual Th) c) Distilled water (25 mℓ) (elution of Pb) ^{212}Pb is determined radiometrically	19
Bronze	Anionite EDE-10P (Cl$^-$ form) a) 2 M HCl sample solution (30—50 mℓ) (adsorption of Pb; Cu passes into the effluent) b) 2 M HCl (50 mℓ) followed by 0.5 M HCl (elution of remaining Cu) c) Hot water (200 mℓ) (elution of Pb) Pb is determined spectrophotometrically Before the separation, the sample (1—5 g) is dissolved in HCl + H$_2$O$_2$	20

[a] See also Tables 20, 25, and 26 in the chapter on Zinc, Volume IV, Table 11 in the chapter on Cadmium, Volume IV; Table 18 in the chapter on Copper, Volume III; and Table 3 in the chapter on Bismuth.

Table 11
DETERMINATION OF Pb IN INDUSTRIAL PRODUCTS AFTER ANION EXCHANGE SEPARATION IN MEDIA CONTAINING ORGANIC ACIDS AND MINERAL ACIDS[a]

Material	Ion exchange resin, separation conditions, and remarks	Ref.
Pb-based alloys	Dowex® 21-K (20-50 mesh; chloride form) Column: 23 × 1.4 cm a) Sample solution (25-mℓ aliquot equivalent to 100 mg of the sample) containing malonic acid (2 g) and adjusted to pH 4.8 (with 0.01 M NaOH and 0.01 M malonic acid) (adsorption of Pb and Sb as anionic malonate complexes) b) 1 M ammonium acetate solution (250 mℓ) (elution of Pb) c) 5% tartaric acid solution (250 mℓ) (elution of Sb) Pb and Sb are determined by conventional methods	40
Metallic Bi	Anionite EDE-10P Column conditioned with 0.06 M HCl a) 0.06 M HCl sample solution (~50 mℓ) containing 20 mℓ of saturated tartaric acid solution (adsorption of Bi; Pb passes into the effluent) (flow rate: 3 mℓ/min) b) Hot water (100 mℓ) (elution of remaining Pb at a rate of 5 mℓ/min) Pb is determined polarographically Before the separation, the sample (0.5 g) is dissolved in conc HCl containing conc HNO$_3$ (few drops)	41
High-purity Bi	Bio-Rad® AG1-X8 (100-200 mesh; chloride form) Column: 15 × 1 cm containing the resin (12 g) conditioned with 6 M HNO$_3$ and operated at a rate of 0.3 mℓ/min a) 5 M HNO$_3$ sample solution (5 mℓ) (adsorption of Pb and Bi as anionic nitrate complexes) b) 8 M HCl (40 mℓ) (elution of Pb) c) 0.5 M HNO$_3$ (300 mℓ) (elution of Bi) Pb is determined by anodic-stripping voltammetry; the precision is ≃ ±5% at the nanogram level; the lower limit of determination is ≃0.01 ppm of Pb Before the separation, the sample (1 g) is dissolved in 1:1 HNO$_3$ (6 mℓ)	35

[a] See also Table 4 in the chapter on Tin and Table 27 in the chapter on Copper, Volume III.

Table 12
ANION EXCHANGE SEPARATION OF ^{210}Pb, ^{210}Bi, and ^{210}Po

Ion exchange resin	Separation conditions and remarks	Ref.
Dowex® 1-X2 (50-100 mesh; chloride form)	Column: 4.2 × 0.85 cm (in a 10-mℓ buret) conditioned with 3 M HCl (for ≤1 week) and operated at flow rates of ∼2 mℓ/min (eluents [a] and [b]) and 1.5 mℓ/min (eluent [c]) a) Sample solution (0.2—0.3 mℓ) (adsorption of Pb, Bi, and Po) b) 3 M HCl (75 mℓ) (elution of Pb) c) 12 M HCl (75 mℓ) (elution of Bi) d) 8 M HNO$_3$ (150 mℓ) (elution of Po) Eluent (c) may be replaced by 1.5 M H$_2$SO$_4$ - 2.3 M Na$_2$SO$_4$ (120 mℓ at a flow rate of 2 mℓ/min) The radionuclides are determined radiometrically The above separation can also be effected on a column of Dowex® 21K using 1 M HCl, conc HCl, and conc HNO$_3$ as eluents for Pb, Bi, and Po, respectively	49—51
Anionite AV-17-X14 (grain size: 20—30 μm; Cl$^-$ form)	Column: 3 × 0.25—0.3 cm a) 1 M HCl sample solution (≥10—15 times the column volume) (adsorption of the elements) b) 0.1 M HCl - 1 M HF (elution of Pb) c) 20—28 M HF (elution of Bi) d) 3 M HNO$_3$ (elution of Po)	7

Table 13
ANION EXCHANGE SEPARATION OF Pb FROM SYNTHETIC MIXTURES WITH OTHER ELEMENTS

Elements separated	Ion exchange resin, separation conditions, and remarks	Ref.
Pb from binary mixtures with Fe(III), Co, Ni, Cu, Mg, Ba, and Sr	Dowex® 1 (20-50 mesh; sulfate form) Column: 6—10 × 0.8 cm conditioned with 50% methanol (40 mℓ) and operated at a flow rate of 2 drops/sec a) 50% methanol solution (50 mℓ, containing 2—150 mg of Pb and ≯2 mℓ HNO₃) (precipitation of Pb as PbSO₄; the other elements pass into the effluent except Ba and Sr, which are also precipitated on the resin) b) 50% Methanol (10—40 mℓ) (elution of residual accompanying elements) c) 3% NaOH solution (25 mℓ) (elution of Pb; Sr and Ba are not coeluted) In the absence of methanol the precipitation of Pb is not complete Pb is determined spectrophotometrically; the method is applicable when the ratios of the Pb to other elements are Fe (1:2), Co (1:35), Ni (1:70), Cu (1:10), Mg (1:500), Ba (1:4), and Sr (1:2); the mean coefficient of variation in 47 determinations was ±0.7%	53
Pb, Cu, Fe(III), Hg(II), and Sn(IV)	Amberlyst® A-26 (macroreticular strong-base resin; 150-200 mesh; Cl⁻ form) Column: 3.8 × 0.63 cm in a liquid chromatograph a) Injection of sample (50 μℓ, containing microgram amounts of the elements) b) 8 M HCl (elution of Pb at a rate of 0.86 mℓ/min) c) 4 M HCl (elution of Cu and residual Pb) d) 1 M HCl (elution of Fe) e) 4 M HCl - 3 M HClO₄ (elution of Hg) f) 5.95 M HClO₄ - 0.1 M HCl (elution of Sn) The entire separation can be completed in 25 min	52

Table 14
DETERMINATION OF Pb IN BIOLOGICAL MATERIALS AND NATURAL WATER AFTER CATION EXCHANGE SEPARATION[a]

Material	Ion exchange resin, separation conditions, and remarks	Ref.
Wine	Dowex® 50 in a column containing 30 g of the resin pretreated with 4 M HCl followed by water a) Wine sample (50 mℓ, freed from solid particles) (adsorption of Pb and other cationic constituents) b) Water (as a rinse) c) 4 M HCl (elution of Pb and coadsorbed elements) Pb is determined colorimetrically with dithizone	56
Urine	Wofatit® P (H$^+$ form) Column containing the resin to a height of 15 cm The sample (3 mℓ, containing >30 μg of Pb per 100 mℓ) is acidified to pH < 1 (with 2 mℓ of 8% HCl) and passed through the resin bed to adsorb interfering substances; the Pb which passes into the effluent (as anionic EDTA complex) is determined polarographically The method can be used for determining Pb in the urine of patients suffering from plumbism and treated with salts of EDTA	68
Natural water	Cationite (grain size: 1 mm; H$^+$ form) Column: 5 × 2 cm operated at a flow rate of 0.6—1.0 ℓ/hr a) Sample (adsorption of Pb and Cu) b) Hot 15% HCl (elution of Pb and Cu) Pb is determined as the sulfide and Cu with dithizone	55

[a] See also Table 20 in the chapter on Cobalt, Volume V, Table 38 in the chapter on Zinc, Volume IV; Tables 41—43, 45, and 46 in the chapter on Copper, Volume III; and Table 7 in the chapter on Strontium, Volume V.

Table 15
DETERMINATION OF Pb IN INDUSTRIAL PRODUCTS AFTER CATION EXCHANGE SEPARATION[a]

Material	Ion exchange resin, separation conditions, and remarks	Ref.
Steel	Zeo-Karb®225 (SRC14) (52—100 mesh; H⁺ form) Column with a resin bed 3.8 cm high and 3 mℓ in volume and operated at a flow rate of ≃2 mℓ/min a) 1 M HF sample solution (≯250 mℓ, containing ≯0.8 mmol of adsorbable metals) (adsorption of Pb, Cu, Co, Mn, and Ni plus a small fraction of Cr[III]; into the effluent pass Fe[III], Ti, Zr, Nb, etc. as anionic fluoride complexes) b) 1 M HF (50 mℓ) (elution of remaining Fe) c) Water (10 mℓ) (as a rinse) d) 2 M HCl (50 mℓ) (elution of Pb and coadsorbed elements) Pb is determined polarographically Before the separation, the sample (1 g) is dissolved in HF-HNO₃ the method could also be applied to the determination of Pb in Nb- and W-base alloys	59
Brass	Column (7 × 0.4 cm) of partially sulfonated cation exchange resin of low capacity (1.89 meq/g; macroreticular resin) operated at a forced flow rate of 2 mℓ/min (liquid chromatograph) a) Injection of sample in 2 M HClO₄ (56.3 μℓ) b) 2 M HClO₄ (8—9 mℓ) (elution of matrix, i.e., Cu, Zn, and Ni) c) 2 M HCl (4 mℓ) (elution of Pb) Pb is determined spectrophotometrically by in-stream addition of PAR reagent The maximum error of the method is ≃2%	63
Ceramic dinnerware	Reeve-Angel® SA-2 ion exchange resin paper (H⁺ form); disk of 1.25 in. diam Solutions (150 mℓ) of 0.01 M HCl, vinegar, or deionized water are placed in the dinnerware which is covered with plastic film and set aside for 24 hr; then the solution is adjusted to pH 2 with 0.1 M HCl and passed seven times through a resin paper disk to adsorb Pb, which is determined directly on the resin by a X-ray fluorescence method; the coefficient of variation for eight successive deterinations is 1.8% and the detection limit is 1.08 μg (= 7.2 ng when 150 mℓ of leaching solution is used)	57
Zn sulfate solutions	A) First column operation (Dowex® 50-X4, 200-400 mesh; H⁺ form) Column: 21.5 × 1.5 cm (removal of sulfate) a) Neutral sample solution (10 mℓ, containing 1.25 g of ZnSO₄) (adsorption of Pb and Zn; sulfate passes into the effluent) b) Deionized water (20 mℓ) (elution of residual sulfate) c) 2 M HCl (elution of Pb and Zn rejecting the first 18 mℓ and collecting the next 20 mℓ; this latter fraction is subjected to the following separation step [B]) On a column (27 × 3.1 cm) of greater capacity it is possible to elute Pb before the Zn using 1 M HCl (240 mℓ) as an eluent; this elution is performed after rinsing the resin with 75 mℓ of eluent (b) B) Second column operation (Dowex® 2-X8, 100-200 mesh; chloride form) Column: 12.5 × 2 cm conditioned with 2 M HCl a) Second fraction (20 mℓ) of eluate (c) obtained by (A) (adsorption of Pb and Zn) b) 2 M HCl (10 mℓ) (as a rinse)	21

Table 15 (continued)
DETERMINATION OF Pb IN INDUSTRIAL PRODUCTS AFTER CATION EXCHANGE SEPARATION[a]

Material	Ion exchange resin, separation conditions, and remarks	Ref.
	c) 8 M HCl (elution of Pb rejecting the first 25 mℓ and collecting the next 55 mℓ) Pb is determined spectrographically	

[a] See also Table 8 in the chapter on Silver, Volume III; Table 11 in the chapter on Sodium, Volume V; Table 19 in the chapter on Cadmium, Volume IV; Table 41 in the chapter on Zinc, Volume IV; Tables 48—52 and 54—56 in the chapter on Copper, Volume III; and Table 7 in the chapter on Bismuth.

Table 16
DETERMINATION OF Pb IN Tl AND Bi MATRICES AFTER CATION EXCHANGE SEPARATION IN ORGANIC SOLVENT MEDIA[a]

Material	Ion exchange resin, separation conditions, and remarks	Ref.
Cyclotron bombarded Tl targets	A) First column operation (Bio-Rad® AG50W-X4, 100-200 mesh, H⁺ form) Column: ~8 × 2 cm conditioned with 0.1 M HCl (50 mℓ) containing 0.05% Br₂ and operated at a flow rate of 3.5 ± 0.3 mℓ/min a) 0.1 M HCl sample solution (200 mℓ) containing saturated Br water (3 mℓ) (adsorption of Pb and Cu; Tl[III] passes into the effluent as anionic chloride complex) b) 0.1 M HCl - 0.05% Br₂ (40 mℓ) (elution of Tl) c) 0.1 M HCl containing 60% acetone (60 mℓ) (elution of residual Tl) d) 0.1 M HCl containing 82% acetone (80 mℓ) (elution of Pb) e) 3 M HCl (80 mℓ) (elution of Cu) To separate Pb from remaining microamounts of Tl and Cu impurities, eluate (d) is evaporated, and after treatment with conc HNO₃ (2 mℓ) and conc HBr (1 mℓ) the following separation is performed B) Second column operation (Bio-Rad® AGl-X4, 200-400 mesh; chloride form) Column: 5.2 × 2 cm conditioned with 0.2 M HBr - 0.1 M HNO₃ (100 mℓ) and operated at a flow rate of 2.4 ± 0.3 mℓ/min a) Sample solution (20 mℓ) which is 0.2 M in HBr and 0.1 M in HNO₃ (adsorption of Pb and Tl[III] as anionic bromide complexes) b) 0.2 M HBr - 0.1 M HNO₃ (80 mℓ) (elution of Cu) c) 2 M HNO₃ - 0.03 M HBr (70 mℓ) (elution of Pb) d) 1 M NH₃ solution (75 mℓ) followed by 0.5 M HNO₃-0.1 M thiourea (elution of Tl) ²⁰³Pb is determined radiometrically Before the first column operation, the deuteron-bombarded target sample (~10 g) is dissolved in 2 M HNO₃ (30 mℓ) and the nitrates are converted into chlorides by repeated evaporation with 10 M HCl + Br₂	29
High-purity Tl	Bio-Rad® AG50W-X4 (100-200 mesh; H⁺ form) Column: ~8.5 × 1.1 cm conditioned with 0.1 M HCl (50 mℓ) containing 0.05% Cl₂ and operated at a flow rate of 2.5 ± 0.3 mℓ/min a) 0.1 M HCl sample solution (~200 mℓ) containing 0.05% Cl₂ (adsorption of Pb, Fe, Cu, Zn, Cd, Mn, Co, and In; Tl[III] passes into the effluent as anionic chloride complex) b) 0.1 M HCl - 0.05% Cl₂ (~100 mℓ) (elution of Tl) c) 0.1 M HCl in 40% acetone (50 mℓ) (elution of residual Tl) d) 3 M HCl (75 mℓ) (elution of Pb and other adsorbed metals) Pb, Fe, Cu, Zn, Cd, Mn, Co, and In are determined by atomic absorption spectrometry Before the separation, the sample (5 g) is dissolved in ~7 M HNO₃ + Br₂ and the nitrates are converted to the chlorides by repeated evaporation with HCl containing some Cl₂	58
Bismuthyl carbonate	Dowex® 50W-X8 (100-200 mesh) Column of 1.1 cm ID containing 4 g of the resin a) Sample solution (5-mℓ aliquot, equivalent to 50 mg of the sample) which is 20% in dioxane, 0.01 M in HClO₄, and 0.001 M in EDTA (adsorption of Pb)	60

Table 16 (continued)
DETERMINATION OF Pb IN Tl AND Bi MATRICES AFTER CATION EXCHANGE SEPARATION IN ORGANIC SOLVENT MEDIA[a]

Material	Ion exchange resin, separation conditions, and remarks	Ref.
	b) 20% Dioxane - 0.01 M HClO$_4$ (80 mℓ) (elution of Bi) c) 2 M HCl (25 mℓ) (elution of Pb) Pb is determined polarographically Before the separation the sample (1 g) is dissolved in HClO$_4$	

[a] See also Reference 83.

Table 17
CATION EXCHANGE SEPARATION OF Pb IN THE PRESENCE OF ORGANIC COMPLEXING AGENTS

Elements separated	Ion exchange resin, separation conditions, and remarks	Ref.
Pb from Sr, Ba, Al, Ce, Zr, Bi, Fe, and Th	Dowex® 50W-X8 (50-100 mesh; H$^+$ and Na$^+$ forms) Column: 14.5 × 1.4 cm A) Separation of Pb from Sr, Ba, and Al After adsorption of the elements the following eluents are used a) Water (50 mℓ) (as a rinse) b) 1 M ammonium acetate (200 mℓ) (elution of Pb) c) 4 M HCl (200 mℓ) (elution of Al, Sr, and Ba) B) Separation of Pb from Ce and Zr a) 5% Citric acid solution of pH 2.7 (adsorption of Pb; Ce and Zr pass into the effluent as anionic citrate complexes) b) Water (50 mℓ) (removal of citric acid) c) 1 M ammonium acetate (200 mℓ) (elution of Pb) C) Separation of Pb from Be, Fe, and Th (Na$^+$-form resin) a) Solution treated with 0.01 M EDTA and adjusted to pH 2.0—2.2 (adsorption of Pb; the anionic EDTA complexes of Bi, Fe, and Th pass into the effluent) b) Water (50 mℓ) (removal of EDTA) c) 1 M ammonium acetate (200 mℓ) (elution of Pb) Recoveries of Pb range from 94—102%	61
Pb from binary mixtures with Sn(IV), Sb(III), Mo(VI), W(VI), Nb(V), and Ta(V)	Bio-Rad® AG50W-X8 (200-400 mesh; H$^+$ form) Column of 2 cm ID containing 60 mℓ of the resin and operated at a flow rate of 3.0 ± 0.3 mℓ/min a) Sample solution (~100 mℓ) which is 0.01 M in HNO$_3$ and 0.25 M in tartaric acid (adsorption of Pb; the other elements pass into the effluent as anionic tartrate complexes) b) 0.01 M HNO$_3$ - 0.1 M tartaric acid (300 mℓ) (elution of remaining amounts of the other elements) c) 0.1 M HNO$_3$ (~100 mℓ) (removal of tartaric acid) d) 3 M HNO$_3$ (300 mℓ) (elution of Pb) Pb is determined by complexometric titration or by atomic absorption spectrometry Precipitates formed in solution (a) between Pb and molybdate, tungstate, niobate, and tantalate can be dissolved by application of the slurry column technique using 5 g of the resin	62
Pb, Mg, Ca, Sr, and Ba	Cationite KU-2 (Na$^+$ form) Column of 1 cm ID containing 10 g of the resin conditioned with 1 M glycine (pH 8.0—8.5) After adsorption of the elements, the following eluents are used a) 1 M glycine (pH 8.3—8.5) (elution of Pb) b) 1 M glycine (pH 9.0—9.5) (elution of Mg) c) 3 M HCl (elution of Ca, Sr, and Ba)	67

Table 18
CATION EXCHANGE SEPARATION OF Pb IN THE PRESENCE OF INORGANIC COMPLEXING AGENTS

Elements separated	Ion exchange resin, separation conditions, and remarks	Ref.
Pb from binary mixtures with Ba, Sr, Ca, Cr, Mg, U(VI), V(IV), Zn, and other elements	Dowex® 50W-X8 (100-200 mesh; H⁺ form) Column: 16 × 1.2 cm operated at 50—60°C to prevent precipitation of $PbBr_2$ After adsorption of the elements the following eluents are used a) 0.6 M HBr (160—240 mℓ) (elution of Pb) b) 3 M HCl (75—300 mℓ) (elution of the other elements) Bi and Cd as well as Sn can be eluted before the Pb with 0.5 M HCl and 0.5 M HF (for Sn), respectively The average recovery is 99.8% with a standard deviation of ± 0.42%	64
Pb from binary mixtures with Tl(I), Cd, and Bi	Amberlite® IR-120 (Na⁺ form) Column containing the resin to a height of 30 cm and operated at a flow rate of 3 mℓ/cm² min After adsorption of the elements the following eluents are used a) 2% $NaNO_2$ solution (300 mℓ) (elution of Cd) b) 5% $NaNO_2$ solution (elution of Pb with the first 280 mℓ; Tl is contained in the 380—880-mℓ fraction) c) 2 M HCl (elution of Bi)	65
Pb from Ba, Sr, Ca, Mg, Fe(III and II)	Amberlite® IR-120 (60-80 mesh; Na⁺ form) Column: 24 × 1.25 cm containing 5 g of the resin and operated at a flow rate of 3 mℓ/min After adsorption, the elements (10—100-mg amounts) are separated using the following eluents a) 0.1 M $Na_2S_2O_3$ (60 mℓ) (elution of Pb) b) 2.5—3 M NaCl or Na-acetate (elution of the other metals except Fe[III], which is further retained as a red-colored cationic thiosulfate complex)	66

Table 19
DETERMINATION OF Pb IN GEOLOGICAL, BIOLOGICAL, AND INDUSTRIAL MATERIALS AFTER SEPARATION ON CHELATING RESINS[a]

Material	Ion exchange resin, separation conditions, and remarks	Ref.
Plants, river sediments, and waters	Dowex® A1 (50-100 mesh; NH_4^+ form) Column: 5 × 0.8 cm operated at a flow rate of 1 mℓ/min a) Sorption solution (50 mℓ) containing 1 M malonic acid (5 mℓ) and which is adjusted to pH 5 with NH_3 solution (adsorption of Pb, Cu, and Zn; matrix elements pass into the effluent) b) 0.1 M ammonium malonate of pH 5 (20 mℓ) (elution of residual matrix elements) c) Water (10 mℓ) (removal of malonate) d) 2 M HNO_3 (15 mℓ) (elution of Pb, Cu, and Zn) Pb, Cu, and Zn are determined by atomic absorption spectrophotometry Before the separation, the plant sample (0.5 g) is wet ashed at 160°C with HNO_3 and $HClO_4$	75
Urine	Dowex® A-1 (50-100 mesh; Na^+ form) Column: 7 × 1.2 cm operated at a flow rate of 4—5 mℓ/min a) Sample (e.g., 50 mℓ) adjusted to pH 3.5 with HNO_3 (adsorption of Pb; the constituents of the matrix pass into the effluent) b) 2 M HCl (18 mℓ) followed by water (10 mℓ) (elution of Pb) Pb is determined spectrophotometrically with dithizone The method shows a precision of ±2.1% in the range of 2—15 µg of Pb per 50-mℓ sample Before the separation the acidified urine is centrifuged or filtered	76
	Chelex® 100 a) Urine sample adjusted to pH 8.0 with 6 M NH_3 (adsorption of Pb) b) Water, methanol, and again water (elution of salts and organic matter) c) 2 M HNO_3 (elution of Pb) Pb is determined by atomic absorption spectroscopy	77
Pb solutions	Bio-Rad Chelex® 100 (<400 mesh; NH_4^+ form) The sample solution (250 mℓ; pH 4) containing ≤0.4 µg of Pb is stirred for 10 min with 100 mg of the resin to adsorb Pb; subsequently, the mixture is filtered through a membrane filter and the resin is suspended in water (5 mℓ); then 10 µℓ (resin: 0.2 mg) of the suspension is injected into an instrument for the electrothermal atomic absorption spectrometric determination of Pb The relative standard deviation was 3.7% at a 0.8 ppb level of Pb	78
Carbonate rocks	Chelating resin based on poly(aminostyrene) and grafted with arsenazo or pyridylazoresorcinol (grain size: 0.2 mm) The sample (0.5 g) is decomposed with 1 M HNO_3, the pH of the solution is adjusted to 6, and after the addition of the resin (100 mg) the mixture is shaken for 2 hr to adsorb Pb, Ga, Cr, Ni, Be, V, rare earths, Mo, Zr, Co, Zn, Ti, Mn, and Nb After filtration, the resin is ashed and Pb and the other elements are determined spectrographically	81
Mixtures of Pb with Zn and Cu	Chelating resin containing a thioglycolloylomethyl functional group (<200 mesh)	82

Table 19 (continued)
DETERMINATION OF Pb IN GEOLOGICAL, BIOLOGICAL, AND INDUSTRIAL MATERIALS AFTER SEPARATION ON CHELATING RESINS[a]

Material	Ion exchange resin, separation conditions, and remarks	Ref.
	Phenyl-CH$_2$-O-C(=O)-CH$_2$SH resin Column: 10 × 0.2 cm in a liquid chromatograph Prior to the introduction of the sample into the column, pH 3.5 acetate buffer (0.01 M) is allowed to flow through for 5 min at a rate of 1 mℓ/min; then the following solutions are passed a) Sample solution-injection (adsorption of Pb) b) 0.01 M acetate buffer of pH 3.5 (elution of Zn and Cd) (flow rate of 2 mℓ/min for 10 min) c) 0.01 M HClO$_4$ (elution of Pb at a flow rate of 1 mℓ/min for 5 min) The method was used for the separation of nanomole quantities of the metals	

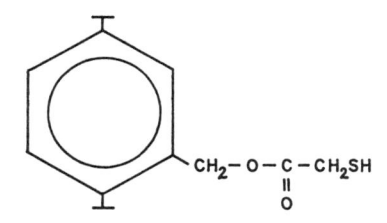

[a] See also Table 25 in the chapter on Cadmium, Volume IV; Table 47 in the chapter on Zinc, Volume IV; Tables 68—74 and 76—78 in the chapter on Copper, Volume III; and Tables 160 and 168 in the chapter on Actinides, Volume II.

REFERENCES

1. **Catanzaro, E. J. and Gast, P. W.**, Isotopic composition of lead in pegmatitic feldspars, *Geochim. Cosmochim. Acta*, 19(2), 113, 1960.
2. **Wampler, J. M. and Kulp, J. L.**, Isotopic study of lead in sedimentary pyrite, *Geochim. Cosmochim. Acta*, 28(9), 1419, 1964.
3. **Johnson, E. I. and Polhill, R. D. A.**, Use of an anion-exchange resin in the determination of traces of lead in food, *Analyst (London)*, 82, 238, 1957.
4. **Ikebuchi, H. and Kametani, K.**, Determination of lead-210 in biological samples by liquid scintillation counting, *Eisei Kagaku*, 23(5), 290, 1977.
5. **Sellers, N. G.**, Ion-exchange separation and atomic-absorption determination of lead in steel, *Anal. Chem.*, 44, 410, 1972.
6. **Kasiura, K. and Meus, M.**, An ion-exchange separation and spectrophotometric determination of traces of lead in cobalt, *Chem. Anal. (Warsaw)*, 23(2), 305, 1978.
7. **Nikitin, M. K. and Katýkhin, G. S.**, Study of ion-exchange in hydrofluoric acid solutions. Separation of lead-210, bismuth-210m and polonium, *At. Energ.*, 14(5), 493, 1963.
8. **Nash, J. R. and Anslow, G. W.**, Polarographic determination of lead in steels and copper-zinc alloys, *Analyst (London)*, 88, 963, 1963.
9. **Joshi, L. U. and Ku, T. L.**, Measurement of lead-210 from a sediment core off the coast of California, *J. Radioanal. Chem.*, 52(2), 329, 1979.
10. **Joshi, S. R. and Durham, R. W.**, Determination of lead-210, radium-226 and caesium-137 in sediments, *Chem. Geol.*, 18(2), 155, 1976.
11. **Khalizova, V. A., Krasyukova, I. G., Donchenko, V. A., Alekseeva, A. Ya., and Smirnova, E. P.**, Complexometric determination of lead in ores after enrichment on an anionite, *Zavod. Lab.*, 33(9), 1064, 1967.
12. **Lyons, H. and Quinn, F. E.**, Measurement of lead in biological materials by combined anion-exchange chromatography and atomic absorption spectrophotometry, *Clin. Chem.*, 17(3), 152, 1971.
13. **Kladnitskaya, M. B. and Khalizova, V. A.**, Determination of traces of lead in "pure" iron and alloys by A.C. polarography, *Zavod. Lab.*, 35(7), 793, 1969.
14. **Lung, W.-H. and Liu, T.-I.**, Micro-determination of lead in zirconium minerals by isotope dilution, *Acta Geol. Sin.*, No. 1, 92, 1975.
15. **Zelenina, T. P.**, Polarographic determination of lead in chromite ores, *Sb. Trud. Vses. Nauchno Issled. Gornometall. Inst. Tsvetn. Met.*, No. 7, 335, 1962.
16. **Marenco, A., Blanc, D., Fontan, J., Lacombe, J. P., and Crozat, G.**, Determination by chemical analysis of Be-7, P-32, Sr-90, Pb-210, and P-210 in air and rain, *Chim. Anal.*, 50(3), 133, 1968.
17. **Shevchuk, I. A., Dovzhenko, N. P., and Kravtsova, Z. N.**, Atomic absorption determination of lead and bismuth in steel using ion-exchange chromatography, *Ukr. Khim. Zh.*, 47(7), 773, 1981.
18. **Stepin, V. V., Ponosov, V. I., Emasheva, G. N., and Zobnina, N. A.**, Micro-determination of zinc, lead and bismuth in chromic oxide using ion exchange chromatography, *Tr. Vses. Nauchno Issled. Inst. Stand. Obratsov*, 4, 100, 1968.
19. **Gorsuch, T. T.**, Separation of lead-212 from thorium, *Analyst (London)*, 85, 225, 1960.
20. **Malakhova, N. M., Olenovich, N. L., and Krainyakova, M. M.**, Use of 4-(2-thiazolylazo)resorcinol for determination of lead, *Zavod. Lab.*, 43(8), 917, 1977.
21. **Leclerca, M. and Duyckaerts, G.**, Chromatographic separation of lead contained in concentrated solutions of zinc sulfate, *Anal. Chim. Acta*, 29, 139, 1963.
22. **Morachevskii, Yu. V., Zvereva, M. N., and Rabinovich, R. Sh.**, Separation of lead from barium by means of anionites, *Zavod. Lab.*, 22(5), 541, 1956.
23. **Barnes, I. L., Murphy, T. J., Gramlich, J. W., and Shields, W. R.**, Separation of lead by anodic deposition and isotope-ratio mass spectrometry of microgram and smaller samples, *Anal. Chem.*, 45, 1881, 1973.
24. **Seymour, M. D. and Fritz, J. S.**, Rapid selective method for lead by forced-flow liquid chromatography, *Anal. Chem.*, 45, 1632, 1973.
25. **Korkisch, J. and Groß, H.**, Atomic-absorption determination of lead in geological materials, *Talanta*, 21, 1025, 1974.
26. **Korkisch, J. and Sorio, A.**, Application of ion-exchange processes to determination of trace elements in natural waters. V. Lead, *Talanta*, 22, 273, 1975.
27. **Strelow, F. W. E. and Toerien, F. von S.**, Separation of lead(II) from bismuth(III), thallium(III), cadmium(II), mercury(II), gold(III), platinum(IV), palladium(II), and other elements by anion-exchange chromatography, *Anal. Chem.*, 38, 545, 1966.
28. **Xu, M., Pan, Z., and Xie, N.**, Spectrophotometric determination of micro-amounts of lead with 5,10,15,20-tetrakis-(4-trimethylammoniophenyl)porphine, *Fenxi Huaxue*, 11(6), 437, 1983.

29. **Van der Walt, T. N., Strelow, F. W. E., and Haasbroek, F. J.**, Separation of lead-203 from cyclotron-bombarded thallium targets by ion-exchange chromatography, *Talanta*, 29, 583, 1982.
30. **Uny, G., Mathien, C., Tardif, J. P., and Van Danh, T.**, Determination of zinc, iron and lead in cobalt of very high purity, *Anal. Chim. Acta*, 53, 109, 1971.
31. **Manton, W. I.**, Sources of lead in blood, *Arch. Environ. Health*, No. 1, 149, July/August 1977.
32. **Gray, C. M. and Oversby, V. M.**, The behavior of lead isotopes during granulite facies metamorphism, *Geochim. Cosmochim. Acta*, 36, 939, 1972.
33. **Korkisch, J. and Groβ, H.**, Contributions to the analysis of nuclear raw materials. VIII. Atomic absorption and spectrophotometric determination of lead in triuranium octaoxide and yellow cake samples, *Mikrochim. Acta*, 2(4-5), 413, 1975.
34. **Strelow, F. W. E.**, Application of ion exchange chromatography to accurate determination of lead, uranium and throrium in tantaloniobates, *Anal. Chem.*, 39, 1454, 1967.
35. **Mizuike, A., Miwa, T., and Oki, S.**, Anodic stripping square-wave voltammetric determination of lead in high-purity bismuth, *Anal. Chim. Acta*, 44, 425, 1969.
36. **Ahluwalia, S. S. and Korkisch, J.**, Anion-exchange separation of bismuth and lead, *Fresenius' Z. Anal. Chem.*, 208, 414, 1965.
37. **Tera, F. and Wasserburg, G. J.**, Precise isotopic analysis of lead in picomole and sub-picomole quantities, *Anal. Chem.*, 47, 2214, 1975.
38. **Ramakumar, K. L., Aggarwal, S. K., Kavimandan, V. D., Raman, V. A., Khodade, P. S., Jain, H. C., and Methews, C. K.**, Separation and purification of magnesium, lead and neodymium from dissolver solution of irradiated fuel, *Sep. Sci. Technol.*, 15(7), 1471, 1980.
39. **Korkisch, J. and Feik, F.**, Separation of lead by anion exchange, *Anal. Chem.*, 36, 1793, 1964.
40. **Rangnekar, A. V. and Khopkar, S. M.**, Anion-exchange studies of lead in chloride and malonate solution: separation from mixtures, *Fresenius' Z. Anal. Chem.*, 232, 432, 1967.
41. **Lyashenko, T. V. and Milaev, S. M.**, Determination of lead in metallic bismuth, *Sb. Nauchn. Tr. Vses. Nauchno Issled. Gornometall. Inst. Tsvetn. Met.*, No. 9, 78, 1965.
42. **Gibalo, I. M., Kamenev, A. I., Agasyan, E. P., Mymrik, N. A., and Chernysheva, A. Yu.**, Analysis of lead tin telluride single crystals, *Zh. Anal. Khim.*, 28(10), 2056, 1973.
43. **Mymrik, N. A., Agasyan, E. P., Gibalo, I. M., and Kamenev, A. I.**, Ion-exchange separation and polarographic determination of the components of the system lead-tin-selenium, *Zavod. Lab.*, 39(11), 1312, 1973.
44. **Waclawik, Z.**, Photometric determination of lead in cast iron after separation by ion exchange, *Pr. Inst. Odlew.*, 20(2), 136, 1970.
45. **Nikulina, I. N., Pastukhova, I. N., Stashkova, N. V., Kurbatova, V. I., Brainina, Kh. Z., and Stepin, V. V.**, Analysis of standard samples by inverse voltammetry of solid phases. Determination of lead in refractory alloys ferrochrome and ferromanganese, *Zavod. Lab.*, 37(10), 1161, 1971.
46. **Vinogradov, L., Komorokhov, B. A., and Kozlova, G. I.**, Determination of small amounts of lead and tin in high temperature nickel-based alloys from one weighing, *Tekhnol. Legk. Splavov. Nauchno Tekh. Byull. VILS*, No. 3, 81, 1971.
47. **Stepin, V. V., Pnosov, V. I., Novikova, E. V., and Zobina, N. A.**, Separation of lead and zinc from bismuth on the Soviet anion-exchange resin AN-31, *Tr. Vses. Nauchno Issled. Inst. Stand. Obraztsov Spektr. Etalonov*, 5, 75, 1969.
48. **Kemula, W., Brajter, K., and Rubel, S.**, Application of anion exchangers in the polarographic determination of lead and zinc on copper ores, *Chem. Anal. (Warsaw)*, 14(6), 1339, 1969.
49. **Pacer, R. A.**, The role of Cherenkov and liquid scintillation counting in evaluating the anion-exchange separation of lead-210, bismuth-210 and polonium-210, *J. Radioanal. Chem.*, 77(1), 17, 1983.
50. **Ishimori, T.**, Separation of Ra-D, Ra-E, and Ra-F by ion-exchange, *Bull. Chem. Soc. Jpn.*, 28(6), 432, 1955.
51. **Fairman, W. D. and Sedlet, J.**, Direct determination of lead-210 by liquid scintillation counting, *Anal. Chem.*, 40, 2004, 1968.
52. **Seymour, M. D. and Fritz, J. S.**, Determination of metals in mixed hydrochloric and perchloric acids by forced-flow anion exchange chromatography, *Anal. Chem.*, 45, 1394, 1973.
53. **Ziegler, M.**, Precipitation of lead with anion exchanger sulfates, *Fresenius' Z. Anal. Chem.*, 180, 1, 1961.
54. **Khristova, R. and Novkirishka, M.**, Sorption and separation of ions by precipitation with ion-exchange resins, *God. Sofii. Univ. Khim. Fak.*, 62, 347, 1967/1968.
55. **Aleskovskii, V. B., Libina, R. I., and Miller, A. D.**, Microdetermination of lead and copper in solutions by a preliminary enrichment on an ion-exchange column, *Tr. Leningr. Tekhnol. Inst. Lensoveta*, No. 48, 5, 1958.
56. **Edge, R. A. and Penny, N.**, Proposed non-ashing technique employing ion exchange resins for the determination of lead in wine, *J. Sci. Food Agric.*, 9(7), 401, 1958.

57. **Tacket, S. L., Bender, G. H., Brunner, T. R., Duncan, D. J., Fedak, M. G., Gentile, R. F., Hiller, J. F., Hooker, K. A., McAuley, A. J., Rollick, K. L., Sandolfini, J. F., Smith, J. L., Vojtko, J. D., Pekala, P. H., and Williams, S. A.**, Determination of lead in dinner-ware by an ion-exchange filter-paper-X-ray fluorescence method, *Anal. Lett.*, 6(4), 355, 1973.
58. **Strelow, F. W. E. and Van der Walt, T. N.**, Quantitative separation of traces of lead, cadmium and many other elements from gram amounts of thallium by cation-exchange chromatography, *Anal. Chim. Acta*, 136, 429, 1982.
59. **Hamza, A. G. and Headridge, J. B.**, Polarographic determination of lead after cation-exchange separation, *Analyst (London)*, 91, 237, 1966.
60. **Sixta, V. and Šulcek, Z.**, Effect of mixed solvents on separation of metal ions by ion exchange chromatography in the presence of EDTA, *Collect. Czech. Chem. Commun.*, 37, 2386, 1972.
61. **Khopkar, S. M. and De, A. K.**, Cation-exchange studies of lead(II) on Dowex 50W-X8. Separation from mixtures, *Talanta*, 7, 7, 1960.
62. **Strelow, F. W. E. and Van der Walt, T.N.**, Separation of lead from tin, antimony, niobium, tantalum, molybdenum, and tungsten by cation-exchange chromatography in tartaric acid-nitric acid mixtures, *Anal. Chem.*, 47, 2272, 1975.
63. **Fritz, J. S. and Story, J. N.**, Chromatographic separation of metal ions on low-capacity microreticular resins, *Anal. Chem.*, 46, 825, 1974.
64. **Fritz, J. S. and Greene, R. G.**, Cation-exchange separation of lead, *Anal. Chem.*, 35, 811, 1963.
65. **Bhatnagar, R. P. and Trivedi, R. G.**, Cation-exchange separation of lead(II) from thallium(I), cadmium(II), and bismuth(III), *Indian J. Chem.*, 5(4), 166, 1967.
66. **Katsura, T.**, Applications of ion-exchange resins in analytical chemistry. IV. Separation of lead from barium, strontium, calcium, magnesium, iron(III), and iron(II), *Jpn. Analyst*, 10(11), 1211, 1961.
67. **Tikhomirov, V. I. and Gornovskaya, N. K.**, Separation of lead and magnesium ions and the sum of alkaline-earth metals on cation-exchange resin in the presence of glycine, *Org. Reagenty Anal. Khim.*, No. 3, 126, 1980.
68. **Chmielowski, J. and Myslak, Z.**, Polarographic determination of lead in urine in the presence of EDTA with use of an ion exchanger to remove interfering substances, *Chem. Anal. (Warsaw)*, 4(1-2), 233, 1959.
69. **Kreshkov, A. P. and Sayushkina**, Separation of copper and lead cations by ion-exchange chromatography, *Tr. Mosk. Khim. Tekhnol. Inst.*, No. 22, 116, 1956.
70. **Khedreyarv, Kh. Kh.**, Possible application of ion-exchange chromatography for the separation of lead from certain accompanying elements, *Tr. Tallin. Politekh. Inst. Ser. A*, No. 215, 121, 1964.
71. **Baetz, R. A. and Kenner, C. T.**, Determination of heavy metals in foods, *J. Agric. Food Chem.*, 21(3), 436, 1973.
72. **Vondenhoff, T.**, Determination of lead, cadmium, copper, and zinc in plant and animal material by atomic absorption in a flame and in a graphite tube after sample decomposition by the Schöniger technique, *Mitteilungsbl. GDCh Fachgruppe Lebensmittelchem. Gerichtl. Chem.*, 29(12), 341, 1975.
73. **Holynska, B.**, Use of chelating ion exchanger in conjunction with radio-isotope X-ray spectrometry for determination of trace amounts of metals in water, *Radiochem. Radioanal. Lett.*, 17(5-6), 313, 1974.
74. **Kuhnhardt, C. and Angermann, W.**, Polarographic determination of lead, cadmium, nickel, and zinc in copper after separation on Wofatit MC 50, *Chem. Anal. (Warsaw)*, 22(1), 37, 1977.
75. **Yano, Y., Odaka, N., Takei, S., and Nagashima, K.**, Determination of trace heavy metals in environmental samples with special reference to lead in plants, *Jpn. Analyst*, 27(8), T25, 1978.
76. **Forman, D. T. and Garvin, J. E.**, Rapid determination of lead in urine by ion exchange, *Clin. Chem.*, 11(1), 1, 1965.
77. **Fiorio, G., Mezzetti, T., and Proietti, G.**, Urinary lead determination by concentration and purification on a chelating resin, *Boll. Chim. Farm.*, 120(6), 330, 1981.
78. **Isozaki, A., Fukuda, Y., and Utsumi, S.**, Electro-thermal atomic absorption spectrometry for lead by direct atomization of lead adsorbed on a chelating resin, *Jpn. Analyst*, 31(7), 404, 1982.
79. **Wynne, E. A., Burdick, R. D., and Fine, L. H.**, Determination of lead using an anion exchanger and sodium chloranilate, *Anal. Chem.*, 33, 807, 1961.
80. **Wynne, E. A. and Burdick, R. D.**, Lead determination using an anion-exchanger and sodium chloranilate, *Anal. Chem.*, 33, 1963, 1961.
81. **Dorokhova, E. M., Shvoeva, O. P., Cherevkov, A. S., and Myasoedova, G. V.**, Spectrochemical determination of trace impurities in carbonate rocks by using chelating sorbents, *Zh. Anal. Khim.*, 34(6), 1140, 1979.
82. **Phillips, R. J. and Fritz, J. S.**, Chromatography of metal ions with a thioglycolate chelating resin, *Anal. Chem.*, 50, 1504, 1978.
83. **Strelow, F. W. E.**, Separation of traces and large amounts of lead from gram amounts of bismuth, tin, cadmium, and indium by cation exchange chromatography in hydrochloric acid-methanol using a macroporous resin, *Anal. Chem.*, 57, 2268, 1985.

NITROGEN

Basic anion exchange resins retain the anionic N species, i.e., nitrate, nitrite, and cyanide, while ammonium ion is adsorbed on resins of the sulfonic acid type. These two adsorption principles have variously been utilized for separations employed in connection with the determination of N in natural waters, biological materials, and industrial products.

Procedures based on ion chromatography of N species are presented in the chapter "Special Analytical Techniques Using Ion Exchange Resins", Volume I.

ANION EXCHANGE RESINS

Nitrate and Nitrite

The anionic species NO_3^- and NO_2^- can be adsorbed on anion exchange resins in the hydroxide (i.e., free base),[1-5] chloride,[6-11] and bromide[12] forms from aqueous solutions of pH 5 to 8 (see Tables 1 to 3*).[1,4,6,8,12,13] Coadsorbed with nitrate and nitrite are other oxyanions such as sulfate and phosphate, but efficient separations are achieved from all cationic constituents. The adsorbed nitrate and nitrite can be eluted by use of the following eluents: 0.9%,[9] 1%,[6] 5%,[1] 20%,[11] 1[10] and 2.5 M NaCl solutions,[8,14] 1 M K bromide,[12,15] 0.2 M Na sulfate,[4] Na sulfate-K-H phosphate solution,[16] and 0.03 M phosphate solution of pH 3,[9] as well as with alkaline solutions such as 1 M ammonia solution,[7] 4 M[2] and 4%[17] NaOH solutions, 1 M Na sulfate - 0.1 M NaOH solution,[3] and 1 M Na bromide - 1 M NaOH.[5]

Adsorption of nitrate and nitrite on basic resins has been applied in connection with the determination of these two N species in natural waters (see Table 1), biological materials (see Table 2), and industrial products (see Table 3). In Table 1 a method is also illustrated which is based on the selective adsorption of the azo dye formed by the diazotization-coupling reaction of nitrite with sulfanilic acid (as substrate) and N-(1-naphtyl)-ethylene-diamine (as coupling agent). It has been shown[18] that nitrate and nitrite can be determined by single-column ion exchange chromatography with use of an Ionosphere A column (25 × 0.46 cm). The mobile phase was 0.02, 0.04, or 0.06 M Na perchlorate or 0.2 M Na methanesulfonate. The method was applied to the analysis of extracts of cooked ham and river water.

Cyanide

Cyanide ion (CN^-), when present as an alkali cyanide or soluble metal cyanide, is readily adsorbed on strongly basic resins in the hydroxide forms. Coadsorbed with cyanide are other anions such as nitrate, nitrite, sulfate, and phosphate, but not metal cations and also not ammonia (see Tables 4 and 5).[19-21] Two applications of this separation principle are illustrated in Table 4 into which a method has been included which is based on reactive ion exchange on Dowex® 50 allowing the indirect determination of cyanide. The resin Dowex® 2-X7.5 has been used as a support for picric acid, which reacts with cyanide to form an orange compound allowing the semiquantitative determination of this anion.

Ammonia

Free ammonia or ammonium ion is not adsorbed on basic anion exchange resins in the hydroxide forms,[19,22-26] a fact which has been utilized for the analytical separations outlined in Table 5. A batch procedure using the resin Bio-Rad® AG1-X8 (OH^- form) has been employed in connection with the determination of ^{15}N ammonia enrichment in blood and urine.[27] Adsorbed on the anion exchangers are virtually all anions contained in the sample solutions, while cations are either precipitated on the resin or coeluted with ammonia.

---

* Tables for this chapter appear at the end of the text.

CATION EXCHANGE RESINS

Nitrate and Nitrite

The nonadsorbability of these oxyanions on strongly acid cation exchange resins has been employed mainly for the removal of cationic constituents of sample solutions prior to the determination of nitrate and/or nitrite in waters,[28-38] biological materials,[5,10,39-43] and industrial products.[11,26,44-55] For this purpose the procedures outlined in Tables 6 to 8 can be used, in which the cation exchange separation step is sometimes combined with the simultaneous or subsequent use of an anion exchange resin. Application of this two-resin technique permits the fractionation and determination of additional N species. Furthermore, when using a batch process to adsorb the azo dye (formed by nitrite in a diazotizing-coupling reaction) on a cation exchange resin, a bulky and, hence, easily filterable precipitate is obtained due to the coagulative action of the anion exchanger (see Table 6).[28] Adsorption of azo dyes on cation exchange resins has also been employed for the microchemical detection of nitrite using the resin spot tests described in Table 9, as well as to separate nitrate from nitrite (present as an azo dye following the diazotizing coupling reaction) using Amberlite® IR-120 in a batch process.[56] To determine nitrite in sea water by ion exchanger colorimetry, diazotization, with sulphanilamide as substrate and 1-naphtylethylene diamine as coupling agent, has been combined with use of Dowex® 50W-X2 (100-200 mesh) as sorbent.[57] In Table 8 a method is included which makes it possible, via reactive ion exchange, to separate nitrate from sulfate.

Ammonia

From neutral to slightly acid solutions (e.g., pH $\simeq$ 4 to 5) ammonium ion (NH_4^+) is adsorbed on cation exchange resins of the sulfonic acid type to an extent comparable to that of K (see chapter on Potassium, Volume V). This fact has variously been utilized in connection with the determination of ammonia in natural waters (see Table 10)[58-60] and biological materials (see Table 11).[61-71] For the quantitative elution of the adsorbed ammonium the following eluents can be employed: 4 M NaCl,[61-64,72] 1 M KCl,[70] 0.1[65,66] and 5 M[58] NaOH, borate buffer of pH $\sim$ 10,[60] and Na phenoxide (phenol-alkaline hypochlorite) solutions.[59,67,68]

CHELATING RESINS

A nitron-poly(vinylbenzyl chloride) polymer has been shown to have strong affinity for certain oxyanions, e.g., nitrate, nitrite, permanganate, perchlorate, chlorate, bichromate, and perrheniate, but not for sulfate, phosphate, carbonate, and halides.[73] This chelating resin may be useful for the selective removal of nitrate and nitrite from contaminated water supplies. It is readily regenerated with ammonia or ammonium chloride solutions and is not affected by the pH of the water over the range 4 to 10.

Table 1
DETERMINATION OF NITRATE AND NITRITE IN NATURAL WATERS AFTER ANION EXCHANGE SEPARATION

Material	Ion exchange resin, separation conditions, and remarks	Ref.
Fresh- and seawater	Dowex® 1-X8 (100-200 mesh; bromide form) Column: ≃10 × 1 cm operated at a flow rate of 5 mℓ/min a) Sample (500 mℓ of freshwater or 50 mℓ of seawater) (adsorption of nitrate, nitrite, sulfate, carbonate, and phosphate; cationic constituents pass into the effluent) b) Distilled water (∼30 mℓ) (as a rinse) c) 0.05 M KBr (70 mℓ) (elution of phosphate and carbonate) d) 1 M KBr (40 mℓ) (elution of nitrate, nitrite, and sulfate) Eluate (d) is freeze dried and nitrate, nitrite, and sulfate are determined simultaneously by infrared spectrophotometry utilizing the KBr-disk technique	12
Seawater	Dowex® 1-X8 (50-100 mesh; Cl$^-$ form) Column of 1 cm ID filled with 4—7 mℓ of the resin pretreated with 60% acetic acid followed by water (50 mℓ) a) Water sample (500 mℓ) to which 20 mℓ of a 0.6% solution of sulfanilic acid in 20% acetic acid and 2 mℓ of a 0.6% aqueous solution of N-1-naphthylethylenediamine dihydrochloride had been added (after 30 min the azo dye formed in presence of nitrite is completely adsorbed on the resin; matrix constituents pass into the effluent) (flow rate: 15—50 mℓ/min) b) Water (≃50 mℓ) (as a rinse) c) 60% Acetic acid (10—25 mℓ) (elution of the azo dye) (flow rate: 1—2 mℓ/min) The concentration of the azo dye in the eluate is determined spectrophotometrically; the method permits the determination of nitrite ion concentrations from nM to 0.1 μm	74
Fresh waters	Weakly basic resin Amberlite® IR-4B (chloride form) Column: 8 × 0.8 cm operated at a flow rate of 1—1.5 mℓ/min a) Sample (500 mℓ) adjusted to pH 5—7 with 0.01 M acetic acid or NaHCO$_3$ and in which the concentration of chloride has been increased to 0.001 M by the addition of NaCl (adsorption of nitrate; cations and organically bound N pass into the effluent) b) Distilled water (25 mℓ) (as a rinse) c) 1% NaCl solution (45—50 mℓ) (elution of nitrate) Nitrate is determined colorimetrically; the lower limit of determination of nitrate in fresh water samples is ≃0.06 ppm	6

Table 2
DETERMINATION OF NITRATE AND NITRITE IN BIOLOGICAL MATERIALS AFTER ANION EXCHANGE SEPARATION

Material	Ion exchange resin, separation conditions, and remarks	Ref.
Whey powder	Dowex® 1-X1 (50-100 mesh; OH⁻ form) Column: 3 × 1 cm operated at a flow rate of 2—4 mℓ/min a) Sample solution (2—25-mℓ aliquot) adjusted to pH 7—8 with 0.2 or 1 M NaOH (adsorption of nitrate and nitrite) b) Water (50 mℓ) (as a rinse) c) 5% NaCl solution (20 mℓ) (elution of nitrate and nitrite) Nitrate and nitrite are determined colorimetrically; the recovery averages 94% for nitrate (30—50 ppm) and 88% for nitrite (10—20 ppm), the detection limit being 2—10 ppm (depending on the sample size) Before the separation, the sample (2—10 g) is extracted at 55°C for 15—20 min with a weakly alkaline-aqueous solution; then the protein is precipitated by the addition of $ZnSO_4$ solution to the extract, and the precipitate is removed by filtration The filtrate thus obtained is the sample solution (a)	1
Fresh and ensiled forages (lucerne and ryegrass)	Amberlite® IR-4B (Cl⁻ form) a) Sample solution containing 15 µg to 3.5 mg of N as NO_3^- (adsorption of NO_3^-) b) 0.05 M HCl (as a rinse) c) 1 M NH_3 solution followed by water (elution of NO_3^-) After reduction of nitrate in a Cd-reductor the nitrite formed is determined colorimetrically; before the separation, the sample is macerated (leached) overnight with 0.5 M HCl and the extract is filtered, thus obtaining solution (a)	7
Tobacco	Weakly basic anion exchange resin Merck II a) Aqueous sample solution (adsorption of NO_3^-) b) Water (as a rinse) c) 0.2 M Na_2SO_4 (elution of NO_3^-)[a] Nitrate is determined by UV spectrophotometry Before the separation, the sample (350 mg) is shaken for 30 min with water (70 mℓ) at 20°C and the mixture is filtered; an aliquot of the filtrate corresponds to solution (a)	4

[a] See also Reference 79.

Table 3
DETERMINATION OF NITRATE AND NITRITE IN INDUSTRIAL PRODUCTS AFTER ANION EXCHANGE SEPARATION

Material	Ion exchange resin, separation conditions, and remarks	Ref.
Aqueous waste from the nuclear fuel fabrication process	Amberlite®IRA-410 (14-52 mesh; chloride form) Column: ~10 × 1 cm operated at a flow rate of 1—1.5 mℓ/min a) Sample (10—15 mℓ) adjusted to pH 5—6 with glacial acetic acid (1 drop) (adsorption of nitrate and nitrite; accompanying cations and organic species pass into the effluent) b) Distilled water (50 mℓ) acidified to pH 5—6 with acetic acid (1 drop) (elution of remaining nonadsorbed elements) c) 2.5 M NaCl (25 mℓ) (elution of nitrate and nitrite) Nitrate and nitrite are determined polarographically; the limits of determination are 5 ng/mℓ of nitrite and 200 ng/mℓ of nitrate Before the separation NH$_3$ is volatilized at 70—80°C and the original volume of the sample is restored with distilled water A 95—98% recovery of both nitrate and nitrite is obtained	8
Stack gas	Dowex®AG1-X4 (200-400 mesh; Br$^-$ form) The gas sample (200 mℓ containing NO and NO$_2$) is shaken with 0.3% H$_2$SO$_4$ containing 0.04% of H$_2$O$_2$ (15 mℓ) and then O$_2$ (200 mℓ) and sufficient pure air is introduced to make the pressure 1 atm; then the mixture is shaken, set aside for 2 hr, diluted with water (5 mℓ), and 1 M Na$_2$SO$_3$ (1 mℓ) is added to remove any excess of H$_2$O$_2$ Subsequently, the mixture, now containing nitrate, is diluted to 25 mℓ with water and the solution is passed at 20°C through a column of the resin to adsorb nitrate, which is then eluted with 1 M KBr using a flow rate of 0.76 mℓ/min; the assay is by thermodetection liquid chromatography	15
Nitrite solutions	Amberlite®IRA-401 Nitrite at ppm levels is oxidized in acidic medium by excess (^{36}Cl)-chloramine T	75

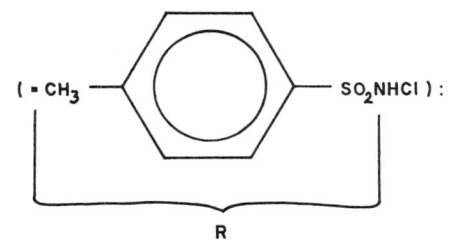

(−CH$_3$ — ⬡ — SO$_2$NHC$\overset{*}{\text{l}}$) :
            └─── R ───┘

$$RNH\overset{*}{Cl} + NO_2^- + H_2O \rightarrow RNH_2 + NO_3^- + H^+ + \overset{*}{Cl}{}^-$$

($\overset{*}{Cl}$ = ^{36}Cl); after 10 min, excess of radiochloramine T is isolated by adsorption on a column of the resin from an alkaline medium using the following solutions

a) NaOH sample solution (10 mℓ) (adsorption of unreacted radiochloramine T; $\overset{*}{Cl}{}^-$ passes into the effluent)

b) 0.3 M NaNO$_3$ (10 mℓ) (elution of $\overset{*}{Cl}{}^-$ formed by the reaction of adsorbed radiochloramine T with NO$_2^-$)

The nitrite originally contained in the sample is determined via measurement of the released radioactive Cl, i.e., $\overset{*}{Cl}{}^-$.

Table 4
DETERMINATION OF CYANIDE AFTER ANION AND CATION EXCHANGE SEPARATION

Material	Ion exchange resin, separation conditions, and remarks	Ref.
Industrial waters	Amberlite® IRA-400 (OH⁻ form) Column of 1 cm ID containing the resin to a height of 7—8 cm and operated at a flow rate of 10 mℓ/min a) Sample (250 mℓ) adjusted to pH 11—12 with NaOH (adsorption of CN^-) b) Water (20—30 mℓ) (as a rinse) c) 1% KNO_3 solution (150 mℓ) (elution of CN^-) Cyanide is determined colorimetrically utilizing the pyridine-pyrazolone reaction	20
Soluble metal cyanides (e.g., $K_2Cd[CN]_4$)	Amberlite® IRA-400 (20-50 mesh; OH⁻ form) a) Aqueous sample solution (2 mℓ) containing ≯100 mg of CN^-) (adsorption of CN^-) b) Water (as a rinse) c) 2 N H_2SO_4 followed by 4.5 N H_2SO_4 (15 mℓ each) and water (elution of CN^- which is collected in 2 M NaOH) CN^- is determined by argentimetry; the recovery of CN^- is complete	21
Cyanide solutions	Dowex® 50W-X8 (100-200 mesh; Hg[II] form) The sample (20 mℓ, containing CN^- in concentrations ranging from 10^{-4}—10^{-3} M) is shaken for 5 min with a 5-mℓ suspension of the resin, the mixture is filtered, and the resin washed with water (10 mℓ); in the combined filtrate and washings $Hg(CN)_2$ (formed by the reaction: $[R_sSO_3]_2Hg + 2 KCN \rightarrow 2RSO_3K + Hg[CN]2$) is determined polarographically; conversion of CN into $Hg(CN)2$ is only ≃80%; Co, Cu, Fe(II), excess of Cl^- or SCN^-, and equimolar concentrations of S^{2-} interfere; when Ni or Fe(III) is present either the shaking time or the temperature must be increased; Ag and Cd are tolerated	76

Table 5
DETERMINATION OF AMMONIA AFTER ANION EXCHANGE SEPARATION[a]

Materials	Ion exchange resin, separation conditions, and remarks	Ref.
Cu amine complexes	Deacidite® FF (100 mesh; OH⁻ form) Column containing 15—20 g of the resin A weighed amount of the complex containing 0.5—3.0 mmol of NH_3 is dissolved in the minimum amount of HCl, and the diluted solution is passed through the resin bed on which Cu is precipitated as $Cu(OH)_2$; the effluent and water washings containing the NH_3 are collected in standard H_2SO_4, and the excess of acid is titrated with 0.1 M NaOH	22
Cyanide plating baths	Anionite EDE-10P (OH⁻ form) Column: 23 × 1 cm operated at a flow rate of 4—5 mℓ/min a) Dilute sample of cyanide electrolyte (50—60 mℓ) (adsorption of CN⁻ and anionic metal cyanides; NH_3 passes into the effluent) b) Water (40—50 mℓ) (elution of remaining NH_3) NH_3 is determined titrimetrically	19
Diverse N-containing organic compounds (e.g., α-alanine, sulfamic acid, and methionine)	Bio-Rad® AGl-X8 (100-200 mesh; OH⁻ form) Column of ~15 cm length operated at 0°C The sample (0.3—0.8 mg) is digested with conc H_2SO_4 (0.01—0.02 mℓ) and the diluted mixture is passed through the resin bed; the ammonium sulfate is converted into NH_3 (SO_4^{2-} is retained by the resin) which is titrated with 2.8 mM $HClO_4$ with use of a pH meter; down to 30 μg of N can be determined	23
Body fluids (urine, plasma, and cerebrospinal fluid) and vegetable matter (e.g., spinach)	Amberlite® IRA-410 (OH⁻ form) Strips of Whatman® No. 1 paper impregnated with the resin (0.1 g/g) Following Kjeldahl decomposition of the sample (e.g., 0.01 mℓ), NH_3 is distilled into 0.001—0.003 M HCl (5 mℓ), the distillate is diluted to 10 mℓ and a portion (0.3 mℓ) is applied to the resin-impregnated paper strip; the areas occupied by the excess of acid are then revealed with suitable indicators and are compared with those produced by standards; the range of the method is 0.04—0.44 μg of N	24,25

[a] See also Table 4 in the chapter on Potassium, Volume V.

Table 6
DETERMINATION OF NITRATE AND NITRITE IN NATURAL WATERS AFTER CATION EXCHANGE SEPARATION

Material	Ion exchange resin, separation conditions, and remarks	Ref.
Natural waters, e.g., riverwater, rain, and snow	Suspensions of the anion exchange resin Amberlyst® A-27 (particle size: <30 μm; 8.97 μeq/mℓ) and the cation exchange resin Amberlyst® 15 (15.4 μeq/mℓ) To the sample (50 mℓ; containing <0.5 μg of NO_2^-), 1 mℓ of p-aminobenzenesulfonamide solution (0.5% in 2 M HCl), 0.5 mℓ of N-(1-naphthyl)-ethylenediamine dihydrochloride solution (0.1% in 2 M HCl), and 2 mℓ of 4 M HCl are added and the mixture is equilibrated for 10 min; subsequently, 3 and 2 mℓ of the anion and cation exchange resin suspensions, respectively, are added and the mixture is stirred for 10 min; during this period, the red-colored azo dye formed by the reaction of NO_2^- with the reagents is adsorbed on the cation exchanger and a bulky precipitate is obtained due to the coagulative action of the anion exchanger Then the mixed resins are filtered by suction onto filter paper (No. 5B) giving a colored thin layer of 17 mm in diameter and ∼0.3 mm in thickness; the thin layer is dipped in 0.2 M HCl for 15 min to stabilize the color and then its absorbance is measured	28
Creek water	The sample and eluent (0.01 M $HClO_4$ at 0.2 mℓ/min) are applied to a two-section glass column (pressure applied by He); the upper section (11 × 0.1 cm) contains Dowex® 50-X8 (80-100 mesh; H^+ form) to remove cations that poison the Cd-electrode of the detector, and the lower section (the analytical column, having the same dimensions) contains Amberlite® IRA-900 (180-250 mesh; ClO_4^- form) Elution is complete in ≃7—9 min In the effluent nitrate and nitrite are determined electrochemically; the detection limit for both anions is ≃0.01 mM The method is also suitable for the analysis of aqueous extracts of maize stalks and of oat hay	29
Lake water	Liquid chromatograph The sample (4 mℓ) is passed through a precolumn (5 × 0.5 cm) of Dowex® 50W-X8 (100-200 mesh; H^+ form) and a Sep-Pak® C_{18} radial cartridge and then through a column (15 × 0.3 cm) of Dowex® 2 (sulfate form) with 50 mℓ mM Na_2SO_4 as the eluent (at a flow rate of 2 mℓ/min) Nitrate is determined by UV spectrometry and an analysis takes ≃10 min	30
Natural waters	Wofatit® K (Na^+ form) Column: 5 × 1.5 cm The sample is passed through the resin bed to adsorb cationic constituents and in the effluent nitrate is determined spectrophotometrically	31

Table 7
DETERMINATION OF NITRATE AND NITRITE IN BIOLOGICAL MATERIALS AFTER CATION EXCHANGE SEPARATION

Material	Ion exchange resin, separation conditions, and remarks	Ref.
Plants	Dowex® 50W-X8 (H$^+$ form) and Dowex® 1-X8 (Cl$^-$ form) a) The dried and powdered sample (125 mg) and the cation exchange resin ($\simeq$50 mg) are shaken with water (25 mℓ) for 15 min to adsorb cationic constituents; then the mixture is filtered (filtrate contains the NO$_3^-$) and the resin is discarded b) Aliquot of filtrate (adsorption of NO$_3^-$ on a column containing the anion exchange resin) c) Water (as a rinse) d) 0.01 M HCl and water (as a rinse) e) 1 M NaCl (10 mℓ) followed by water (14 mℓ) (elution of NO$_3^-$) Nitrate is determined spectrophotometrically with brucine	10
Plants	Mixture of Dowex® 50 (200-400 mesh; H$^+$ form) and Dowex® 1 (200-400 mesh; OH$^-$ form) containing a slight excess of the latter resin Column of 2 cm length consisting of 2 mℓ of the resin mixture a) Aqueous plant extract (from 1—5 g of tissue) (adsorption of the N species; into the effluent pass glucose and other interfering substances) b) Water (10 mℓ) (removal of residual glucose etc.) c) 1 M NaF followed by water (10 mℓ each) (elution of NH$_4^+$ and amide-N from the cation exchanger) d) 1 M NaBr - 1 M NaOH (15 mℓ) followed by water (10 mℓ) (elution of NO$_2^-$ and NO$_3^-$ from the anion exchange resin) The various N species are determined titrimetrically following distillation	5
Cheese	Dowex® 50W-X8 (50-100 mesh; Ag$^+$ and Al^{3+} forms) To the filtered sample solution ($\sim$50 mℓ) 3 g each of the two resin forms are added and the mixture is stirred for 5 min to remove chloride and protein (chloride is precipitated as AgCl); after removal of the resin by filtration, NO$_3^-$ is determined in the filtrate using an ion selective electrode; the limit of detection is 5 mg nitrate per kilogram Before this batch separation of nitrate from chloride and protein, the fat of the sample (20 g) is removed by homogenizing it with water (70 mℓ) which involves heating for 1 hr on a boiling water bath and then cooling at 5°C for 1 hr followed by filtration	39
Plants	Dowex® 50-X8 (H$^+$ form) mixed in a slurry with Al$_2$(SO$_4$)$_3$ solution The dried, ground plant sample (0.4 g containing $\simeq$250—5000 ppm of N) is mixed with water (50 mℓ) and the resin-Al sulfate slurry (1 mℓ) is shaken for 10—15 min The mixture is filtered and nitrate and nitrite in the filtrate are determined by using a NO$_3^-$-sensitive electrode and a pH meter	40

Table 8
DETERMINATION OF NITRATE,[a] NITRITE, AND AMMONIA IN INDUSTRIAL PRODUCTS AFTER CATION EXCHANGE SEPARATION

Material	Ion exchange resin, separation conditions, and remarks	Ref.
Fertilizers	A) Separation of nitrate (cation exchange resin MK-2, grain size: 0.5—0.8 mm, H^+ form) Column of 0.7—1.0 cm ID containing 50 g of the resin a) Aqueous sample solution (5—10 mℓ) (adsorption of cations; HNO_3 passes into the effluent) (flow rate: 2—3 mℓ/min) b) Distilled water (300—500 mℓ) (elution of residual HNO_3 until eluate is neutral using a flow rate of 30—50 mℓ/min) HNO_3 is determined titrimetrically B) Separation of ammonium (anion exchange resin Amberlite® IRA 410 [~50 g; OH^- form] in a column of 0.7—1.0 cm ID) a) Aqueous sample solution (5—10 mℓ) containing ammonium sulfate (adsorption of sulfate; NH_3 passes into the effluent) (flow rate: 2—3 mℓ/min) b) Distilled water (300—500 mℓ) (elution of remaining NH_3 at a rate of 30—50 mℓ/hr) NH_3 is determined titrimetrically	26
	Dowex® 50-X8 (Na^+ form) and Dowex® 21K (Cl^- form) The sample solution (pH 4.1) is passed through columns of the cation and anion exchange resins connected in series; the NH_4^+ is retained by the first column and NO_3^- and PO_4^{3-} by the second column and amide passes through both; then the ions are eluted from the separated columns with 20% NaCl solution and water	11
Cigarettes	Rexyn® AG50 (16-50 mesh; H^+ form) Column: 8 × 1.3 cm containing 10 g of the resin and operated at a flow rate of 3—4 mℓ/min a) 5% NaOH solution (10 mℓ) (adsorption of Na; nitrate and nitrite pass into the effluent) b) Deionized water (30 mℓ) (elution of residual nitrate and nitrite) N is determined colorimetrically; the coefficient of variation is 5.2% The average recovery is 97.2% Before the separation, the cigarette smoke (containing oxides of N) is collected on a column of charcoal which is extracted with solution (a) and the extract is passed directly through the ion exchange column	44
Industrial waters	Cation exchange resin of the sulfonic acid type (H^+ form) Column of 2.5 cm ID containing 60 mℓ of the resin and operated at a flow rate of ~30 mℓ/min a) Water sample (adsorption of cations; nitrates, sulfates, and chlorides pass into the effluent) b) Distilled water (until effluent is neutral; elution of residual anions) After removal of CO_2 by passing air through the eluates, NO_3^-, Cl^-, and SO_4^{2-} are determined potentiometrically	45
Sulfate solutions	Cationite KU-2 (Pb^{2+} form) When passing the sample solution through a column of the resin, SO_4^{2-} reacts with Pb(II) to form insoluble $PbSO_4$ which remains in the resin bed; the NO_3^- combines with Pb(II) form-	46

Table 8 (continued)
DETERMINATION OF NITRATE,[a] NITRITE, AND AMMONIA IN INDUSTRIAL PRODUCTS AFTER CATION EXCHANGE SEPARATION

Material	Ion exchange resin, separation conditions, and remarks	Ref.
	ing soluble Pb(NO$_3$)$_2$ which passes into the effluent so that separation from sulfate is achieved	
	The method was used for the separation of milligram amounts of the elements	

[a] See also Table 9 in the chapter on Sulfur.

Table 9
MICROCHEMICAL DETECTION OF NITRITE BY RESIN SPOT TESTS[a]

Ion exchange resin	Experimental conditions and remarks	Ref.
Strongly acidic cation exchange resin, e.g., Dowex® 50W-X1 (~30 mesh; H⁺ form)	Several beads of the resin are placed on a spot plate with 1 drop of a 0.5% solution of m-phenylenediamine in either 0.3 M HCl or 10% acetic acid and thoroughly mixed and set aside for several minutes; then 1 drop of the test solution containing nitrite is added, and after a few minutes the surface of the resin is examined with a lens for an orange to brown coloration (azo dye: Bismarck brown) A detection limit of 0.039 μg and a limiting concentration of 1 in 1×10^6 for nitrites is obtained; the method can also be used for nitrates that can be readily reduced to nitrite with Zn	77
Dowex® 50W-X2 (H⁺ form)	To 1 drop of the sample solution (containing nitrite) on a spot-test plate are added a few beads of the resin and 1 drop of 1% Griess-Romijn reagent; in the presence of nitrite an azo dye is formed which appears as a red-purple color on the resin particles The limit of detection is 3 ng of NO_2^- (dilution limit = $1:10^7$) At the 30-ng level, there is interference from 30 μg of Cu, Fe, Ni, or sulfide and by 300 μg of Al, F^-, I^-, or SCN^-	78

[a] See also Table 17 in the chapter on Sulfur.

Table 10
DETERMINATION OF AMMONIA IN NATURAL WATERS AFTER CATION EXCHANGE SEPARATION

Material	Ion exchange resin, separation conditions, and remarks	Ref.
Rainwater	Amberlite® IR-120 (Na⁺ form) Column: 6.5×0.3 cm inserted in a flow-injection analysis system in a sample loop with an electronically operated injection valve a) Neutral or slightly acid sample (adsorption of NH_4^+) b) Carrier stream of 5 M NaOH (elution of NH_4^+) NH_3 is determined spectrophotometrically with Nessler reagent In samples containing 10—200 ppb of NH_4^+ the coefficient of variation was ≃1%	58
Seawater	Bio-Rad® AG50-X8 (50-100 mesh) Column: 4×0.7 cm a) Sample solution of pH 4—5 (adsorption of NH_4^+) b) Distilled water (1 mℓ) (as a rinse) c) ≃0.3 M Na phenoxide solution (elution of NH_4^+) N is determined spectrophotometrically; from 10—150 μg/ℓ of ammonia-N could be determined with a coefficient of variation of ±4.2% (seven determinations); before the separation, the sample (10 mℓ) is centrifuged at 4°C after adding 1 M NaOH (1 mℓ) to precipitate $Mg(OH)_2$ etc., and the supernatant is decanted into 0.6 M acetic acid to prepare solution (a)	59
Natural waters (e.g., river- and seawater)	Durrum® DC-4A cation exchange resin PTFE column: 5.5×0.3 cm operated at a flow rate of 0.2 mℓ/min a) Water sample (<0.5 mℓ) (adsorption of NH_4^+; into the effluent pass amino compounds) b) Borate buffer solution of pH 10.15 (elution of NH_4^+) NH_4^+ is determined fluorometrically	60

Table 11
DETERMINATION OF AMMONIA IN BIOLOGICAL MATERIALS AFTER CATION EXCHANGE SEPARATION

Material	Ion exchange resin, separation conditions, and remarks	Ref.
Blood plasma	Dowex® 50-X8 (100-200 mesh; Na$^+$ form) Column: 2 × ≃0.18 cm a) Plasma (100 µℓ) immediately followed by eluent (d) (500 µℓ) (adsorption of NH$_4^+$) (this mixture is passed in ∼5—7 min) b) Water (200 µℓ) (as a rinse) c) Water (500 µℓ) (prewarmed to 37°C) d) 4 M NaCl (4 × 200 µℓ) (elution of NH$_4^+$) NH$_3$ is determined spectrophotometrically	61
Plasma	Dowex® 50-X12 (50-100 mesh; Na$^+$ form) Column of the resin conditioned with 0.24 M sucrose a) Plasma sample (0.5 mℓ) (adsorption of NH$_4^+$) b) 0.24 M sucrose followed by water (as rinses) c) 4 M NaCl (elution of NH$_4^+$) N is determined colorimetrically with indophenol	62
	Amberlite® CG-120 (100-200 mesh) a) Plasma sample at neutral pH (adsorption of NH$_4^+$) b) 4 M NaCl (elution of NH$_4^+$) After removal of any protein in the eluate by coprecipitation with Al(OH)$_3$, NH$_3$ is determined colorimetrically with indophenol blue	63
	Permutit®-Folin-W resin (cation exchange resin which is more specific for NH$_4^+$ than, for example, Dowex® 50W-X12, 200-400 mesh) The sample (0.5 mℓ) and 50 mg of the resin are mixed for 15 min to adsorb NH$_4^+$; after washing the resin with water (10 mℓ), the adsorbed NH$_4^+$ is eluted by shaking it for 10 min with 1 mℓ of a 3 + 1 mixture of 4 M NaCl and ethanol; in the eluate NH$_4^+$ is determined colorimetrically	64
	Dowex® 50W-X8 (Na$^+$ form) packed in a 20-mℓ syringe and equilibrated with phosphate buffer solution (pH 7.4) a) Plasma sample (adsorption of NH$_4^+$) b) Water (elution of plasma proteins) c) 0.1 M NaOH (elution of NH$_4^+$) N is determined colorimetrically with phenoxide	65
	Dowex® 50W-X16 (20-50 mesh; Na$^+$ form) Column containing the resin conditioned with 0.2 M phosphate buffer of pH 7.4 a) Sample (3 mℓ) (adsorption of NH$_4^+$) b) Water (as a rinse) c) 0.1 M NaOH (5 mℓ) (elution of NH$_4^+$) N is determined spectrophotometrically	66
Whole blood	Dowex® R50-X8 (200-400 mesh) The sample is added to the resin and water and shaken for 1—2 min to adsorb NH$_4^+$; then the resin is filtered off, washed with water, and the color reaction (with phenol-alkaline hypochlorite) is applied directly to the resin	67
Plasma	Permutit® Q resin is equilibrated with the plasma sample to adsorb NH$_4^+$	68

Table 11 (continued)
DETERMINATION OF AMMONIA IN BIOLOGICAL MATERIALS AFTER CATION EXCHANGE SEPARATION

Material	Ion exchange resin, separation conditions, and remarks	Ref.
	Subsequently, the resin is treated with the Na phenoxide solution in the presence of OCl$^-$ to effect simultaneous elution and reaction to give a blue color	
Urine	Dowex® 50W-X8 (50–100 mesh; H$^+$ form)	69
	Column: 5 × 0.5 cm conditioned with 30 mM citric acid of pH 2	
	a) Sample solution containing 1 M HCl (0.5 mℓ) (adsorption of (^{14}C)-glutamic acid which has a net positive charge; uncharged 2-oxo-(^{14}C)-glutarate passes into the effluent)	
	b) 30 mM citrate buffer (50 mℓ) of pH ∼ 2.0 (elution of residual glutarate)	
	c) 1 M NH$_3$ solution ([^{14}C]-glutamate is eluted with the 3rd or 4th mℓ of the eluent and then determined radiometrically)	
	Before the separation, NH$_4^+$ (at the picamole or nanomole level) is converted into (^{14}C)-glutamate in the presence of an excess of 2-oxo-(^{14}C) glutarate and NADH on addition of glutamate dehydrogenase; then this enzymatic reaction is stopped with HCl, thus obtaining solution (a)	
Plant extracts	Dowex® 50-X8 (200–400 mesh; Na$^+$-K$^+$ form)	70
	a) Sample solution (1—7 mℓ) (adsorption of NH$_4^+$)	
	b) Water (18—24 mℓ) (elution of amides, i.e., asparagine and glutamine)	
	c) 1 M KCl (20 mℓ) (elution of NH$_4^+$)	
	N is determined spectrophotometrically	

REFERENCES

1. **Sen, N. P. and Lee, Y. C.**, Determination of nitrate and nitrite in whey powder, *J. Agric. Food Chem.*, 27(6), 1277, 1979.
2. **Barbieri, G., Sala, G., Gavioli, E., and Beneventi, G.**, Determination of nitrite and nitrate in meat, *Boll. Chim. Unione Ital. Lab. Prov. Parte Sci.*, 5(4), 611, 1979.
3. **Guertler, O. and Holzapfel, H.**, Separation of nitrate, nitrite, hyponitrite, and N-nitrohydroxylaminate anions on the anion-exchanger Wofatit SBW, *Angew. Makromol. Chem.*, 7, 194, 1969.
4. **Barkemeyer, H.**, Determination of nitrates in tobacco by ultraviolet spectrophotometry, *Beitr. Tabakforsch.*, 3(7), 455, 1966.
5. **Varner, J. E., Bulen, W. A., Vanecko, S., and Burrell, R. C.**, Determination of ammonium, amide, nitrite, and nitrate nitrogen in plant extracts, *Anal. Chem.*, 25, 1528, 1953.
6. **Westland, A. D. and Langford, R. R.**, Determination of nitrate in fresh water. Concentration of samples by an ion-exchange procedure, *Anal. Chem.*, 28, 1996, 1956.
7. **Bousset-Fatianoff, N. and Gouet, P.**, Research on nitrates in fresh and ensiled forages. I. Determination of nitrates in green and ensiled forages, *Ann. Biol. Anim. Biochem. Biophys.*, 11(4), 705, 1971.
8. **Buldini, P. L., Ferri, D., Pauluzzi, E., and Zambianchi, M.**, Differential pulse polarographic determination of nitrate and nitrite in nuclear wastes, *Mikrochim. Acta*, 1(1-2), 43, 1984.
9. **Chasko, J. H. and Thayer, J. R.**, Rapid concentration and purification of nitrogen-13-labelled anions on a high-performance anion-exchanger, *Int. J. Appl. Radiat. Isot.*, 32(9), 645, 1981.
10. **Baker, A. S.**, Colorimetric determination of nitrate in soil and plant extracts with brucine, *J. Agric. Food Chem.*, 15(5), 802, 1967.
11. **O'Neal, J. M. and Clark, K. G.**, Separation of various forms of nitrogen in fertilizers, *J. Assoc. Off. Agric. Chem.*, 47(6), 1054, 1964.
12. **Citron, I., Tai, H., Day, R. A., and Underwood, A. L.**, Simultaneous infrared determination of sulfate, nitrate and nitrite in water samples, *Talanta*, 8, 798, 1961.
13. **Tanaka, A., Nose, N., and Watanabe, A.**, Gas-chromatographic determination of nitrite in foods as trimethylsilyl derivative of 1H-benzotriazole, *J. Chromatogr.*, 194(1), 21, 1980.
14. **Matsumoto, R., Ono, A., Ibata, T., and Uemura, T.**, Analytical systems for continuously measuring trace amounts of nitrite and nitrate in the activated-sludge aeration bath for ammonia liquor treatment, *Tetsu To Hagane*, 67(6), 809, 1981.
15. **Kuroda, D.**, Analytical methods for nitrogen oxides. V. Determination by thermal detection liquid chromatography, *Jpn. Analyst*, 22(9), 1186, 1973.
16. **Schultz, F. A. and Mathis, D. E.**, Ion-selective electrode detector for ion exchange liquid chromatography, *Anal. Chem.*, 46, 2253, 1974.
17. **Collet, P.**, Determination of nitrite and nitrate in foods, *Dtsch. Lebensm. Rundsch.*, 79(11), 370, 1983.
18. **Eek, L. and Ferrer, N.**, Sensitive determination of nitrite and nitrate by ion exchange chromatography, *J. Chromatogr.*, 322(3), 491, 1985.
19. **Degtyarenko, Ya.A. and Kogut, L. N.**, Determination of ammonia in cyanide plating baths, *Ukr. Khim. Zh.*, 29(1), 94, 1963.
20. **Hissel, J. and Cadot-Dethier, M.**, Determination of cyanide in water, *Trib. CEBEDEAU*, 18, 272, 1965.
21. **Gilath, I.**, Breakdown, of alkaline complex cyanide by ion exchange, *Anal. Chem.*, 49, 516, 1977.
22. **Khorasani, S. S. M., Das, P. K., and Khundkar, M. H.**, Determination of ammonia in copper-amine complexes by employing ion-exchange, *Chemist Analyst*, 54(4), 125, 1965.
23. **Griepink, B. F. A. and Terlouw, J. K.**, Determination of centimilligram amounts of nitrogen in organic compounds, *Mikrochim. Acta*, No.3, 624, 1968.
24. **Madrowa, M.**, Chromatographic ultra-micro determination of nitrogen in body fluids by using anion-exchange paper, *Chem. Anal. (Warsaw)*, 22(2), 319, 1977.
25. **Lewandowski, A., Madrowa, M., and Skirbiszewski, J.**, Determination of microgram amounts of nitrogen in vegetable matter by means of ion-exchange paper, *Chem. Anal. (Warsaw)*, 12(5), 1043, 1967.
26. **Kowalska, E. and Sollorz, J.**, Determination of nitrate-nitrogen and ammonia-nitrogen in nitrogenous fertilizers by application of ion-exchange methods, *Fresenius, Z. Anal. Chem.*, 210, 271, 1965.
27. **Nissim, I., Yudkoff, M., Yang, W., Terwilliger, T., and Segal, S.**, Gas chromatography-mass spectrometry determination of ammonia enrichment in blood and urine, *Anal. Biochem.*, 114(1), 125, 1981.
28. **Matsuhisa, K. and Ohzeki, K.**, Determination of trace amounts of nitrite in water by coagulated ion exchanger colorimetry, *J. Chem. Soc. Jpn.*, No.(11), 1593, 1983.
29. **Davenport, R. J. and Johnson, D. C.**, Determination of nitrate and nitrite by forced-flow liquid chromatography with electrochemical detection, *Anal. Chem.*, 46, 1971, 1974.
30. **Wakida, S., Tanaka, T., Kawahara, A., and Hiiro, K.**, Rapid and highly sensitive analytical method for determination of nitrate in lake water by liquid chromatography, *Jpn. Analyst*, 32(10), 615, 1983.
31. **Procházková, L.**, Determination of nitrates in water, *Fresenius, Z. Anal. Chem.*, 167, 254, 1959.

32. **Hale, D. R.,** Performance evaluation of an automated batch analyzer, *Int. Lab.*, 10(1), 79, 1980.
33. **Fishman, M. J., Skougstad, M. W., and Scarbro, G. F.,** Diazotisation method for nitrate and nitrite, *J. Am. Water Work. Assoc.*, 56(5), 633, 1964.
34. **Pappenhagen, J. M.,** Colorimetric determination of nitrates, *Anal. Chem.*, 30, 282, 1958.
35. **Aruga, R., Baiocchi, C., Campi, E., Gennaro, M. C., and Mentasi, E.,** Nephelometric determination of nitrates in aqueous solution, *Ann. Chim. (Rome)*, 74(1-2), 87, 1984.
36. **Ceauşescu, D.,** Rapid determination of sulfate, chloride and nitrate in water, *Fresenius' Z. Anal. Chem.*, 165, 424, 1959.
37. **Ohno, S. and Tsutsui, T.,** Rapid determination of nitrogen-13 and fluorine-18 in reactor cooling water by an ion-exchange method, *Analyst (London)*, 95, 396, 1970.
38. **Luk'yanov, V. F., Duderova, E. P., Novak, E. F., Barabanova, T. E., and Polyakova, I. A.,** Spectrophotometric determination of nitrate in waste waters with 2,6-diacetamino-pyridine, *Zh. Anal. Khim.*, 39(2), 2175, 1984.
39. **Kaneda, Y., Kanamori, T., and Iwaida, M.,** Determination of nitrate in cheese by use of an ion-selective electrode, *Eisei Kagaku*, 23(5), 301, 1977.
40. **Paul, J. L. and Carlson, R. M.,** Nitrate determination in plant extracts by the nitrate electrode, *J. Agric. Food Chem.*, 16(5), 766, 1968.
41. **Eipeson, W. E., Mahadeviah, M., Gowramma, R. V., and Sastry, L. V. L.,** Improved method for the determination of nitrate and nitrite in fresh and canned fruit and vegetable products, *J. Food Sci. Technol.*, 11(5), 209, 1974.
42. **Green, L. C., Wagner, D. A., Glogowski, J., Skipper, P. L., Wishnok, J. S., and Tannenbaum, S. R.,** Analysis of nitrate, nitrite and nitrogen-15 nitrate in biological fluids, *Anal. Biochem.*, 126(1), 131, 1982.
43. **Hoshikawa, G. and Fudano, Y.,** Atomic absorption determination of nitrate in plant material, *Kagawa Daigaku Nogakuba Gakujutsu Hokoku*, 27, 111, 1976.
44. **Scherbak, M. P. and Smith, T. A.,** Colorimetric method for the determination of total oxides of nitrogen in cigarette smoke, *Analyst (London)*, 95, 964, 1970.
45. **Wagner, A.,** Rapid potentiometric determination of sulfates, chlorides and nitrates in water, *Cent. Belg. Etud. Document. Eaux.*, 37, 164, 1957.
46. **Ryabinin, A. I. and Bogatyrev, V. L.,** Use of cation-exchange resins for separation and titrimetric determination of anions. I. Separations of sulfate from nitrate by means of a cation exchange resin in the lead form, and indirect titration of nitrate, *Zh. Anal. Khim.*, 23(6), 894, 1968.
47. **Kozák, M. and Krajči, P.,** Use of ion exchange resins in the determination of nitrate nitrogen, *Chem. Prum.*, 13(5), 246, 1963.
48. **Gabrielson, G. and Andersson, B.,** Determination of nitrate and chloride in sodium hydroxyde solutions by means of cation exchangers in the manufacturing control of alkaline batteries, *Anal. Chim. Acta*, 16, 425, 1957.
49. **Petrova, S. Yu,** Methods for determining nitrate content using methyl orange, *Energetik*, No.(10), 18, 1980.
50. **Tsitovich, I. K. and Lapina, T. A.,** Use of cationites in the salt form for the removal of accompanying anions in the determination of nitrate, *Zh. Vses. Khim. Ova.*, 7(5), 579, 1962.
51. **Chromniak, E.,** Determination of nitrogen in nitro-lime and ammonium nitrate, *Chem. Anal. (Warsaw)*, 15(4), 789, 1970.
52. **Kreshkov, A. P., Kazaryan, N. A., and Rubtsova, E. S.,** Potentiometric analysis of mixtures of phosphoric and nitric acids in the presence of trioctylamine, *Tr. Mosk. Khim. Tekhnol. Inst.*, No.(58), 265, 1968.
53. **Ungar, J.,** Determination of nitrates in aqueous solution, *J. Appl. Chem.*, 6(6), 245, 1956.
54. **Kosina, F., Kupec, J., Mladek, M., Podesvova, E., Zamazalova, J., and Maslanova, L.,** Determination of nitrate and nitrite in tannery effluent, *Collect. Czech. Chem. Commun.*, 40(1), 278, 1975.
55. **Bastian, R., Weberling, R., and Palilla, F.,** Ultra-violet spectrophotometric determination of nitrate. Application to analysis of alkaline-earth carbonates, *Anal. Chem.*, 29, 1795, 1957.
56. **Lambert, J. L. and Zitomer, F.,** Differential colorimetric determination of nitrite and nitrate ions, *Anal. Chem.*, 32, 1684, 1960.
57. **Capitan Garcia, F., Valencia, M. C., and Capitan-Vallvey, L. F.,** Determination of nitrite in seawater by ion-exchanger colorimetry, *Mikrochim. Acta*, 3(3-4), 303, 1984.
58. **Bergamin, Filho, H., Reis, B. F., Jacintho, A. O., and Zagatto, E. A.,** Ion-exchange in flow injection analysis. Determination of ammonium ions at the $\mu g/\ell^{-1}$ level in natural waters with pulsed Nessler reagent, *Anal. Chim. Acta*, 117, 81, 1980.
59. **Riley, R. T. and Mix, M. C.,** Ion-exchange technique for concentrating ammonia from small volumes of seawater, *Mar. Chem.*, 10(2), 159, 1981.
60. **Gardner, W. S.,** Micro-fluorimetric method to measure ammonium in natural waters, *Limnol. Oceanogr.*, 23(5), 1069, 1978.

61. **Oberholzer, V. G., Schwarz, K. B., Smith, C. H., Dietzler, D. N., and Hanna, T. L.**, Micro-scale modification of a cation-exchange column procedure for plasma ammonia, *Clin. Chem.*, 22(12), 1976, 1976.
62. **Kurahasi, K., Ishihara, A., and Uehara, H.**, Determination of ammonia in blood plasma by an ion-exchanger method, *Clin. Chim. Acta*, 42(1), 141, 1972.
63. **Fenton, J. C. B.**, Estimation of plasma ammonia by ion exchange, *Clin. Chim. Acta*, 7(2), 163, 1962.
64. **Wenzel, E., Oppolzer, R., and Schuster, D.**, Micro-determination of ammonia. Modified simple hatch procedure with Permutit Folin-W resin for the isolation of ammonia from blood plasma, *Mikrochim. Acta*, No.(4), 680, 1971.
65. **Dienst, S. G. and Morris, B.**, Plasma ammonia determination by ion exchange, *J. Lab. Clin. Med.*, 64(3), 495, 1964.
66. **Miller, G. E. and Rice, J. D.**, Determination of the concentration of ammonia nitrogen in plasma by means of a simple ion-exchange method, *Am. J. Clin. Pathol.*, 39(1), 97, 1963.
67. **Rahiala, E. L. and Kekomaki, M. P.**, Simplified one-stage cation exchange technique for the determination of whole blood ammonia, *Clin. Chim. Acta*, 30(3), 761, 1970.
68. **Forman, D. T.**, Rapid determination of plasma ammonia by an ion-exchange technique, *Clin. Chem.*, 10(6), 497, 1964.
69. **Cheema-Dhadli, S., Hamat, R., Sonnenberg, H., and Halperin, M.**, Micro-method to measure ammonia, *Kidney Int.*, 19(1), 80, 1981.
70. **Henderlong, P. R. and Schmidt, R. R.**, Determination of free ammonium and asparagine and glutamine amide-nitrogen in extracts of plant tissue, *Plant Physiol.*, 41(7), 1102, 1966.
71. **Burczynska-Niedzialek, A., Tyczkowska, K., and Wojcik, J.**, Procedure for estimating ammonia in blood serum, *Biul. Inf. Przem. Paszow.*, 19(1), 50, 1980.
72. **Vujovic, A.**, Critical evaluation of blood-ammonia determination with cation-exchange resin, *Rass. Int. Clin. Ter.*, 54(15), 940, 1974.
73. **Chiou, S. J., Gran, T., Meloan, C. E., and Danen, W. C.**, Nitron-poly(vinylbenzyl chloride)polymer to selectively remove oxidizing anions from non-oxidizing anions. Removal of nitrate and nitrite from polluted waters, *Anal. Lett. Part A*, 14(11), 865, 1981.
74. **Wada, E. and Hattori, A.**, Spectrophotometric determination of traces of nitrite by concentration of azo-dye on an anion-exchange resin. Application to seawater, *Anal. Chim. Acta*, 56, 233, 1971.
75. **Narayanan, S. S. and Rao, V. R. S.**, A radio-release method for the determination of nitrite, *Radiochem. Radioanal. Lett.*, 58(2), 69, 1983.
76. **Hanagata, G., Ohzeki, K., and Kambara, T.**, Alternating current polarographic determination of cyanide ion after conversion into mercury(II) cyanide by use of a Hg-form resin, *Bull. Chem. Soc. Jpn.*, 53(10), 3025, 1980.
77. **Fujimoto, M.**, Micro-analyses with ion-exchange resins. VI. The detection of small quantities of nitrites with m-phenylenediamine, *Bull. Chem. Soc. Jpn.*, 29(5), 600, 1956.
78. **Ichikawa, T., Kato, K., and Kakihana, H.**, Microchemical detection of nitrite ion using ion-exchange-resin particles, *J. Chem. Soc. Jpn.*, No.(7), 1186, 1981.
79. **Unger, M. and Heumann, K. G.**, New calibration method for the trace determination of nitrate and nitrite in food samples, *Fresenius' Z. Anal. Chem.*, 320, 525, 1985.

PHOSPHORUS

Adsorption of orthophosphate and other oxyanions of P on anion exchange resins makes it possible to isolate this element from complex matrices such as geological, biological, and industrial materials for which purpose cation exchange resins can also be successfully employed. On the latter resins accompanying cationic constituents of the samples are retained, while phosphate is not adsorbed. Anion exchange resins have also been utilized for the chromatographic separation of phosphate species of varying degrees of condensation.

ANION EXCHANGE RESINS

General

The relative affinity of strongly basic resins, e.g., Amberlite® IRA 400 for orthophosphate, is given as $PO_4^{3-} > SO_4^{2-} > Cl^- > OH^-$. Consequently, this anion is usually adsorbed on the chloride forms of the resins, although occasionally the free base forms are also used for the adsorption, e.g., from neutral sample solutions[1,2] and natural waters (see Table 1*).[3-5] Furthermore, retention of phosphate on the anion exchangers also occurs from 0.1 M chloride solutions as well as from 0.05 M ethylenediaminetetraacetic acid (EDTA) at pH 5 (see Table 2),[6] or directly from aqueous suspenions of very finely ground samples of soils (see Table 1)[7,8] or phosphate rocks,[9] using batch procedures. Coadsorbed with phosphate from all these systems are other anions, as, for example, sulfate, fluoride, arsenate, vanadate, and nitrate, but efficient separations of phosphate are obtained from virtually all accompanying cationic constituents of the samples. This is also the case when adsorbing phosphate as the heteropolyacid anion molybdophosphate which is a very effective means for the isolation of phosphate from natural waters (see Table 1).[10-12] Following reduction of the adsorbed molybdophosphate, the determination of phosphate is possible directly on the resin using ion exchanger colorimetry.

For the elution of adsorbed orthophosphate the following eluents have been employed: 0.02,[13] 0.05,[1,14-16] 0.1,[13,17] and 0.4 M HCl,[4,18] 1 M HNO$_3$,[19] 1 N Na$_2$SO$_4$,[7,8] 1 M NaCl,[9] and 1 M NaOH.[2] With most of these eluents all other adsorbed anions will be eluted together with the phosphate. Coadsorbed fluoride can be removed from the resin before elution of the phosphate, using 0.1 M NaCl[9] or 0.25 M NaOH[2] as eluent.

The fact that phosphate is not adsorbed on anionic resins from the acid solutions mentioned above or at higher acid concentrations has also been utilized for analytical separations. Thus, with the separation procedure described in Table 3,[20] utilizing 2 M hydrofluoric acid (HF) as the sorption medium, phosphate can be separated from a very large number of metals by using coupled columns containing anion and cation exchange resins. An example in which the nonadsorbability of phosphate from strong hydrochloric acid (HCl) medium is employed for a separation is shown in Table 5.[21,22] Separations of this type are very common and occur whenever anionic metal complexes are adsorbed on anion exchange resins from mineral acid solutions at acid concentrations comparable to or higher than those used for the elution of the adsorbed orthophosphate (see above).

In addition to the applications illustrated in Tables 1 to 3 and 5, strongly basic anion exchanges have also been successfully used in the separation of acids containing P in various oxidation states as well as in the separation of mixtures of condensed phosphates.[1,14-16,23-55] Examples for separations of this type are shown in Tables 2 to 4.[6,14,15,53-55] For the separation of mixtures of condensed phosphates by ion exchange chromatography on columns of anion exchange resins various eluents have been employed, among which solutions containing K

---

* Tables for this chapter appear at the end of the text.

chloride of varying concentrations are the most widely used.[25,26,29,30,32-37,39-42,49-52,56,57] In most cases these solutions are buffered to pH 5.0 by addition of acetate, formate, or borate and in order to accelerate the chromatographic elution a concentration gradient is often used. Linear polymers of phosphoric acid containing up to 14 P atoms per molecule can be separated by gradient elution. With 1 M KCl at pH 5 the order of elution on Dowex® 1-X10 is ortho-, pyro-, tri-, tetra-, meta-, and trimeta-phosphate.[34] Successful chromatographic fractionations of oxyanion species of P can also be effected by the use of eluents containing Na chloride of varying concentrations.[28,46,58,59] When separating the individual linear phosphates (ortho- to octa-) on BioRad® AGl-X8 (chloride form) by an exponential gradient elution with 0.12 to 0.32 M NaCl at pH 7.0, it is possible to prevent hydrolysis of the P-O-P linkages, due to the presence of heavy metal traces, by addition of 0.005 M EDTA. The effectiveness of EDTA masking of the heavy metals present is better than that, for example, of citric acid.[24] Therefore, the NaCl eluents frequently contain EDTA,[5,24,43,44] which is also effective in KCl solutions (see Table 2).[6] Acetate eluents, e.g., ammonium acetate solutions, have been employed for the fractionation of the lower oxyacids of P.[60,61] Thus, separation of peroxymonophosphoric acid (H_3PO_5) from orthophosphoric acid can be effected on Deacidite® FF using an acetate buffer of pH 5 as the eluent.[61] Dilute HCl solutions of varying concentrations can be used for the separations of ortho- from pyro-phosphate (see Table 4).[15]

High molecular weight polymeric anions (for example, polymetaphosphate) are also adsorbed on strongly basic resins. However, their elution with salt solutions is difficult. The polymers should be hydrolyzed in an acid medium at slightly increased temperature and then eluted. It has been shown that Na ions are retained by anionic resins in the polymetaphosphate form.[62]

Applications

The separation principles discussed in the preceding paragraphs have been used in connection with the determination of orthophosphate and other oxyanions of P in natural waters (see Table 1),[3-5,10-12,21] soils (see Table 1),[7,8,63,64] sediments,[65,66] natural phosphates,[67] food and diets (see Table 2),[6,68] bone (see Table 2),[13] yeast,[69] feces,[68] other biological materials,[70] cola drinks,[71] and in industrial products such as Nb (see Table 3),[20] high-purity Zr,[17,72] fertilizers,[57] washing powder,[73] P smokes (aerosols formed by burning red P impregnated in butyl rubber),[59] and commercial tripolyphosphate (see Table 3).[53]

For the mutual separation of phosphate species of varying degrees of condensation the procedures outlined in Table 4 have been employed.[7,8,63,64]

To separate orthophosphate from synthetic mixtures with other elements the procedures presented in Table 5 can be employed.[18,22,74] A resin spot test that can be used for the microchemical detection of phosphate is outlined in Table 80 (see in chapter on Copper, Volume III).

Procedures based on ion chromatography of phosphate are presented in the chapter 2 "Special Analytical Techniques Using Ion Exchange Resins", Volume I.

CATION EXCHANGE RESINS

General

Orthophosphate and other oxyanions of P are not retained by strongly acid cation exchange resins, e.g., Dowex® 50, from acid, neutral, and alkaline solutions. Therefore, ready separation of phosphate is achieved from virtually all cationic constituents of sample solutions provided that their adsorption is performed under suitable conditions, as, for example, from very dilute mineral acid solutions. Among these 0.1 M HCl is by far the most frequently used sorption medium for accompanying elements such as alkalies, alkaline earth elements, rare earths, Al, Fe, other transition metals, etc. (for examples see Tables 6 to 12).[75-90]

Other HCl systems from which this adsorption can be performed include: 0.015,[91] 0.05,[92,93] and 0.2 M HCl,[89,94] 90% ethanol - 0.2 M HCl,[95] and 0.6 M HCl.[96] It has been shown[83,84] that Fe(III), which has been adsorbed on cationic resins from, e.g., 0.05 or 0.1 M HCl media, interacts with the phosphate so that a portion of the latter is retained in the resin bed. Consequently, incomplete separation results, especially when a large amount of Fe(III) is present. This coadsorption of phosphate with Fe can be avoided by performing the adsorption in the presence of holding reductants such as SO_2,[84,90] hydroxylamine hydrochloride,[97] and ascorbic acid.[98,99] These agents reduce Fe(III) to Fe(II) which is also strongly retained by cation exchange resins, but does not react with phosphate. The use of the reductants is also recommended for separations of phosphate from V[92,93,100] and Cr.[92,93]

Separations of phosphate from the cationic constituents of sample solutions can also be effected by their adsorption from dilute solutions of nitric acid,[101-104] HF,[20] sulfuric acid,[98-100] perchloric acid,[105] trichloroacetic acid,[106] EDTA,[107] and K hydroxide.[108,109]

By application of the batch technique it is also possible to adsorb accompanying cations simply by contacting the ion exchange resin with an aqueous suspension of the finely ground sample. This separation of phosphate can be effected by equilibration of the sample with a cation exchange resin only (see Table 8)[110-112] or with a mixture of cationic and anionic resins (see Table 6).[9]

After adsorption of the metal cations from the systems mentioned above residual phosphate remaining in the resin is usually recovered by rinsing it with water (e.g., until the effluent is neutral, i.e., acid free.[1,2,75-77,79-81,84,87,89,94,100,101,104,106,108,109,113-122] For the same purpose the following eluents have also been used: 0.015,[123] 0.2,[92] and 0.5 M HCl[96] as well as 1:10 ammonia solution.[98,99]

Applications

Cation exchange separations of phosphate from accompanying elements have been employed in connection with the determination of P in natural phosphates (see Tables 6 to 8),[9,75,76,84,88,89,94,96,110,111,121,123-127] U ores and concentrates (see Table 6),[113] waters,[128-131] biological materials (see Table 9),[77-79,95,103,106,119,132] and in industrial products such as fertilizers (see Table 10),[80,122,133,134] metals and alloys (see Table 10)[114] which include Fe and steel,[135] ferrosilicon,[98,99] and V,[136] as well as in organic compounds (see Tables 10 and 11),[1,81,108,109,137] detergents (see Table 11),[82,116] Cu-phosphate electrolytic baths,[87] and compounds and salts of complex composition (see Tables 10 and 11).[92,100-102,112,115] To separate phosphate from synthetic mixtures with other elements the procedures outlined in Table 12 can be used.[2,117,118,138,139] In this table two methods are also illustrated utilizing reactive ion exchange as a means for separating phosphate.[138,139]

Table 1
DETERMINATION OF P IN GEOLOGICAL MATERIALS AFTER ANION EXCHANGE SEPARATION[a]

Material	Ion exchange resin, separation conditions, and remarks	Ref.
Rain, snow, lake, and seawater[b]	Suspensions of the finely divided macroreticular resins Amberlyst® 15 (sulfonic acid type; H^+ form) and Amberlyst® A-27 (strongly basic type; Cl^- form)	10, 11
	To the sample (50 mℓ, containing <0.3 µg of phosphate) reagent solution (6 mℓ) (2.5 M H_2SO_4 [100 mℓ] + 1% ammonium molybdate solution [30 mℓ] + K-Sb-tartrate solution [10 mℓ; 1 mg Sb per milliliter]) and 0.1 M ascorbic acid (2 mℓ) are added and the mixture is stirred mechanically for 5 min; following addition of 2 and 4 mℓ, respectively, of the cation and anion exchange resin suspensions, stirring is continued for 15 min; then the coagulated material is collected on one end of a filter paper strip and a disk (17 mm diameter and ≃0.3 mm thick) of colored resin (adsorbed and reduced molybdophosphate = Mo-blue complex) is formed, the adsorbance of which is measured by ion exchanger colorimetry at 700 nm; preceding the measurement, the filter strip is dipped for 20 min in a solution prepared by dilution of 2.5 M H_2SO_4 (5 mℓ), 0.1 M ascorbic acid (6 mℓ), and K-Sb-tartrate solution (1 mℓ) with water to 50 mℓ.	
	Before the separation, the sample is filtered (if orthophosphate is to be determined); if total P is to be assayed the sample is first digested with hot H_2SO_4 and HNO_3	
Riverwater[b]	Dowex® 1-X8 (100-200 mesh; molybdate form); this molybdate form is prepared by stirring the chloride form in ammonium molybdate solution, then washing it with water and drying in vacuo at ambient temperature	12
	To the sample is added 0.2 M HCl (2.5 mℓ), if required, and the volume is adjusted to 25 mℓ with water before the resin (0.5 g) is added, the mixture being stirred for 20 min to adsorb P (as molybdophosphate); the resin is then washed with water, 10 mℓ of reducing solution (0.3% in ascorbic acid, 0.01% in Sb-K-tartrate, and 0.25 M in H_2SO_4) is added, and the mixture is stirred at 25°C for 10 min; the supernatant solution is decanted and the blue-colored resin is washed with water before subjecting it to ion exchanger colorimetry at 705 nm	
	Most anions at the 100—1000-ppm level do not interfere; by using larger sample volumes ppb levels of P can be determined	
	Dowex® 1-X4 (50-100 mesh; OH^- form)	3
	Column of 1.3 cm ID containing 10 mℓ of the resin and operated at a flow rate of ≯12 mℓ/cm² min	
	a) Sample (1 ℓ) containing 0.01% V solution (0.3 mℓ) and 1 M HCl (0.1 mℓ) (adsorption of phosphate, arsenate, and vanadate; matrix elements pass into the effluent)	
	b) Distilled water (100 mℓ) (as a rinse)	
	c) 1 M HNO_3 (35 mℓ) (elution of phosphate, arsenate, and vanadate)	
	P and As are determined spectrographically with V as the internal standard	
	Concentrations as low as 0.001 ppm were determined with a precision of ±0.001 ppm (95% confidence level for a single measurement)	

Table 1 (continued)
DETERMINATION OF P IN GEOLOGICAL MATERIALS AFTER ANION EXCHANGE SEPARATION[a]

Material	Ion exchange resin, separation conditions, and remarks	Ref.
Aqueous environmental samples	Dowex® 1-X8 (100-200 mesh; chloride form) Column of 1 in. I.D. containing ~5 g of the resin conditioned with conc HCl (25 mℓ) and operated at a flow rate of ~3 mℓ/min a) Sample solution (~50 mℓ) saturated with HCl-gas (adsorption of ^{95}Zr/^{95}Nb; P passes into the effluent) b) Conc HCl (2 × 20 mℓ) (elution of residual P) Subsequently, P is extracted as the molybdophosphate and ^{32}P is determined radiometrically; the sensitivity is 14 pCi/ℓ, and the precision is 11% when a 50-mℓ sample is analyzed for ^{32}P; interference from ^{131}I is prevented by CHCl$_3$ extraction; decontamination of ^{32}P from interferences is ≃10^4 and recovery is ≮85%	21
Soils	Deacidite® FF-510 (particle size: >0.5 mm; Cl$^-$ form) The sample (2 mℓ, dried, and ground to pass a 0.5-mm sieve) is mixed with the resin (2.9 g) in a volume of water, depending on the expected PO$_4^{3-}$ content of the soil (100 mℓ for normal agricultural soil; 1 ℓ for glasshouse soil), and the mixture is shaken for 16 hr at ≃20°C to adsorb available phosphate; the suspension is then poured onto stretched Terylene® net (mesh ≃0.5 mm) to separate the resin from soil and water; the resin retained by the net is washed with water to remove adhering soil particles and the adsorbed phosphate is eluted with 1 N Na$_2$SO$_4$ Phosphate is determined spectrophotometrically	7, 8

[a] See also Table 1 in the chapter on Titanium, Volume IV; Table 1 in the chapter on Silicon; and Table 6 in the chapter on Lead.
[b] See also Reference 141.

Table 2
DETERMINATION OF P IN BIOLOGICAL MATERIALS AFTER ANION EXCHANGE SEPARATION[a]

Material	Ion exchange resin, separation conditions, and remarks	Ref.
Food (cheese, ham, and sausage)	Dowex® AG1-X4 (100-200 mesh; chloride form) Column: 60 × 1 cm a) Sorption solution (∼300 mℓ) which is 5 mM in EDTA and adjusted to pH 5 with 1 M NaOH (adsorption of phosphates) b) 5 mM EDTA solution of pH 5 (100 mℓ) (as a rinse) c) 5 mM EDTA (pH 5) - 0.15 M KCl (200 mℓ) (elution of orthophosphate) d) 5 mM EDTA (pH 5) - 0.26 M KCl (200 mℓ) (elution of pyrophosphate) e) 5 mM EDTA (pH 5) - 0.38 M KCl (200 mℓ) (elution of tripolyphosphates) f) 2 M HCl (100 mℓ) (elution of other polyphosphates) Polyphosphates are hydrolyzed by heating with HClO$_4$ and the phosphates are converted into molybdophosphate, which is extracted into butyl acetate for spectrophotometric measurement Before the ion exchange separation step, the sample (5 g) is homogenized with ice-cold 20% trichloroacetic acid (45 mℓ) and after dilution to 50 mℓ, a 1-mℓ aliquot is added to the sorption solution (a)	6
Bone tissue	Deacidite® (120-200 mesh; Cl$^-$ form) After adsorption of phosphates on a column of the resin, the following eluents are used a) 0.02 M HCl (elution of orthophosphate and other phosphates except pyrophosphate) b) 0.1 M HCl (elution of P$_2$O$_7^{4-}$; ≃0.5% of the total P in bone)	13

[a] See also Table 4 in the chapter on Iron, Volume V and Table 15 in the chapter on Zinc, Volume IV.

Table 3
DETERMINATION OF P IN INDUSTRIAL PRODUCTS AFTER ANION EXCHANGE SEPARATION[a]

Material	Ion exchange resin, separation conditions, and remarks	Ref.
Nb	Coupled columns of Dowex® 1-X8 (10 × 0.7 cm; upper column) and Dowex® 50W-X8 (5 × 0.7 cm) (both 100-200 mesh) conditioned with 2 M HF (20 mℓ) and operated at a flow rate of 0.5—0.7 mℓ/min a) Dilute HF sample solution (~8 mℓ) (adsorption on the anion exchange resin of the Nb matrix together with impurities such as Zr, Hf, Mo, W, and Ta [all these metals are adsorbed as anionic fluoride complexes]; on the lower column containing the cation exchanger the alkali metals, alkaline earth metals, rare earths, Co, Cu, Mn, Ni, Zn, and others are retained; P passes into the effluent from both columns) b) 2 M HF (20 mℓ) (elution of P) After evaporation of the eluent, P is extracted substoichiometrically (as molybdophosphate) and determined radiometrically; the detection limit was 2 ng/g	20
Commercial tripolyphosphate	Before the ion exchange separation step, the sample (≃100 mg) is irradiated with neutrons and then decomposed with conc HF (0.5 mℓ) and conc HNO$_3$ (~0.3 mℓ) Following the dissolution, P-carrier (100 μg) is added and the solution is diluted with water to prepare solution (a) After adsorption of the phosphates on a column of an anion exchange resin the following eluents are used a) 0.05 M HCl - 0.1 M KCl (elution of orthophosphate) b) 0.2 M KCl at pH 5 (elution of pyrophosphate) c) 0.4 M KCl at pH 5 (elution of triphosphate) d) 0.75 M KCl at pH 5 (elution of trimetaphosphate) Pyrophosphate in eluate (b) is determined spectrophotometrically after acid hydrolysis with H$_2$SO$_4$ to orthophosphate	53

[a] See also Table 5 in the chapter on Zirconium and Hafnium, Volume IV; Table 2 in the chapter on Tungsten, Volume IV; Table 19 in the chapter on Copper, Volume III; and Table 13 in the chapter on Sulfur.

Table 4
MUTUAL ANION EXCHANGE SEPARATION OF PHOSPHATE SPECIES

Ions separated	Ion exchange resin, separation conditions, and remarks	Ref.
$H_2PO_2^-$, PO_3^{3-}, PO_4^{3-}, $P_2O_7^{4-}$, and $P_3O_{10}^{5-}$	Amberlite® IRA-410 (30-50 mesh; Cl^- form) Column: 53 × 0.8 cm operated at a flow rate of 60 mℓ/hr After adsorption of the ions the following eluents are used for their fractionation a) 0.1 M KCl - 0.025 M glycine - 0.0265 M KOH (150 mℓ) (elution of hypophosphite) b) 0.05 M HCl (sequential elution of phosphate and phosphite with 70 mℓ each) c) 0.3 M HCl (90 mℓ) (elution of pyrophosphate) d) 0.5 M HCl (90 mℓ) (elution of triphosphate)	14
PO_4^{3-} from $P_2O_7^{4-}$	Amberlite® IRA-400 (20-50 mesh; Cl^- form) Column: 24 × 1.2 cm operated at a flow rate of 3 mℓ/min a) Sample solution (150 mℓ) containing 80 mg of the phosphates and 2 mℓ of 1 M HCl (adsorption of ortho- and pyrophosphates) b) 0.05 M HCl (250 mℓ) (elution of orthophosphate) c) 0.5 M HCl (150 mℓ) (elution of pyrophosphate)	15
PO_2^{3-}, PO_3^{3-}, and PO_4^{3-}	Dowex® 1-X8 (100-200 mesh; Cl^- form) Column containing the resin to a height of 50 cm a) Sample solution (1 mℓ, containing up to 2 mg of P per anion) b) KCl solution buffered at pH 6.8 (elution in the order: PO_2^{3-}, PO_4^{3-}, PO_3^{3-}) c) KCl solution buffered at pH 11.4 (elution in the order: PO_2^{3-}, PO_3^{3-}, PO_4^{3-})	55
Traces of polyphosphates from orthophosphate	Dowex® 1-X8 (100-200 mesh) Column operated at a flow rate of 15—18 mℓ/min After addition of the sample solution (25-mg sample) to the ion exchange resin column the following eluents are passed a) 0.1 M KCl - 0.05 M HCl (300 mℓ) (elution of PO_4^{3-}) b) 1 M KCl (100 mℓ) (elution of polyphosphates)	54

Table 5
ANION EXCHANGE SEPARATION OF P FROM SYNTHETIC MIXTURES WITH OTHER ELEMENTS

Ions separated	Ion exchange resin, separation conditions, and remarks	Ref.
PO_4^{3-} from Ca, Fe(III), and Al	Anionite PE-9 or EDE-10 (Cl^- form) Column: 15—18 × 1 cm operated at a flow rate of 1—2 mℓ/min a) Sample solution (50 mℓ, 0.1 M with respect to Cl^-) containing ⪢0.1 g of the ions (adsorption of phosphate; Ca, Fe, and Al pass into the effluent) b) Water (200 mℓ) (elution of remaining cations) c) 2 M HCl (100 mℓ) (elution of phosphate)	74
^{32}P from ^{95}Zr/^{95}Nb	From 12 M HCl sample solution Zr and Nb are adsorbed on a column of Dowex® 1, while P passes into the effluent where it is determined radiometrically The method can be used in connection with the determination of P in hair samples	22
Phosphate from sulfate	Dowex® 1-X2 (200-400 mesh; chloride form) Column: 6 × 0.8 cm With 0.4 M HCl as eluent the separated anions pass (phosphate is eluted first) to the burner of an atomic absorption spectrometer wherein they reduce the atomic absorption of Ca at 422.7 nm by amounts related to their amounts (thereby allowing their determination)	18

Table 6
TITRIMETRIC DETERMINATION OF P IN GEOLOGICAL MATERIALS AFTER CATION EXCHANGE SEPARATION[a]

Material	Ion exchange resin, separation conditions, and remarks	Ref.
Natural phosphates	Lewatit® S100 (H⁺ form) in a column operated at a flow rate of 5 drops/min a) Aqueous sample solution (100—130 mℓ, containing 1 mℓ of conc HCl) (adsorption of cationic constituents; phosphate passes into the effluent) b) Water (150—180 mℓ) (elution of remaining phosphate) Phosphate (H_3PO_4) is determined titrimetrically Before the separation, the sample (300 mg) is dissolved in conc HCl (5 mℓ) and a few drops of conc HNO_3	75
Apatite	Phenolsulfonic acid resin, e.g., cationite KU-1 (H⁺ form) Column containing 10 g of the resin a) ≃0.1—0.2 M HCl sample solution (20-mℓ aliquot, containing 200 mg of the sample) (adsorption of Ca and other cationic constituents; phosphate passes into the effluent) b) Water (80—100 mℓ) (elution of remaining phosphate) Phosphate is determined titrimetrically Before the separation, the sample (1 g) is decomposed with HNO_3-HCl	89
Crude phosphate	Wofatit® FN (H⁺ form) Column containing the resin to a height of 20 cm and operated at a flow rate of 10 mℓ/min a) ≃0.2 M HCl sample solution (25 mℓ, containing 150 mg of sample) (adsorption of Fe and Al; phosphate passes into the effluent) b) Water (5 × 25 mℓ) (elution of remaining phosphate) Phosphate is determined titrimetrically Before the separation, the sample is decomposed with HNO_3 and HCl	94
Natural phosphate (vivianite) and ferric oxide deposit	Amberlite® IR-120 or Amberlite® IR-112 (A.G.) (50-100 mesh; Na⁺ form) Column: 11 cm × 0.9 cm² operated at a flow rate of 1—2 mℓ/min a) 0.1 M HCl sample solution through which SO_2 was passed for ∼30 min to reduce Fe to Fe(II) (adsorption of Fe[II]; phosphate passes into the effluent) b) Distilled water (30 mℓ) (elution of residual phosphate) P is determined titrimetrically Before the separation, the sample (1 g) is decomposed with aqua regia (20 mℓ) and nitrate is removed by repeated evaporation with conc HCl (5 × 5 mℓ)	84
Phosphate rock	Mixture of cationite KU-1 (grain size <1.2 mm; H⁺ form) and anionite AN-1 (grain size >1.2 mm; OH⁻ form) The sample (50—100 mg) is stirred for 30 min under water with the resin mixture to adsorb Ca on the cation and phosphate plus fluoride on the anion exchange resin, respectively; subsequently, the slurry is passed through a sieve to separate the two exchangers from one another; the fluoride retained by the anionite remaining on the sieve is eluted with 0.1 M NaCl (30 mℓ) and then phosphate is desorbed with 1 M NaCl (50 mℓ)	9
U ores, concentrates, and liquors	Amberlite® IR-120 (H⁺ form) Column: 8 × 1 in. a) Dilute acid sample solution (100 mℓ) (adsorption of U[VI] and other cations; phosphate passes into the effluent) b) Water until the effluent is neutral (elution of remaining phosphate)	113

Table 6 (continued)
TITRIMETRIC DETERMINATION OF P IN GEOLOGICAL MATERIALS AFTER CATION EXCHANGE SEPARATION[a]

Material	Ion exchange resin, separation conditions, and remarks	Ref.
	Phosphate is determined titrimetrically Before the separation, the sample (0.1—0.2 g) is decomposed with HBr and HNO$_3$	

[a] See also Table 2 in the chapter on Sodium, Volume V; Table 3 in the chapter on Silicon; and Reference 142.

Table 7
GRAVIMETRIC DETERMINATION OF P IN NATURAL PHOSPHATES AFTER CATION EXCHANGE SEPARATION[a]

Material	Ion exchange resin, separation conditions, and remarks	Ref.
Natural Fe phosphates	Amberlite® IR-120 (grain size: 0.15—0.3 mm; H$^+$ form) Column: 15 cm × 1 cm^2 conditioned with 0.6 M HCl (50 mℓ) and operated at a flow rate of 5 mℓ/min a) 0.6 M HCl sample solution (150—200 mℓ) (adsorption of Fe[III], Mn, Mg, and Ca; phosphate passes into the effluent) b) 0.6 M HCl (20 mℓ) (as a rinse) c) 0.5 M HCl (100 mℓ) (elution of remaining phosphate) d) 4 M HCl (250 mℓ) (elution of Fe [III] and other adsorbed metals) Phosphate and the other elements are determined gravimetrically Before the separation, the sample (0.3—0.4 g) is dissolved in HCl and HNO$_3$	96
Rock phosphates	Amberlite® IR-120 (H$^+$ form) Column of 1.5 cm I.D. containing an amount of resin corresponding to 60-meq capacity and operated at a flow rate of ~10 mℓ/min a) Dilute HCl sample solution (~80 mℓ) (adsorption of Mg; phosphate passes into the effluent) b) 0.015 M HCl (250 mℓ) (elution of remaining phosphate) Subsequently, phosphate is isolated as NH$_4$MgPO$_4$ and ignited to Mg$_2$P$_2$O$_7$ The specific activity of this material is then compared with the standard Before this ion exchange purification step, samples of standard Mg$_2$P$_2$O$_7$ labeled with ^{32}P are digested with the rock sample in acid solution	123

[a] See also Table 4 in the chapter on Rare Earth Elements, Volume I.

Table 8
DETERMINATION OF P BY OPTICAL METHODS IN PHOSPHATE ROCK AFTER CATION EXCHANGE SEPARATION

Material	Ion exchange resin, separation conditions, and remarks	Ref.
Phosphate rock	Dowex® 50W-X8 (e.g., 200-400 mesh; H⁺ form)	110
	To an aqueous suspension (100 mℓ) of the sample (containing $\simeq$22 mg of P) the resin ($\sim$8.5 g) is added and the mixture is shaken for $\sim$15 min to adsorb the cations	
	The insoluble residue is allowed to settle and in an aliquot of the clear solution, P is determined by cool flame emission spectrometry; the average relative error of the method is 4.2%	
	A similar technique has been described[111] using Amberlite® IR-120 (10 g) to isolate P from 30 mg of phosphate rock suspended in 30 mℓ of water; the mixture is shaken for 2—3 hr at 80°C and P is determined by atomic absorption spectroscopy via measurement of added Ca	
	The batch process using Dowex® 50W (2 g) can also be applied after decomposition of the sample (2 g) with HNO₃ and HCl;[124] in this case the average recovery from samples containing 17.3—32.8% of P₂O₅ is 100.2%	
Phosphate rocks (e.g., apatite)	Dowex® 50 (20-50 mesh; H⁺ form)	76
	Column: 50-mℓ buret containing the resin (10 g) to a height of 8—9 in. and operated at a flow rate of $\sim$85 mℓ/10 min	
	a) Sample solution (20 mℓ) containing 2 drops of conc HCl (adsorption of Ca and other cationic constituents; phosphate passes into the effluent)	
	b) Water (65 mℓ) (elution of residual phosphate)	
	Phosphate is determined flame photometrically	
	Before the separation, the sample (containing 3.6—8.5 mg of P₂O₅) is decomposed with HCl and HNO₃	

Table 9
DETERMINATION OF P IN BIOLOGICAL MATERIALS AFTER CATION EXCHANGE SEPARATION[a]

Material	Ion exchange resin, separation conditions, and remarks	Ref.
Liver, bone, and cherry leaves	Dowex® 50W-X8 (H⁺ form) The sample solution (2—3 mℓ 2 M HCl diluted with ethanol to a final solution 90% in ethanol) is shaken for 30 min with the dry resin (10 g) to adsorb interfering cations such as Ca, Ba, Ag, Fe, Co, Sb, Sn, and Pb (solution of bone samples have to be equilibrated with the resin for 60 min to remove all the Ca) Subsequently, the resin is filtered off and in the filtrate, phosphate is determined flame photometrically Before the separation, the sample (1 g) is wet ashed with HNO_3 -$HClO_4$	95
Orange juice	Amberlite® IR-120 or Dowex® 50 (H⁺ form) The sample solution (few milliliters, containing the ash of 25 mℓ of orange juice dissolved in 0.5 mℓ of 2 M HCl) is equilibrated with the resin (2—3 g) to adsorb Fe and Al; subsequently, the resin is filtered off, washed with water, and in the filtrate H_3PO_4 is determined titrimetrically	79
Wine	Amberlite® IR-120 or Dowex® 50 (H⁺ form) A solution in dilute acid of ash (from 25 or 50 mℓ of wine) is boiled for a short time with the resin (0.3 or 0.5 g, respectively) to adsorb Fe(III) and Al After removal of the exchanger by filtration, phosphate (H_3PO_4) is determined titrimetrically in the filtrate	78
Lysimeter leachates	Lewatit® S 100 (H⁺ form) Column: 15 × 0.5 cm operated at a flow rate of 30 mℓ/10 — 20 min a) Sample aliquot (30 mℓ of a solution containing 2 mℓ of 25% HCl per 100 mℓ) (adsorption of cationic constituents; phosphate passes into the effluent) b) Water (elution of residual phosphate) Phosphate is determined colorimetrically Before the separation, the sample (300 mℓ) is evaporated to dryness, the residue is dry ashed, and the residue is dissolved in HCl; final sample volume: 100 mℓ	77
Rubber	Amberlite® IR-112 (H⁺ form) Column: 10 × 0.8 cm a) Sample solution containing trichloroacetic acid (adsorption of cationic constituents; phosphate passes into the effluent) b) Water (elution of remaining phosphate) Phosphate is determined colorimetrically Before isolation of the free PO_4^{3-} by the ion exchange method, NH_3 is removed by air blowing 0.5—3 g of latex, which is coagulated by dilution to 5 mℓ and addition of 10% trichloroacetic acid	106

[a] See also Table 3 in the chapter on Calcium, Volume V and Table 5 in the chapter on Sulfur.

Table 10
TITRIMETRIC AND GRAVIMETRIC DETERMINATION OF P IN INDUSTRIAL PRODUCTS AFTER CATION EXCHANGE SEPARATION[a]

Material	Ion exchange resin, separation conditions, and remarks	Ref.
Superphosphate	Wofatit® (Na$^+$ form) (60 g) contained in a column operated at a flow rate of 15—20 mℓ/min a) $\simeq$0.1 M HCl sample aliquot (100 mℓ containing $\sim$0.5 g of the sample) (adsorption of cationic constituents; phosphate passes into the effluent) b) Water (300 mℓ) (elution of residual phosphate) Phosphate is determined titrimetrically Before the separation, the sample (2.5 g) is dissolved in 1 M HCl; a similar separation method has been described using the cationite SBS (Na$^+$ form)[140]	80
Fertilizers	Wofatit® F or KPS 200 (H$^+$ form) Column operated at a flow rate of 8—10 mℓ/min An aliquot (50 mℓ) of the sample solution is passed through the resin bed to adsorb alkali and alkaline earth metals and in the effluent P is determined gravimetrically (as MgNH$_4$PO$_4$) Before the separation, the sample (2 g) is shaken with 2% citric acid solution (200 mℓ) for 30 min and the solution is filtered	133
Phosphor, Cu and allied brazer filler	Dowex® 50W-X8 (20-25 mesh; H$^+$ form) Column: 14.5 cm $\times$ 4.2 cm^2 operated at a flow rate of $\geqslant$6 mℓ/min a) Very dilute acid sample solution (50 mℓ) (adsorption of Cu and Fe; phosphate passes into the effluent) b) Water (100 mℓ) (elution of remaining phosphate) c) 2 M HNO$_3$ (150 mℓ) (elution of Cu and Fe) Phosphate is determined titrimetrically Before the separation, the sample (0.2 g) is dissolved in HNO$_3$-HCl	114
U phosphates	Dowex® 50 (H$^+$ form) Column: 25 $\times$ 2 cm a) Sorption solution (300 mℓ) (adsorption of U; phosphate passes into the effluent) b) Water (200 mℓ) (elution of remaining phosphate) Phosphate is determined by amperometric titration Before the separation, the phosphate solution (5 mℓ) containing UO$_2$(II), sulfate, perchlorate, and $\leqslant$25 mg of orthophosphate is neutralized with NH$_3$, acidified with conc HNO$_3$ (5 mℓ), and diluted with water to obtain solution (a)	101
Organic P compounds	Dowex® 50-X8 (100-200 mesh; H$^+$ form) Column of 0.5 cm ID containing 1.5 mℓ of the resin a) Sample solution of pH $\simeq$ 1 ($\sim$0.5 mℓ) (adsorption of metals; phosphate passes into the effluent) b) Water (2 $\times$ 1.5 mℓ) (elution of residual phosphate) P is determined titrimetrically Before the separation, the sample is dry ashed	81
Na and Ba hypophosphites	Cationite SBS or SBSR (H$^+$ form) Column containing the resin to a height of 40 cm and operated at a flow rate of 5—15 mℓ/min a) Aqueous solution (50 mℓ, containing 30 mg of the sample) (adsorption of Na and Ba; hypophosphite passes into the effluent) b) Water (200 mℓ) (elution of remaining hypophosphite) Hypophosphite (H$_3$PO$_2$) is determined titrimetrically with 0.1 M NaOH	115

[a] See also Table 3 in the chapter on Calcium, Volume V and Table 13 in the chapter on Molybdenum, Volume IV.

Table 11
DETERMINATION OF P IN INDUSTRIAL PRODUCTS BY OPTICAL METHODS AFTER CATION EXCHANGE SEPARATION

Material	Ion exchange resin, separation conditions, and remarks	Ref.
V compounds (V_2O_5, NH_4VO_3, and Na_3VO_4)	Wofatit® F (H^+ form) Column: 11 × 3 cm a) Sample solution (∼ 130 mℓ) (adsorption of blue V[IV]; P passes into the effluent) b) Water (elution of residual P; until effluent is acid free) c) 5 M HCl (elution of V) P is determined colorimetrically Before separation, the sample (400 mg) is dissolved in 2 M NaOH, diluted with water (120 mℓ), and after addition of Na_2SO_3 (350 mg; to reduce V) the solution is acidified with 2—3 mℓ of H_2SO_4 (1:1) and heated until it acquires a pure blue color	100
Detergents	Dowex® 50 (H^+ form) Column: buret containing ∼ 15 mℓ of the resin and operated at a flow rate of 1—2 drops/sec a) Aqueous sample solution (10-mℓ aliquot) (adsorption of cations; phosphate passes into the effluent) b) Deionized water (90 mℓ) (elution of phosphate) P is determined by flame emission spectrometry Before the cation exchange separation step, the sample is heated for 15 min at 500—520°C and a solution of the ash is prepared containing ≃60 mg of P in 100 mℓ of water	116
	Dowex® 50-X8 Column: 20 × 1.5 cm a) Aqueous solution (25-mℓ aliquot containing 62.5 mg of the sample) (adsorption of cationic constituents; phosphate passes into the effluent) b) Water (225 mℓ) (elution of remaining phosphate) Phosphate is determined by emission spectroscopy	82
Pharmaceuticals	Coupled columns containing Dowex® 50-X4 (50-100 mesh; H^+ form; ∼10 × 1.7 cm, upper column) and Dowex® 1-X2 (50-100 mesh; Cl^- form; 17 × 1.7 cm) a) Neutralized sample solution (∼ 6 mℓ) (adsorption of cationic constituents on upper column; the phosphates are retained by the anion exchange resin) b) Water (∼150 mℓ) (as a rinse for both columns) c) Disconnection of the columns and fractionation of the adsorbed phosphates by using the following eluents (d) and (e) d) 0.05 M HCl (195 mℓ) (elution of orthophosphate) e) 3 M HCl (195 mℓ) (elution of polyphosphates, i.e., pyro- and hexametaphosphate) P is determined colorimetrically by the Mo-blue method	1
Lubricating oils	Dowex® 50W-X8 (H^+ form) Column: 25 × 2 cm The sample (1 g containing 0.009—0.2% of P) is ashed in the presence of solid KOH (0.5 g) and the residue is dissolved in water (10 mℓ); this solution is equilibrated for 10 min (or until it loses all turbidity) with 5 g of the resin to adsorb interfering cations; then the resin slurry is transferred to the column and in the effluent plus successive water washings (250 mℓ total volume), phosphate is determined by emission spectrography; the precision is 5% at the 0.04% level (ten determinations)	108

Table 11 (continued)
DETERMINATION OF P IN INDUSTRIAL PRODUCTS BY OPTICAL METHODS AFTER CATION EXCHANGE SEPARATION

Material	Ion exchange resin, separation conditions, and remarks	Ref.
Fire-resistant textiles	Dowex® 50W-X8 (H⁺ form) a) The alkaline sample solution (25-mℓ aliquot) is equilibrated with the resin (10 g) for 30 min to adsorb cationic constituents b) The slurry (a) is transferred to a column of the same resin and the effluent and water washings containing the phosphate are diluted to 250 mℓ Phosphate is determined by emission spectroscopy Before the separation, the sample (2.5g) is fused with KOH (10 g), the melt is dissolved in water, the solution is filtered, and the filtrate diluted with water to 250 mℓ to prepare solution (a)	109
Cr concentrate and V pentoxide	Cationite KU-2 (H⁺ form) a) Sample solution (130 mℓ) containing ∼6 M HCl (1—1.5 mℓ) and 10% NH$_2$OH-HCl solution (10 mℓ) as a holding reductant for Cr and V (adsorption of Cr[III] and V[IV]; phosphate passes into the effluent) b) 0.2 M HCl (40 mℓ) (elution of remaining phosphate) Phosphate is determined spectrophotometrically Before the separation, the sample (0.2 g of Cr concentrate) is decomposed with HNO$_3$-HClO$_4$; the V$_2$O$_5$ sample (01.1 g) is dissolved in 5% alkali solution	92
Mg and Ca polyphosphates	Dowex® 50W-X12 (H⁺ form) The sample (150 mg) is mixed with the resin (5 g) and water 15 mℓ) and then stirred to absorb Mg and Ca; for polyphosphates obtained at 100, 250, 360, or 450°C (by heating of Mg[H$_2$PO$_4$]$_2$· 2H$_2$O or [H$_2$PO$_4$]$_2$· H$_2$O) stirring is carried out for 10 min, 20—30 min, 1 hr, or 3 hr, respectively; in 0.08 mℓ of the supernate the polyacids are analyzed by paper chromatography	112

Table 12
CATION EXCHANGE SEPARATION OF P FROM SYNTHETIC MIXTURES WITH OTHER ELEMENTS

Elements separated	Ion exchange resin, separation conditions, and remarks	Ref.

Phosphate from cations and anions	A) First column operation (Dowex® 50-X8, H⁺ form)
	Column containing 40 g of the resin
	  a) Aqueous sample solution (containing ≯50 mg of PO_4^{3-}) (adsorption of cations; phosphate and other anions pass into the effluent)
	  b) Water (elution of residual anions)
	B) Second column operation (Dowex® 1, OH⁻ form)
	Column containing 20 g of the resin
	  a) Combined effluents (a) and (b) obtained by (A) (adsorption of phosphate, fluoride, and other anions)
	  b) 0.25 M NaOH (elution of F⁻)
	  c) Water (as a rinse)
	  d) 1 M NaOH (elution of phosphate)
	After removal of Na by passing eluate (d) through a column of Dowex® 50, the P contained in the effluent and water washings is determined spectrophotometrically
		2

PO_4^{3-} from Fe(III)	Wofatit® KPS-200 (grain size: 0.1—0.3 mm; H⁺ form)
	Column: 12 cm × 1 cm² operated at a flow rate of 4—5 mℓ/min
	  a) Sample solution of pH ≃ 5 (adsorption of Fe[III])
	  b) Water (same volume as [a]) (as a rinse)
	  c) 0.5 M HCl (30 mℓ) (elution of phosphate)
		117

P from Mg, Ca, Mn, and Fe	Cation exchange resin Merck I (H⁺ form)
	Column: 10 × 1 cm
	  a) Dilute acid sample solution (10—20 mℓ) (adsorption of cations; phosphate passes into the effluent)
	  b) Water (~20 mℓ) (elution of residual phosphate)
	Phosphate is determined titrimetrically
		118

PO_4^{2-} from SO_4^{2-}	Diaion® SK-1 (100-200 mesh; La³⁺ or Fe³⁺ form)
	Column: 10 × 0.6 cm
	  a) 0.5% NH₄Cl solution (pH 9) containing carrier-free ³²PO_4^{3-} and ³⁵SO_4^{2-} (precipitation of La- or Fe-phosphate in the resin bed; SO_4^{2-} passes into the effluent)
	  b) 2 M HCl (elution of phosphate)
		138

Phosphate	Weakly acidic cation exchange resin Amberlite® IRC 50 (30-50 mesh; loaded with Zr[IV])
	Column of 0.55 cm ID containing 3.8 g of the resin
	  a) Sample solution of pH 2.0 (adsorption of phosphate) (flow rate: 10—20 mℓ/hr)
	  b) 1 M NaOH (40 mℓ) (elution of phosphate at a rate of 12 mℓ/hr)
		139

REFERENCES

1. **Beckstead, H. D., French, W. N., and Truelove, J. F.,** Determination of phosphate in tetracycline preparations, *Can. J. Pharm. Sci.,* 2(1), 9, 1967.
2. **Wynne, E. A., Burdick, R. D., and Fine, L. H.,** Determination of microgram amounts of phosphates. A new general technique, *Mikrochem. J.,* 5(2), 185, 1961.
3. **Ko, R.,** Ion-exchange-spectrographic determination of arsenic and phosphorus in river water, Report HW-59008, U.S. Atomic Energy Commission, 1959.
4. **Gunz, D. and Schnell, E.,** Determination of phosphate in lake and waste waters with ion exchange chromatography combined with atomic absorption spectroscopy, *Mikrochim. Acta,* 3 (5-6), 409, 1984.
5. **Kabeya, H.,** Determination of condensed phosphates at low concentration in water by liquid chromatography, *Nippon Kagaku Kaishi,* No. 1, 65, 1983.
6. **Toyoda, M. and Iwaida, M.,** Determination of ortho- and polyphosphates in processed cheese, ham and sausage, *Nippon Shokuhin Kogyo Gakkaishi,* 30(8), 462, 1983.
7. **Cooke, I. J. and Hislop, J.,** Use of anion-exchange resin for assessment of available soil phosphate, *Soil Sci.,* 96(5), 308, 1963.
8. **Hislop, J. and Cooke, I. J.,** Anion-exchange resin as a means of assessing soil phosphate status: a laboratory technique, *Soil Sci.,* 105(1), 8, 1968.
9. **Taneeva, G. G. and Freze, N. A.,** Decomposition of phosphate rock by means of ion-exchange resins, *Zavod. Lab.,* 36(6), 665, 1970.
10. **Matsuhisa, K., Ohzeki, K., and Kambara, T.,** Application of coagulated ion-exchanger colorimetry to the determination of trace amounts of phosphate in water, *Bull. Chem. Soc. Jpn.,* 55(10), 3335, 1982.
11. **Matsuhisa, K., Ohzeki, K., and Kambara, T.,** Improved ion-exchanger colorimetry for determination of trace amount of phosphate, *Bull. Chem. Soc. Jpn.,* 54(9), 2675, 1981.
12. **Tanaka, T., Hiiro K., and Kawahara, A.,** Colorimetric determination of phosphate ion using anion-exchange resin in molybdate form, *Jpn. Analyst,* 28(1), 43, 1979.
13. **Cartier, P.,** Separation and identification of pyrophosphates in bone tissue, *C. R.,* 243(14), 982, 1956.
14. **Beremzhanov, B. A., Burkitbaev, M. M., Muratbekov, M. B., and Makhmeeva, G. Kh.,** Ion-exchange separation of hypophosphite, phosphite, phosphate, pyrophosphate, and triphosphate, *Zh. Anal. Khim.,* 32(9), 1694, 1977.
15. **Winand, L.,** Separation of ortho- and pyro-phosphate mixtures on an anion exchanger, *J. Chromatogr.,* 7(3), 400, 1962.
16. **Beremzhanov, B. A., Muratbekov, M. B., Burkitbaev, M. M., Ergozhin, E. E., and Prdius, L. N.,** Use of macroreticular anion exchanger of the Ional MA-9 type for separation of phosphite, phosphate, pyrophosphate and triphosphate, *Izv. Akad. Nauk Kaz. SSR Ser. Khim.,* 28(5), 67, 1978.
17. **Mignonsin, E. P. and Albert, P.,** Analysis of purified zirconium by neutron activation, *Bull. Soc. Chim. Fr.,* No. 2, 553, 1965.
18. **Gunz, D. and Schnell, E.,** Use of anion-exchange in combination with atomic absorption spectrophotometry for determination of phosphate and sulfate, *Mikrochim. Acta,* 3 (1-2), 125, 1983.
19. **Waki, H., Yoshimura, K., and Ohashi, S.,** The stabilities of complexes of linear phosphate ions with tetramine copper (II) ions in aqueous solution, *J. Inorg. Nucl. Chem.,* 36, 1337, 1974.
20. **Caletka, R., Vorwalter, C., and Krivan, V.,** Determination of phosphorus at the ng g^{-1} level in niobium by a radiochemical neutron-activation method, *Anal. Chim. Acta,* 141, 393, 1982.
21. **Milham, R. C.,** Determination of phosphorus-32 in aqueous environmental samples, Report DP-MS-73-35, U.S. Atomic Energy Commission, 1973.
22. **Taylor, R. W. and Boulogne, S.,** Rapid determination of phosphorus-32, *Health Phys.,* 13, 657, 1967.
23. **Pollard, F. H., Nickless, G., and Rothwell, M. T.,** Chromatographic studies on the hydrolysis of phosphorus compounds. VII. Hydrolysis of phosphorus trihalides and pseudohalides, *J. Chromatogr.,* 11(3), 383, 1963.
24. **Nakamura, T., Yano, T., Fujita, A., and Ohashi, S.,** Anion-exchange chromatographic separation of linear phosphates with eluent containing a chelating agent, *J. Chromatogr.,* 130, 384, 1977.
25. **Nickless, G. and Peacock, C. J.,** Chromatography of phosphorus anions, *J. Chromatogr.,* 40(1), 192, 1969.
26. **Deswaef, R.,** Study of polyphosphates: separation, identification and determination. I. Sodium mono-, pyro- and tri-phosphates, *Trib. CEBEDEAU,* 21, 312, 1968.
27. **Beukenkamp, J., Rieman, W., III, and Lindenbaum, S.,** Behavior of the condensed phosphates in anion-exchange chromatography, *Anal. Chem.,* 26, 505, 1954.
28. **Tominaga, M., Nakamura, T., and Ohashi, S.,** Improved method for the ion-exchange chromatographic separation of phosphorus-32-labelled oxoanions of phosphorus produced in neutron-irradiated orthophosphates, *J. Inorg. Nucl. Chem.,* 34(4), 1409, 1972.
29. **Rothbart, H. L. and Rieman, W.,** Determination of trimetaphosphate ion in the presence of linear phosphates, *Talanta,* 11(1), 43, 1964.

30. **Rothbart, H. L., Weymouth, H. W., and Rieman, W.**, Separation of the oligophosphates, *Talanta*, 11, 33, 1964.
31. **Kroschwitz, H., Pungor, E., and Ferenczi, S.**, Separation of inorganic phosphates on ready-made thin layers of anion-exchange resin, *Talanta*, 19, 695, 1972.
32. **British Standards Institution**, Methods of test for sodium tripolyphosphate (penta sodium triphosphate) and sodium (tetra-sodium) pyrophosphate for industrial use. XI. Separation by column chromatography and determination of the different phosphate forms, British Standards Institution Publ. BS 4427, 12, 1983.
33. **Benz, C. and Paixao, L. M.**, Rapid determination of polyphosphates by an improved chromatographic method using ion exchange resin, *Chim. Anal.*, 50(5), 247, 1968.
34. **Kolloff, R. H.**, Analysis of commercial sodium tripolyphosphate by reverse-flow ion-exchange chromatography, *Bull. A.S.T.M.*, No. 237, 74, 1959.
35. **Spangler, W. G., Howes, D. E., and Kish, J. A.**, Application of ion-exchange chromatography to the analysis of commercial triphosphate, *Bull. A.S.T.M.*, No. 228, 61, 1958.
36. **Wieker, W.**, Chemistry of condensed phosphates and arsenates. XXVII. Separation on anion-exchange resins and quantitative determination of condensed phosphates having a degree of condensation up to three, *Z. Chem.*, 1(1), 19, 1960.
37. **Peters, T. V. and Rieman, W., III**, Analysis of mixtures of the condensed phosphates by ion-exchange chromatography. II. Mixtures of ortho-, pyro-, tri-, tetra-, trimeta-, and tetrameta-phosphates and Graham's salt, *Anal. Chim. Acta*, 14, 131, 1956.
38. **Lindenbaum, S., Peters, T. V., and Rieman, W., III**, Analysis of mixtures of the condensed phosphates by ion-exchange chromatography, *Anal. Chim. Acta*, 11, 530, 1954.
39. **Shiraishi, N. and Iba, T.**, Separation and determination of various phosphates in commercial sodium triphosphate by ion-exchange chromatography, *Jpn. Analyst*, 13(9), 883, 1964.
40. **Ukrainskaya, L. M. and Kuznetsov-Fetisov, L. I.**, Separation of sodium polyphosphates on AV-17 anionite, *Izv. Vyssh. Uchebn. Zaved. Khim. Khim. Tekhnol.*, 13(11), 1694, 1970.
41. **Peters, T. V.**, Analysis of mixtures of the condensed phosphates by ion-exchange chromatography, *Diss. Abstr.*, 16(10), 1809. 1956.
42. **Uhlíř, Z. and Šolcová, V.**, Determination of condensed phosphates by ion-exchange chromatography, *Prum. Potravin*, 15(6), 305, 1964.
43. **Nakamura, T., Yano, T., Nunokawa, T., and Ohashi, S.**, Anion-exchange chromatographic separation of phosphorus oxo-anions, *J. Chromatogr.*, 161, 421, 1978.
44. **Yamaguchi, H., Nakamura, T., Hirai, Y., and Ohashi, S.**, High-performance liquid-chromatographic separation of linear and cyclic condensed phosphates, *J. Chromatogr.*, 172, 131, 1979.
45. **Wiersma, J. H. and Sandoval, A. A.**, Hydrolysis products of diphosphorus tetrachloride and diphosphorus tetraiodide, *J. Chromatogr.*, 20(2), 374, 1965.
46. **Yoza, N., Ishibashi, K., and Ohashi, S.**, Gel and ion-exchange chromatographic purification of reaction products between pyrophosphate and pyrophosphonate, *J. Chromatogr.*, 134(2), 497, 1977.
47. **Pollard, F. H., Nickless, G., and Rothwell, M. T.**, Separation of some lower oxo-acids of phosphorus by anion-exchange chromatography, *J. Chromatogr.*, 10(2), 212, 1963.
48. **Pollard, F. H., Nickless, G., Rogers, D. E., and Rothwell, M. T.**, Quantitative inorganic chromatography. XII. Automatic analysis of phosphorus-anion mixtures by anion-exchange chromatography, *J. Chromatogr.*, 17(1), 157, 1965.
49. **Kobayashi, E.**, Studies on nitrogen-phosphorus compounds. VII. Ion-exchange chromatography of condensed phosphates, *J. Chem. Soc. Jpn. Pure Chem. Sect.*, 85(5), 317, 1964.
50. **Ohashi, S. and Takada, S.**, Determination of ortho-, pyro- and triphosphate by the use of a small column of ion-exchange resin, *Bull. Chem. Soc. Jpn.*, 34(10), 1516, 1961.
51. **Ohashi, S., Tsuji, N., Ueno, Y., Takeshita, M., and Muto, M.**, Elution-peak positions of linear phosphates in gradient elution chromatography with an anion-exchange resin, *J. Chromatogr.*, 50(2), 349, 1970.
52. **Nakamura, T., Kimura, M., Waki, H., and Ohashi, S.**, The pH dependence of anion-exchange chromatographic separation of tri- and tetraphosphate anions, *Bull. Chem. Soc. Jpn.*, 44(5), 1302, 1971.
53. **Weiser, H. J.**, Determination of pyrophosphate in commercial tripolyphosphate, *J. Am. Oil Chem. Soc.*, 34(3), 124, 1957.
54. **Chess, W. B. and Bernhart, D. N.**, Quantitative determination of traces of pyrophosphate in orthophosphates, *Anal. Chem.*, 31, 1116, 1959.
55. **Pollard, F. H., Rogers, D. E., Rothwell, M. T., and Nickless, G.**, Separation of hypophosphite, phosphite and phosphate by anion-exchange chromatography, *J. Chromatogr.*, 9(2), 227, 1962.
56. **Pollard, F. H., Nickless, G., and Murray, J. D.**, Chromatographic studies of phosphorus compounds. XI. The reaction of sodium hydroxide with phosphoryl chloride, *J. Chromatogr.*, 22(1), 139, 1966.
57. **Lucansky, D.**, Separation of condensed phosphates by ion-exchange chromatography, and their determination in liquid fertilizer based on ammonium phosphate, *Chem. Prum.*, 26(10), 514, 1976.
58. **Fukuda, T., Nakamura, T., and Ohashi, S.**, Anomalous behavior of pyrophosphate anions in ion-exchange chromatography, *J. Chromatogr.*, 128(1), 212, 1976.

59. **Brazell, R. S., Holmberg, R. W., and Moneyhun, J. H.**, Application of high-performance liquid chromatography-flow injection analysis for the determination of polyphosphoric acids in phosphorus smokes, *J. Chromatogr.*, 290, 163, 1984.
60. **Koguchi, K., Waki, H., and Ohashi, S.**, Separation of lower oxo-acids of phosphorus by gradient-elution chromatography with use of an anion-exchange resin, *J. Chromatogr.*, 25(2), 398, 1966.
61. **Heslop, R. B. and Lethbridge, J. W.**, Isolation of peroxymonophosphoric acid by anion-exchange chromatography, *J. Chromatogr.*, 13(1), 199, 1964.
62. **Grundelius, R. and Samuelson, O.**, Sorption of polymetaphosphate on anion-exchange resins, *Anal. Chim. Acta*, 27, 67, 1962.
63. **Muriuki, S. K. and Barber, R. G.**, Study on the merits of separating topical soils into groups using different chemical extractants for different groups in the routine measurement of available soil phosphorus, *Commun. Soil Sci. Plant Anal.*, 14(6), 521, 1983.
64. **du Plessis, S. F. and Burger, R. du T.**, Comparison of chemical extraction methods for the evaluation of phosphate availability of top soils, *S. Afr. J. Agric. Sci.*, 8(4), 1113, 1965.
65. **Blanchar, R. W. and Riego, D. C.**, Tripolyphosphate and pyrophosphate hydrolysis in sediments, *J. Am. Soil Sci. Soc.*, 40(2), 225, 1976.
66. **Blanchar, R. W. and Riego, D. C.**, Pyro- and triphosphate contents of sediments, *Water Air Soil Pollut.*, 7(1), 27, 1977.
67. **De Bussetti, S. G., Ferreiro, E. A., Natale, I. M., and Helmy, A. K.**, Extraction of phosphate from some phosphatic compounds, *Commun. Soil Sci. Plant Anal.*, 11(7), 741, 1980.
68. **Santini, R. and de Jesus, J. M.**, Rapid method for the determination of phosphorus, nitrogen, magnesium, calcium, sodium, and potassium in faeces and diets, *Am. J. Clin. Pathol.*, 31(2), 181, 1959.
69. **Matsuhashi, M.**, Separation of polyphosphates by ion-exchange chromatography. Application to yeast polyphosphates, *Z. Physiol. Chem.*, 333, 28, 1963.
70. **Samsahl**, An anion-exchange method for the separation of P-32 activity in neutron-irradiated biological material, Report AE-149, Aktiebolaget Atomenergi Stockholm, 1964.
71. **Gjerde, D. T., Schmuckler, G., and Fritz, J. S.**, Anion chromatography with low-conductivity eluents. II, *J. Chromatogr.*, 187(1), 35, 1980.
72. **Rigler, F. H.**, Further observations inconsistent with the hypothesis that the molybdenum blue method measures orthophosphate in lake water, *Limnol. Oceanogr.*, 13(1), 7, 1968.
73. **Gohla, W., Nielen, H. D., and Sorbe, G.**, Automatic, rapid determination of phosphates in washing powder production and quality control, *GIT Fachz. Lab.*, 23(2), 89, 1979.
74. **Morachevskii, Yu. V., Zvereva, M. N., and Kuznetsova, A. A.**, Separation of the phosphate ion from certain cations by means of anionites, *Zavod. Lab.*, 22(10), 1170, 1956.
75. **Skřivánek, V. and Klein, P.**, Determination of phosphoric oxide in natural phosphates, *Rudy*, 9(10) (Addendum 6), 51, 1961.
76. **Dippel, W. A., Bricker, C. E., and Furman, N. H.**, Flame-photometric determination of phosphate, *Anal. Chem.*, 26, 553, 1954.
77. **Möller, G.**, Photometric determination of phosphorus in lysimeter waters with vanadate-molybdate, *Z. Pflanzenhernaehr. Dueng. Bodenkd.*, 106(1), 20, 1964.
78. **Schneyder, J.**, Routine determination of phosphoric acid in wine ash, *Mitt. Wein Obstbau Wien A*, 6(6), 309, 1956.
79. **Benk, E.**, Natural phosphate content of orange juice, *Riechst. Aromen Koerperpflegem.*, 18(6), 244, 1968.
80. **Senger, A. I.**, Volumetric determination of total phosphoric oxide in superphosphate, *Nauchn. Zap. Odess. Politekh. Inst.*, 20, 55, 1960.
81. **Griepink, B. and Römer, F. G.**, Determination of phosphate using electrical end-point detection. IV, *Mikrochim. Acta* p. 1192, 1968.
82. **Elliott, W. N. and Mostyn, R. A.**, Determination of phosphate in detergents by cool-flame emission spectroscopy, *Analyst (London)*, 96, 452, 1971.
83. **Yoshino, Y.**, Separation of phosphoric acid and arsenic acid from cations by means of a cation-exchange resin, *Jpn. Analyst*, 3(2), 121, 1954.
84. **Yoshino, Y.**, The separation of phosphorus and iron by means of cation exchange resin, *Bull. Chem. Soc. Jpn.*, 26(7), 401, 1953.
85. **Negina, V. R., Zamyatnina, V. N., and Egorova, A. A.**, Determination of chlorine, arsenic and phosphorus present as impurities in some organic materials by the activation method, *Radiokhimiya*, 5(2), 270, 1963.
86. **Püschel, R. and Lassner, E.**, Application of metal-specific indicators to precipitation titrations. VI. The rapid direct volumetric determination of phosphate ion with standard cerium(III) solution after the separation of interfering ions with the aid of ion-exchange resins, *Fresenius' Z. Anal. Chem.*, 174, 1, 1960.
87. **Lakomkin, I. G.**, Determination of the pyrophosphate ion in copper-phosphate electrolytic baths, *Zavod. Lab.*, 21(5), 540, 1955.

88. **Gohda, S., Yamazaki, H., and Nishikawa, Y.**, Indirect determination of phosphoric acid with zinc-65, *Jpn. Analyst*, 28(2), 102, 1979.
89. **Lewandowski, A. and Witkowski, H.**, Titrimetric determination of phosphorus in apatite by means of ion exchange resins, *Pr. Kom. Mat. Przyr. Poznań. Tow. Przyj. Nauk*, 7(9), 3, 1959.
90. **Fischer, W., Paul, R., and Abendroth, H. J.**, Separation of small amounts of phosphoric acid from iron and vanadium by extraction and ion exchange, *Anal. Chim. Acta*, 13, 38, 1955.
91. **Singhal, K. C., Banerji, A. C., and Banerjee, B. K.**, Indirect method of estimation of phosphorus by atomic-absorption spectrophotometry, *Technology (Sindri India)*, 5(2), 117, 1968.
92. **Verbitskaya, V. A., Stepin, V. V., and Kamaeva, L. V.**, Determination of phosphorus in chromium concentrate and vanadium pentoxide by using chromatographic separation, *Tr. Vses. Nauchno Issled. Inst. Stand. Obraztsov*, 3, 133, 1967.
93. **Fleps, V. and Inczédy, J.**, The determination of the phosphoric oxide content of polyvanadates, *Magy. Kem. Foly.*, 62(9), 322, 1956.
94. **Kadič, K.**, Alkalimetric determination of phosphoric oxide, *Chem. Prum.*, 10(12), 627, 1960.
95. **Skogerboe, R. K., Gravatt, A. S., and Morrison, G. H.**, Flame-spectrophotometric determination of phosphorus, *Anal. Chem.*, 39, 1602, 1967.
96. **Povandra, P. and Roubalová, D.**, Rapid analytical methods for the examination of metals and inorganic raw materials. XI. Determination of phosphoric acid in naturally occurring iron phosphates by ion-exchange separation, *Collect. Czech. Chem. Commun.*, 25(7), 1890, 1960.
97. **Vigh, K., Inczédy, J., and Erdey, L.**, Determination of phosphorus in steel, crude iron and ferrovanadium by using ion-exchange resins, *Magy. Kem. Foly.*, 69(2), 73, 1963.
98. **Borisova, R. and Mitropolitska, E.**, Spectrophotometric determination of phosphate ions with the system cerium(III)-arsenazo III, *Talanta*, 26(7), 543, 1979.
99. **Mitropolitska, E. and Borisova, R.**, Indirect spectrophotometric determination of phosphate using the reaction between cerium(III) and xylenol orange, *Fresenius' Z. Anal. Chem.*, 294, 285, 1979.
100. **Hartmann, S.**, Determination of small amounts of phosphate in the presence of large amounts of vanadate, *Fresenius' Z. Anal. Chem.*, 151, 332, 1956.
101. **Cogbill, E. G., White, J. C., and Susano, C. D.**, Determination of phosphate in perchloric and sulfuric acid solutions of uranium phosphates. Ion-exchange separation and amperometric determination, *Anal. Chem.*, 27, 455, 1955.
102. **Timofeeva, V. I. and Soldatova, G. V.**, Determination of phosphates in salts of complex composition, *Khim. Promst. Ser. Metody Anal. Kontrolya Kach. Prod. Khim. Promsti.*, No. 4, 30, 1979.
103. **Asada, T.**, Determination of phosphate and sulfate in urine or blood by an ion-exchange method, *Jpn. Analyst*, 6(2), 100, 1957.
104. **Bassett, J.**, Volumetric method for determining phosphate and arsenate in a mixture, *Analyst (London)*, 88, 238, 1963.
105. **Anbar, M., Halmann, M., and Silver, B.**, Determination of oxygen-18 in phosphate ion, *Anal. Chem.*, 32, 841, 1960.
106. **Tunnicliffe, M. E.**, Determination of magnesium, total phosphorus and free phosphate in rubber, *I.R.I. Trans.*, 31(5), 141, 1955.
107. **Prodan, E. A., Shashkova, I. L., and Galkova, T. N.**, Effect of cations on the separation of phosphates by thin-layer chromatography, *Zh. Anal. Khim.*, 33(12), 2304, 1978.
108. **Elliott, W. N., Heathcote, C., and Mostyn, R. A.**, Determination of phosphorus in lubricating oils by cool-flame emission spectroscopy, *Talanta*, 19, 359, 1972.
109. **Elliott, W. N., Heathcote, C., and Mostyn, R. A.**, Determination of phosphorus in fire-resistant textiles by cool-flame emission spectroscopy, *Text. Res. J.*, 42(2), 86, 1972.
110. **Syty, A.**, Determination of phosphorus in phosphate rock by cool flame emission spectrometry, *At. Absorpt. Newsl.*, 12(1), 1, 1973.
111. **Singhal, K. C. and Banerjee, B. K.**, Indirect estimation of phosphorus in rock phosphate by atomic-absorption spectrometry with an ion-exchange decomposition technique, *Technology (Sindri India)*, 5(3), 239, 1968.
112. **Pechkovskii, V. V., Shchegrov, L. N., and Shul'man, A. S.**, Analysis of poorly-soluble magnesium and calcium polyphosphates, *Izv. Akad. Nauk Kazakh. SSR Ser. Khim.*, No. 3, 13, 1968.
113. **Sporek, K. F.**, Determination of phosphate in uranium ores, concentrates and liquors via an EDTA titration, *Chemist Analyst*, 47(1), 12, 1958.
114. **Dutta, R. K. and Rao, P. K.**, Rapid analysis for phosphorus in phosphor copper and allied brazing filter metals, *Indian J. Technol.*, 3(1), 24, 1965.
115. **Kutarkina, A. K. and Sirina, A. M.**, Chromatographic determination of total phosphorus in hypophosphites, *Tr. Ural. Nauchno Issled. Khim. Inst.*, No. 11, 93, 1964.
116. **Syty, A.**, Determination of phosphorus in detergents by flame emission spectrometry, *Anal. Lett.*, 4(8), 531, 1971.

117. **Kozhukharov, M. and Gudev, N.**, Ion-exchange separation of ferric and phosphate ions, *Zh. Anal. Khim.*, 18(2), 280, 1963.
118. **Püschel, R.**, Application of metal-specific indicators to precipitation titrations. VII. A new rapid direct method for the determination of microgram quantities of orthophosphate, *Mikrochim. Acta*, No. 3, 352, 1960.
119. **Pickup, J. F.**, Accurate method for determination of inorganic phosphate in serum, *Ann. Clin. Biochem.*, 13(1), 306, 1976.
120. **Flaschka, H. A. and Wolfram, W. E.**, Assay of reagent-grade "metaphosphoric acid", *Chemist Analyst*, 48(3), 65, 1959.
121. **Talipov, Sh. T. and Fedorova, T. I.**, Use of a cationite in the analysis of phosphorites, *Tr. Sredneaziat. Gos. Univ. (Tashkent) Khim. Nauk.*, 55(7), 135, 1954.
122. **Mitropolitska, E. K. and Borisova-Pangarova, R. G.**, Indirect spectrophotometric determination of phosphorus in ferro-silicon, *Dokl. Bolg. Akad. Nauk*, 31(8), 1019, 1978.
123. **Rijkheer, J.**, Isotope dilution procedure for the determination of the true phosphate content of rock phosphates, *J. S. Afr. Chem. Inst.*, 13(1), 1, 1960.
124. **Calokerinos, A. C. and Hadjicoannou, T. P.**, The determination of inorganic phosphorus compounds by using molecular emission cavity analysis, *Anal. Chim. Acta*, 157(1), 171, 1984.
125. **Babachev, G.**, Method for analyzing phosphorite and apatite, *Stroit. Mater. Silik. Promst.*, 11(5 and 6), 22, 1970.
126. **Nikolina, E. S., Frolov, I. V., Baimuratov, V. I., and Aziev, R. G.**, Quantitative analysis of intermediate products of reduction and dissociation of natural calcium phosphates, *Zh. Anal. Khim.*, 32(9), 1739, 1977.
127. **Skřivánek, V. and Klein, P.**, Determination of P_2O_5 in natural phosphates, *Rudy*, 6, 51, 1961.
128. **Banks, J. and Robson, J.**, Rapid methods for boiler-water analysis, *Analyst (London)*, 83, 98, 1958.
129. **Sugimae, A.**, Determination of phosphorus in river water, sewer water and domestic waste water by inductively coupled plasma optical emission spectrometry, *Jpn. Analyst*, 29(8), 502, 1980.
130. **Prager, M. J. and Seitz, W. R.**, Flame-emission photometer for determining phosphorus in air and natural waters, *Anal. Chem.*, 47, 148, 1975.
131. **Lin, C. I. and Huber, C. O.**, Determination of phosphate, silicate and sulphate in natural and waste water by atomic absorption inhibition titration, *Anal. Chem.*, 44, 2200, 1972.
132. **Kramer, G. H. and Joseph, S.**, Simultaneous analysis of chlorine-36, sulfur-35 and phosphorus-32 in urine, *Can. J. Chem.*, 62(11), 2344, 1984.
133. **Riedel, V.**, Determination of phosphate in fertilizers, in the presence of soluble silicic acid with ion exchangers, *Chem. Tech. Berlin*, 18(9), 567, 1966.
134. **Ceausescu, D. and Aşteleanu, M.**, Micro-determination of components of some mixed nitrogen-phosphorus-potassium fertilizers in a sample, *Rev. Roum. Chem.*, 13(3), 325, 1968.
135. **Cartier, P. H. and Thuillier, L.**, Measurement of inorganic pyrophosphate in biological fluids and bone tissues, *Anal. Biochem.*, 61(2), 416, 1974.
136. **Stepin, V. V., Pliss, A. M., and Silaeva, E. V.**, Determination of impurities in metallic vanadium, *Byull. Nauchno Tekh. Inf. Ural. Nauchno Inst. Chern. Met.*, No. 4, 103, 1958.
137. **Hunter, T. L.**, Determination of phosphorus in organic compounds. Titration with lead nitrate, *Anal. Chim. Acta*, 35(3), 398, 1966.
138. **Shikata, E. and Yamaguchi, C.**, Separation of carrier-free sulfuric and phosphoric acid with anion-exchange resin, *Jpn. Analyst*, 8(11), 753, 1959.
139. **Yoshida, I., Nishimura, M., Matsuo, K., and Ueno, K.**, Studies on the selective adsorption of anion by metal-ion loaded ion-exchange resin. V. Adsorption of phosphate ion on ion-exchange resin loaded with zirconium(IV), IRC 50-Zr(IV), *Sep. Sci. Technol.*, 18(1), 73, 1983.
140. **Lakomkin, I. G.**, Determination of phosphoric acid in superphosphates by means of ion exchangers, *Zavod. Lab.*, 24(6), 679, 1958.
141. **Matsuhisa, K. and Ohzeki, K.**, Determination of trace amounts of phosphate in water samples by ion exchange resin thin layer spectrophotometry, *Analyst (London)*, 111, 685, 1986.
142. **Fukushi, K. and Hiiro, K.**, Determination of various phosphate ions in seawater by capillary isotachophoresis after coprecipitation enrichment, *Jpn. Analyst*, 34(1), 21, 1985.

ARSENIC

The four As species commonly reported to be present in environmental materials are arsenite, arsenate, monomethylarsonate, and dimethylarsinate. For the isolation and fractionation of these species ion exchange procedures are frequently utilized, employing both strongly and weakly basic anion and strongly acidic cation exchange resins. Ion exchangers of this type are also used in connection with the determination of arsenic in industrial products. Chelating polymers have found but limited application for separations involving As.

ANION EXCHANGE RESINS

General
As (V)

This As species, which in solution is predominantly present as arsenate, i.e., AsO_4^{3-}, shows an ion exchange behavior very much resembling that of orthophosphate PO_4^{3-} (see chapter on Phosphorus). Thus, arsenate is adsorbed on strongly[1-4] and weakly[5] basic anion exchange resins (e.g., chloride or acetate forms) from solutions of pH 4 to 10 (e.g., from natural waters)[6] and is readily eluted with, for example, 2 M NaCl[5] or with acid eluents, even with very dilute mineral acids, e.g., 0.12 M hydrochloric acid (HCl) (see Table 1).*[3] At higher HCl concentrations As(V) is weakly retained by, for example, Dowex® 1, and in the 2, 4, 6, 8, and 10 M acid distribution coefficients of ~2, ~3, ~3.5, 4, and ~4, respectively, have been measured.[7] This slight adsorption is probably due to extraction of As acid (H_3AsO_4) by the resin and not caused by adsorption of a negatively charged chloro complex. Coadsorbed with arsenate from systems of pH 4 to 10 are not only anions of other elements such as phosphate, sulfate, and nitrate, but also organic forms of As(V), i.e., dimethylarsinate (anion of dimethyl arsinic acid (= cacodylic acid): $(CH_3)_2AsOOH$; DMAA for short) and monomethylarsonate (anion of monomethylarsonic acid: $CH_3AsO(OH)_2$; MMAA for short). The adsorption of these two compounds is lower than that of arsenate. Not or only negligibly retained under these conditions are arsenite, borate, and silicate as well as the cationic constituents of the sample solutions so that rather selective separations of As(V) can be accomplished. To ensure that all As is present in the adsorbable pentavalent state, K permanganate or K bichromate can be added to the sorption solution (see Tables 1 and 6).[1,2,8] Quantitative reduction of As(V) to As(III) can be effected with iodide (see Table 2).[9]

The fact, that the organic forms of As(V) are less strongly adsorbed on anionic resins than arsenate (see above) has variously been employed for the fractionation by sequential elution of As species using the procedures illustrated in Tables 1 and 4.[4,6,10,11] In all cases, the order of elution with the various eluents employed is arsenite (which is not adsorbed on anionic resins from the systems used), DMAA, MMAA, and arsenate. Carbonate eluents other than that used in the procedure described in Table 1[10] have been employed in connection with an HPLC technique which allows the As species to be separated within 36 min.[12] If trimethylarsine oxide ($[CH_3]_3AsO$) is also present it is eluted before the arsenite (see Table 1).

On Dowex® 1-X8 in the molybdate form, arsenate was found to be adsorbable from mineral acid solutions such as 1 N sulfuric acid or 2 M HCl.[13] Under these conditions As is retained as a heteropolyacid anion; if present, phosphate is coadsorbed as molybdophosphate (see chapter on Phosphorus).

As(V) but not As(III) is appreciably adsorbed on strongly basic anion exchange resins

---

* Tables for this chapter appear at the end of the text.

from hydrofluoric acid (HF) solutions.[14-16] In the $< 2\ M$ acid the distribution coefficient of As has a value of $>10^2$ which then decreases to values of ~ 80 and ~ 40 to 60 in the 4 and 8 to 24 M acid, respectively.[16]

As(III)

Arsenite, i.e., AsO_3^{3-} is only negligibly retained on strongly basic resins from very weakly acid to neutral solutions and absolutely no adsorption of this As species is observed on weakly basic resins such as Dowex® 3 (see Tables 2 and 7).[5,9] Consequently, on the latter resins separations of As(III) from As(V) are readily achieved since the pentavalent element is appreciably adsorbed under these conditions. No separation of arsenite from arsenate can be obtained in dilute alkaline solutions from which both species are retained by basic resins (see Table 3).[17-19] Maximum sorption of arsenite is at pH $\simeq 9.2$.[20] In HCl media of the concentrations 0.1 to 2, 2 to 6, 6 to 9, and 9 to 12 M, the predominating As(III) species are H_3AsO_3, $As(OH)_2Cl$, $AsOHCl_2$, and $AsCl_3$, respectively. From these solutions As is moderately adsorbed on strongly basic resins, e.g., Dowex® 1 (probably as an anionic chloro complex) showing distribution values of < 1, ~ 2, 10, ~ 20, and ~ 25 in the 2, 4, 6, 8, and 10 to 12 M acid, respectively.[7] Since As(V) is only slightly retained by the resin from 8 to 12 M HCl (see preceding section) it can be separated from As(III) at these acidities. For example, As(V) is eluted with 9 M HCl and then As(III) is desorbed with the 0.1 M acid.[7] This sequence of elution is reversed when using very dilute HCl solutions as eluents as for example the 0.12 M acid (see Table 1)[3] or a phosphate-ammonium acetate-acetic acid solution of pH 4.6 (see Table 3).[21] From these systems arsenate is more strongly retained than arsenite so that the latter is eluted first.

The relatively weak adsorption of As(III) from, for example, 1 to 7 M HCl can be utilized for its separation from elements which are strongly retained by basic resins from these media. For example, in 4 M HCl As can be separated from Sb(III and V) and Bi (see Table 3),[22] and at higher acid concentrations, e.g., in 6 to 7 M HCl its separation from Sb(V)[23] and many other elements including Fe(III), UO_2(II), Mo(VI), Cu, Sn, Zn, Cd, Hg, and Bi can readily be effected. At lower acid molarities, e.g., in the 1 to 2 M acid, As can be separated from Zn, Cd, Hg, Cu(I), Ag, Au, and many other elements.[24] In all these cases, however, As(III) will be accompanied by arsenate, phosphate, and other anions as well as by numerous cationic constituents including the alkalies, alkaline earth elements, rare earths, Al, Ni, and Mn.

Adsorption of As(III) on basic resins from both hydrobromic and hydroiodic acid media approaches completeness with increasing acidity.[25] It is higher than the sorption from HCl media of comparable acidity.

While As(V) is essentially nonadsorbed on, for example, Dowex® 1 from oxalic acid media,[26] appreciable retention of arsenite is observed from 17.5 M acetic acid ($K_d = >10^2$).[27]

Applications

The anion exchange separation principles discussed above have been employed in connection with the determination of As species in natural waters (see Tables 1 and 4),[1-4,6,11,24,28] biological materials (see Table 1),[10,29,30] and industrial products such as glasses (see Table 2),[9] high-purity Se,[23] bronze,[31] arsenical-pesticide residues,[12] and mixtures of arsenates (or arsenites) and phosphates.[32] To separate As from synthetic mixtures with other elements procedures of the type outlined in Table 3 can be used.[17,21,22]

To determine As in meteorites, rocks, and atmospheric pollutants, the procedures described in Tables 1 in the chapter on Tin, 2 in the chapter on Copper (Volume III), and 3 in the chapter on Mercury (Volume IV) can be used.

Procedures based on the ion chromatographic determination of As are described in the chapter "Special Analytical Techniques Using Ion Exchange Resins", Volume I.

If total As is to be determined in solid materials the samples must be mineralized to convert all As to inorganic forms.[33] This can be accomplished by wet ashing, dry ashing, fusions, or O combustion. Wet ashing of the samples can be achieved by digestion with sulfuric acid, perchloric acid,[4] and nitric acid plus K bromate.[34]

CATION EXCHANGE RESINS

General

Like phosphate, arsenate is not adsorbed on strongly acid cation exchange resins from acid, neutral, and alkaline solutions so that excellent separations of this As species can be achieved from essentially all cationic constituents of sample solutions provided that the adsorption of the latter is performed from very dilute mineral acid media, e.g., 0.1 to 0.3 M hydrochloric or nitric acids or at pH values in the range of 2 to 10 (see Tables 4 to 8).[4,5,8,11,35-47] Very slightly coadsorbed with the metal cations, such as those of the alkalies, alkaline earth elements, Al, Fe, Cu, Ni, and other heavy metals, are other As species, i.e., As(III)), MMAA, and DMAA. Separation of As(III) from the other two As species can be effected by washing the resin bed with a few milliliters of 0.5 M HCl (see Table 6)[44-46] or with 0.1 M orthophosphoric acid (see Table 5),[29] which also elute any remaining arsenate. If water is used as the eluent not only As(III) is eluted but this is followed in this order by MMAA and DMAA (see Table 7)[38,39] indicating that the adsorbability of the As species on cation exchange resins increases in the order: $AsO_4^{3-} < AsO_3^{3-} <$ MMAA $<$ DMAA. From anion exchange resins DMAA is eluted before the MMAA (see preceding section).

Although water has been employed most frequently to elute MMAA from cation exchange resins[8,38,39,44-46] this separation from DMAA can also be achieved by the use of 0.02 M acetic acid,[4] 0.006 M trichloroacetic acid,[11] 1 M ammonium acetate,[35,36] and 0.02 M ammonia solution[37] as eluents (see Tables 4 and 5). Subsequent elution of the adsorbed DMAA can be effected with 1,[4,37] 1.5,[8,11,38] and 4 M[46] and 20%[44,45] ammonia solutions or with 1 M NaOH[4] (see Tables 4 to 6). With the above eluents effective separations of the organic As species from each other and from the inorganic arsenic is obtained. To separate As(III) from As(V) the method of coupled columns containing cation and anion exchange resins can be used (see Table 7)[5] on which accompanying cations, arsenate, and other elements are adsorbed while arsenite passes into the deionized effluent.

In the chapter on Phosphorus it has been mentioned that complete separation of phosphate from Fe(III) on cation exchange resins is only achieved in the presence of a holding reductant. This is also the case when separating As from Fe (see Table 8)[47] Thus, quantitative separation is only obtained in the presence of SO_2 which reduces Fe to the divalent and As to the trivalent states, respectively. Although the nature of this retention of As (or phosphate) by the ferric ion adsorbed on the cation exchanger is not yet fully elucidated, it is very likely that the adsorption of As may occur on the unoccupied coordination sites of the resin-bound ferric ion.

No significant retention of As(III and V) is observed from 1 to 24 M HF.

Applications

Separations of As species by means of cation exchange, which is often combined with the subsequent use of anionic resins, have variously been employed in connection with the determination of As in natural waters (see Tables 4 and 5),[4,11,35-37,40,47-49] biological materials (see Tables 6 and 7),[5,8,38,39,43-45,50] industrial products (see Table 8),[41,42,51] and synthetic mixtures.[52,53]

CHELATING RESINS

Separations of As with chelating polymers from accompanying elements contained in natural waters and Cu can be effected by application of the procedures described in Table 9. Interfering ions can also be removed from natural waters, e.g., riverwater, and from solutions of biological standard materials by the use of Chelex® 100.[54]* The chelating resin RGS-KSH containing mercapto groups can be employed for the removal and recovery of As(III) from aqueous solutions.[55]

RESIN SPOT TESTS

A method for the microchemical detection of As is outlined in Table 80 in the chapter on Copper (Volume III).

---

* See also Reference 60.

Table 1
DETERMINATION OF As IN NATURAL WATERS[a] AND BIOLOGICAL MATERIALS[b] AFTER ANION EXCHANGE SEPARATION

Material	Ion exchange resin, separation conditions, and remarks	Ref.
Riverwater	Amberlite® IRA-400 ISC.P. (strongly basic resin in the chloride form) Column: 10 mℓ buret containing 5 g of the resin and operated at a flow rate of 0.75 mℓ/min a) Water sample (adsorption of As[V] as AsO_4^{3-}; into the effluent pass matrix cations and interfering metals such as Cu, Co, Hg, Ni, and Sb) b) 9 M HCl (80 mℓ) (elution of As) As is determined spectrophotometrically The method allows the quantitative determination of As-concentrations below 0.005 mg/ℓ Before the ion exchange separation, the water sample (e.g., 2 ℓ) is digested with $KMnO_4$ and the excess of this reagent is destroyed by addition of $NH_2OH \cdot HCl$	1, 2
Groundwaters	Dowex® 1-X8 (100-200 mesh; acetate form) Column of 0.7 cm ID containing the resin (2.3 g) to a height of ~10 cm a) Acidified sample (5 mℓ) (adsorption of As[V]; As [III] passes into the effluent) b) 0.12 M HCl (3 × 5 mℓ) (elution of residual As[III] with the first 5 mℓ and desorption of As [V] with the next 10 mℓ) As(III) and As(V) are determined by atomic absorption spectrophotometry The detection limit for each species is 1 µg/ℓ and the recovery is 80—120% Before the separation, the sample (50 mℓ) is filtered through a 0.45 µm membrane filter and then acidified with conc HCl (0.5 mℓ)	3
Sediment interstitial waters and aquatic plants	Zerolit® FF SRA 70 (HCO_3^--form) Column: 25 mℓ buret containing the resin to a height of 25 cm and operated at a flow rate of 4 mℓ/min (at this rate the ion exchange separation takes 75 min) After adsorption of the As species, their fractionation is effected by using the following eluents: a) CO_2-HCO_3^--buffer solution at pH 5.5 (150 mℓ) (elution of As[III] and DMAA) b) CO_2-NH_4Cl-solution at pH 4.0 to 4.2 (150 mℓ) (elution of MMAA and inorganic As[V]) As is determined by atomic absorption spectroscopy Mean recoveries of As species range from 94.8% for salts of DMAA to 98.8% for salts of MMAA	6
Vegetables	Bio-Rad Aminex® A-27 anion exchange resin Radial compression column After injection of the sample the solvent program begins with water and becomes 100% 0.2 M ammonium carbonate after 15 min on a convex program; the various As species are eluted in the following order: trimethylarsine oxide, arsenite, dimethylarsinate, methylarsonate, and arsenate, and then determined by graphite-furnace atomic absorption spectroscopy	10

[a] See also Table 1 in the chapter on Phosphorus and Table 10 in the chapter on Zinc, Volume IV.
[b] See also Table 5 in the chapter on Mercury, Volume IV; Table 6 in the chapter on Cadmium, Volume IV; and Tables 10 and 11 in the chapter on Copper, Volume III.

Table 2
DETERMINATION OF As IN INDUSTRIAL PRODUCTS AFTER ANION EXCHANGE SEPARATION[a]

Material	Ion exchange resin, separation conditions, and remarks	Ref.
Glasses	Dowex® 3 (weakly basic resin; Cl⁻-form) Column of 15-cm length and operated at a flow rate of ~6 mℓ/min a) Sample solution (adsorption of the anionic iodide or chloride complex of Sb[III] together with iodine [probably as triiodide]; As[III] passes into the effluent together with the matrix elements) b) HCl-Borax wash solution (50 mℓ) (elution of remaining As and accompanying elements) As is determined by coulometric titration Solution (a) is prepared by dissolution of the sample (~0.1 g, containing 0.05—1.0% As_2O_5) in a mixture consisting of KI (5 mg), 6 M HCl (2 mℓ), 5.8 M HI (1 mℓ), and 14 M HF (2 mℓ), and finally 6 M HCl (3 mℓ) followed by water (5 mℓ) and borax (1 g, to complex F⁻) are added (I⁻ reduces all As[V] to As[III]) For the preparation of eluent (b), borax (30 g) is mixed with conc HCl (140 mℓ) and the mixture is diluted with water to 1500 mℓ	9
W and Ta	Dowex® 1 (100-200 mesh) Column: ~25 × 0.6 cm conditioned with 15 mℓ of 40% HF-methanol (1 + 2) (procedure A) or 10 M HF (15 mℓ) (procedure B) A) W and Ta a) HF-HNO_3-sample solution (5 mℓ) (adsorption of W and Ta as anionic fluoride complexes; Se passes into the effluent) b) 40% HF-methanol (1 + 2) (15 mℓ) (elution of Se) B) W a) 10 M HF-sample solution (5 mℓ) (adsorption of W as anionic fluoride complex; Se passes into the effluent) b) 10 M HF (15 mℓ) (elution of Se) As is determined via radiometric measurement of ^{75}Se; the Se-recovery is ~97—99% in both procedures; the detection limit for As is ≃10 ng/g Before separation (A) or (B) the sample is irradiated with protons (^{75}As [p,n]^{75}Se) or deuterons (^{75}As[d,2n]^{75}Se) and then dissolved in HF + HNO_3 (in the presence of Se-carrier)	56

[a] See also Table 3 in the chapter on Tin; Table 7 in the chapter on Vanadium, Volume IV; Tables 9 and 10 in the chapter on Gold, Volume III; Tables 16, 17, 19, 26, and 49 in the chapter on Copper, Volume III; Table 27 in the chapter on Zinc, Volume IV; Table 96 in the chapter on Rare Earth Elements, Volume I, and Table 3 in the chapter on Selenium, Tellurium, and Polonium.

Table 3
ANION EXCHANGE SEPARATION OF As FROM SYNTHETIC MIXTURES WITH OTHER ELEMENTS

Elements separated	Ion exchange resin, separation conditions, and remarks	Ref.
As(III), Sb(III), and Bi(III)	Amberlyst® A-26 macroreticular strong base anion exchange resin (150-200 mesh) Column: 9.15 × 0.63 cm in a liquid chromatograph a) Injection of 4 M HCl-sample solution (50 µℓ) b) 4 M HCl (elution of As at a forced flow rate of 1 mℓ/min) c) 1 M HCl - 4.5 M HClO$_4$ (elution of Bi before Sb)	22
AsO_3^{3-}, AsO_4^{3-}, SeO_3^{2-}, and SeO_4^{2-}	Column (25 cm length) of Nucleosil® Dimethylamine operated at a forced flow rate of 1.4 mℓ/min (HPLC system equipped with ICP a.e.s. detection) a) 0.002 M NH$_4$H$_2$PO$_4$ - 0.005 M ammonium acetate-acetic acid (pH 4.6) (elution in the order of AsO_3^{3-}, SeO_3^{2-}, and AsO_4^{3-}) b) 0.08 M NH$_4$H$_2$PO$_4$-NH$_3$-solution (pH 6.9) (elution of SeO_4^{2-})	21
As from Sb	Dowex® 21K (100-200 mesh; Cl$^-$-form) a) 1 M NaOH (adsorption of arsenite; Sb passes into the effluent) b) 2 N H$_2$SO$_4$ (elution of As)	17

Table 4
DETERMINATION OF As IN NATURAL WATERS AFTER CATION AND ANION EXCHANGE SEPARATIONS

Material	Ion exchange resin, separation conditions, and remarks	Ref.
Natural waters (e.g., pond water)	A) Cation exchange separation (Dowex® 50W-X8, 50-100 mesh; H$^+$-form; column: 16 × 1 cm) a) Sample solution (200 mℓ aliquot of pH 4—10) acidified with 1.74 M acetic acid (2 mℓ) (adsorption of dimethylarsinate and monomethylarsonate at a flow rate of 5 mℓ/min) b) 0.02 M acetic acid (70 mℓ) (elution of As[III], As[V], and monomethylarsonate at a similar flow rate) c) 1 M NaOH (elution of dimethylarsinate which is contained in the fraction from 31—42 mℓ; flow rate: 1 mℓ/min) B) Anion exchange separation (Bio-Rad® AG1-X8, 50-100 mesh; acetate form) Column: 11 × 0.8 cm conditioned with a solution of pH 4.7 which is 0.1 M in total acetate concentration and operated at a flow rate of 1—2 mℓ/min a) Sample solution (50 mℓ aliquot of pH 4—10) mixed with Na acetate/acetic acid buffer to obtain a total acetate concentration of 0.01 M and a pH of 4.7 (adsorption of monomethylarsonate and As[V]; into the effluent pass As[III] and dimethylarsinate) b) Eluent of similar composition as solution (a) (50 mℓ) (elution of residual As[III] and dimethylarsinate) c) Eluent of similar composition as solution (a) but of a total acetate concentration of 0.1 M (monomethylarsonate and As[V] are collected between 20—40 mℓ and after 75 mℓ, respectively) In the various fractions, As is determined by differential pulse polarography (following digestion with HClO$_4$); recoveries of	4

Table 4 (continued)
DETERMINATION OF As IN NATURAL WATERS AFTER CATION AND ANION EXCHANGE SEPARATIONS

Material	Ion exchange resin, separation conditions, and remarks	Ref.
	monomethylarsonate and dimethylarsinate average 98 and 100%, and the detection limits are 18 and 8 ppb, respectively Before the ion exchange separations, the sample solution (pH 4—10) is divided into 4 aliquots; 2 of these are used to determine As(III) and total inorganic As (As[V] is calculated by difference) A similar separation scheme as described above has been utilized in connection with the determination of the As-species in other environmental and biological samples using graphite-furnace atomic absorption spectrophotometry[48] In this method the time delays caused by the $HClO_4$-digestions of the eluates are avoided and also better detection limits (ng/mℓ) for DMAA, MMAA, As(V), and total As of 0.02, 2.0, 0.4, and 4.0, respectively, are obtained (As[III] is determined by difference); in place of eluent (c) (A) 1 M ammonia solution is used to elute DMAA (collected in the fraction from 40—50 mℓ); monomethylarsonate and As(V) adsorbed on the anion exchange resin column (10 × 1 cm) are eluted with 0.5 M acetate buffer of pH 4.7 (the fraction of effluent between 10 and 30 mℓ contains the methylarsonate while As[V] begins to elute until 65 mℓ of this eluent have passed and is collected in the next 30 mℓ [flow rates of 1.8—2.0 mℓ/min are required])	
Natural waters (e.g., lake water)	Composite column of 1 cm ID packed with a 9-cm layer of Bio-Rad® AG1-X8 (100-200 mesh; chloride form) surmounted by a 26-cm layer of Bio-Rad® AG50W-X8 (100-200 mesh; H$^+$-form) a) Sample (≯2 mℓ) at pH 2.5—7 containing 80—4000 ng of total As (adsorption of As species together with alkalies, alkaline earth elements, and other constituents of the matrix) b) 0.006 M trichloroacetic acid (pH 2.5) (55 mℓ) (elution of As[III] followed by monomethylarsonate) (flow rate: 2 mℓ/min) c) 0.2 M trichloroacetic acid (8 mℓ) (elution of As[V] at the same flow rate) d) 1.5 M Ammonia solution (55 mℓ) followed by 0.2 M trichloroacetic acid (50 mℓ) (elution of dimethylarsinate at a flow rate of 6 mℓ/min; with the first eluent [ammonia solution] this As species is desorbed from the cation exchange resin and then readsorbed on the anion exchanger; from the latter it is eluted with the second eluent) In the various fractions As is determined by atomic absorption spectroscopy; recoveries range from 97—104% for lake water; the detection limit is 10 ppb	11

Table 5
DETERMINATION OF As IN NATURAL WATERS AFTER CATION EXCHANGE SEPARATION

Material	Ion exchange resin, separation conditions, and remarks	Ref.
River-, lake-, and pore-water and sediment extracts	Bio-Rad® AG50W-X8 (100-200 mesh; H$^+$-form) Column: 25—30 × 1 cm a) Sample solution of pH > 2 ($\ngtr$2 mℓ, containing 1—5 µg of As) (adsorption of MMAA, DMAA, and matrix elements such as Na, K, Mg, and Ca) b) 0.2 M trichloroacetic acid (30 mℓ) (elution of inorganic As [III and V]) c) 1 M ammonium acetate (70 mℓ) (elution of MMAA and DMAA) In the eluates As is determined by atomic absorption spectrometry	35, 36
Seawater	Dowex® AG50W-X8 (100-200 mesh; H$^+$-form) Column containing 5 g of the resin a) Seawater sample (e.g., 3.5 mℓ) (adsorption of DMAA + MMAA) b) 0.1 M H$_3$PO$_4$ (elution of inorganic As) c) 0.02 M NH$_3$-solution (elution of MMAA) d) 1 M NH$_3$-solution (elution of DMAA) As is determined by atomic absorption spectroscopy	37
Natural waters and effluents (containing > 0.1 mg/ℓ of heavy metals)	Zerolit® 225 (52-100 mesh; Na$^+$-form) Column of 0.8 cm^2 (cross-sectional area) containing ∼5 mℓ of the resin a) Sample solution (25 mℓ) adjusted to pH 3.0—3.5 with 2 M NH$_3$-solution (adsorption of Cu, Ni, Ag, Cd, Bi, and other heavy metals; As passes into the effluent) b) Distilled water (20 mℓ) (elution of remaining As) As is determined spectrophotometrically Before the separation, the sample (50 mℓ) is evaporated in the presence of conc HNO$_3$ (2 mℓ) and conc H$_2$SO$_4$ (1 mℓ)	40

Table 6
DETERMINATION OF As IN BIOLOGICAL MATERIALS AFTER CATION EXCHANGE SEPARATION

Material	Ion exchange resin, separation conditions, and remarks	Ref.
Marine organisms (algae and clam tissue) and estuarine sediments	Dowex® AG50-X8 (100-200 mesh; H$^+$-form) Column: 18 × 1.25 cm conditioned with 0.5 M HCl (50 mℓ) a) Dilute HCl-sample solution (aliquot of <1.5 mℓ) (adsorption of organic As species and coextracted cations) b) 0.5 M HCl (10 mℓ) (elution of inorganic As[V]) c) Water (16 mℓ) (elution of methylarsonic acid) d) 1.5 M ammonia solution (25 mℓ) (elution of DMAA) In the various fractions As is determined by atomic absorption spectroscopy; recoveries for As range from 92—97% Before the ion exchange separation step, extracts of biological samples with 0.1 M NaOH are filtered and evaporated in vacuo, and each residue is dissolved in 8.5 M HCl; extracts of dried sediments with 6 M HCl are centrifuged, and the supernatant solutions are retained; the residual sediments are twice extracted under reflux with 0.1 M NaOH - 1 M NaCl, the combined extracts are evaporated to dryness in vacuo, and each residue is dissolved in 8.5 M HCl; an aliquot of any of these acid solutions is reduced by treatment with 1 M KI saturated with ascorbic acid, and the As species are extracted into toluene and then back-extracted into water (2 × 4 mℓ); conc HCl (1 mℓ) is added to this aqueous extract and after the addition of 5% K$_2$Cr$_2$O$_7$-solution (50 $\mu\ell$; to oxidize As to As[V]) the solution is diluted to 10 mℓ thus obtaining solution (a)	8
Dog plasma and urine	Bio-Rad® AG50W-X8 (100-200 mesh; H$^+$-form) Column: 10-mℓ disposable glass pipette packed with 7.5 g of the resin conditioned with 0.5 M HCl After passage of 1 mℓ of the sample the following eluents are used a) 0.5 M HCl (10 mℓ) (elution of inorganic As[III and V]) b) Water (15 mℓ) (elution of methylarsonate) c) 5% NH$_3$-solution (15 mℓ) (as a rinse) d) 20% NH$_3$-solution (15 mℓ) (elution of DMAA) ^{74}As is determined radiometrically Before the separation of the above As metabolites, ^{74}As is injected intravenously	44
Urine	Bio-Rad® AG50W-X8 After adsorption of the various As-species on a column of the resin, their fractionation is achieved by using the following eluents: a) 0.5 M HCl (elution of inorganic As) b) Water (elution of methylarsonic acid) c) 5% NH$_3$-solution (elution of As in other forms from dietary sources) d) 20% NH$_3$-solution (elution of DMAA) As is determined by atomic absorption spectroscopy; limits of detection were 0.5 μg/ℓ for each chemical form of As; recovery was 95—102% for samples with As concentrations from 40—400 μg/ℓ with coefficients of variation from 3—6% Before the separation, the sample is acidified with HCl and mineralized by dry oxidation with MgO-Mg(NO$_3$)$_2$	45
Diverse biological materials (e.g., hair)	Dowex® 50-X8 (100-200 mesh; H$^+$-form) Column: 1 × 0.5 cm operated at a flow rate of 15 mℓ/cm^2 min	43

Table 6 (continued)
DETERMINATION OF As IN BIOLOGICAL MATERIALS AFTER CATION EXCHANGE SEPARATION

Material	Ion exchange resin, separation conditions, and remarks	Ref.
	a) 0.0005 M H_2SO_4-sample solution (1 mℓ) (adsorption of Na, Cu, and other cations; arsenate passes into the effluent) b) 0.0005 M H_2SO_4 (5 mℓ) (elution of residual As) ^{76}As is determined radiometrically Before the separation, the sample is irradiated with neutrons and then wet ashed with H_2SO_4-HNO_3-$HClO_4$ in the presence of As-carrier (10 μg)	

Table 7
DETERMINATION OF As IN BIOLOGICAL MATERIALS AFTER SEPARATION BY TWO-COLUMN CATION AND ANION EXCHANGE

Material	Ion exchange resin, separation conditions, and remarks	Ref.
Marine organisms (e.g., seaweed)	Dowex® 50-X8 (50-100 mesh; H$^+$-form) The sample solution at pH 2 is passed through a short column (12 × 1.5 cm) connected to a long column (11.8 × 1.9 cm) using a flow rate of 9—11 mℓ/cm^2 hr (DMAA is adsorbed on the short column); subsequently, the columns are disconnected and water (300 mℓ) is employed to elute As(V), As(III), and methylarsonate (in that order) from the longer column (the As-species being contained in the eluate fractions of 100—150, 150—220, and ~240—300 mℓ, respectively) From the short column DMAA is eluted with 1.5 M ammonia solution (100 mℓ) In the various eluates As is determined by atomic absorption spectroscopy; the recovery of the As-species is ~100%	38, 39
Plants	The dried sample (500 mg) is extracted with 0.5 M HClO$_4$ (2 × 10 mℓ) and the centrifuged extracts are adjusted to pH 5.0 with solid K_2CO_3; a portion of the solution containing ≃1.5 μg of As is applied to a column of Bio-Rad® AG50W-X4 (100-200 mesh; H$^+$-form) surmounting a column of the weakly basic anion exchange resin Bio-Rad® AG3-X4A (100-200 mesh; chloride form) and water (100 mℓ) is passed through both columns to elute As(III); the As(V) is then eluted from the disconnected anion exchange column using 2 M NaCl (100 mℓ) In the two eluates As is determined by atomic absorption spectrometry; recoveries were 93—104% at levels of ~5, 10, and 20 ppm of As	5

Table 8
DETERMINATION OF As IN INDUSTRIAL PRODUCTS AFTER CATION EXCHANGE SEPARATION[a]

Material	Ion exchange resin, separation conditions, and remarks	Ref.
As-Fe alloys	Dowex® 50W-X8 (50-100 mesh; H$^+$-form) Column of 1 cm ID containing a 10-mℓ resin bed and operated at a flow rate of 5 mℓ/min a) 0.3 M HCl-sample solution (50 mℓ) saturated with SO$_2$-gas (as a reductant for As[V] and Fe[III]) (adsorption of Fe[II], Co, Cu, and Ni; As[III] passes into the effluent) b) 0.3 M HCl (50 mℓ) (elution of remaining As) As is determined titrimetrically Before the separation, the sample (containing ≃125 mg of As) is dissolved in 1:1 HNO$_3$	41
High-As alloys	Cationite KU-2 Column containing 40 g of the resin and operated at a flow rate of 10 mℓ/min a) Sample solution (60 mℓ) containing 2 mℓ of conc HNO$_3$ (adsorption of cations; As passes into the effluent) b) Water (100 mℓ containing 1 mℓ of conc HNO$_3$) (elution of residual As) As is determined by complexometric titration Before the separation, the As-Fe alloy (0.1 g, containing up to 30% of As) is dissolved in HNO$_3$ (1:1)	51

[a] See also Table 2 in the chapter on Tungsten, Volume IV; Tables 55 and 56 in the chapter on Copper, Volume III; Table 143 in the chapter on Actinides, Volume II; and Table 10 in the chapter on Gallium.

Table 9
DETERMINATION OF As IN NATURAL WATERS AND Cu AFTER SEPARATION ON CHELATING RESINS[a]

Material	Ion exchange resin, separation conditions, and remarks	Ref.
Riverwater	Spheron Thiol resin (hydrophilic methacrylate gel containing thiol groups; grain size: 40—63 μm, average pore diameter 37 mm) (Lachema, Brno, Czechoslovakia)	57
	The sample (50 mℓ) made 1 M in H_2SO_4 is shaken for 2 hr with 50 mg of the resin to adsorb As; after centrifugation, the resin is injected into a graphite tube furnace for the determination of As by atomic absorption spectroscopy; the coefficient of variation for 10 ng of As is 4%; good recoveries of As are reported from 5% KCl and $MgCl_2$ solutions and from distilled water and riverwater	
	Completely coadsorbed with the As are Sb and Bi; the detection limits for these 3 elements are 0.32, 1.3, and 0.28 ng/mℓ, respectively	
Seawater	Chelating resin Uniselec® UR-10 containing (2-hydroxybenzyl) nitrilodiacetic acid as functional groups (100-200 mesh; Fe^{3+}-form)	58
	Column: 11.5 × 1.05 cm containing a 10-mℓ resin bed and operated at a flow rate of 40 mℓ/hr	
	a) Water sample (10—20 ℓ) (adsorption of As[III and V]; the matrix constituents pass into the effluent)	
	b) 2 M HCl (80 mℓ) (elution of As and Fe)	
	As is determined spectrophotometrically	
Electrolyte Cu	Chelex® 100 (100-200 mesh; NH_4^+-form)	59
	Column: 15 × 1.5 cm containing 30 mℓ of the resin	
	a) Sample solution (∼7 mℓ) containing 3 mℓ of 15 M NH_3-solution (adsorption of Cu as cationic tetramine complex; As and Sb pass into the effluent)	
	b) 0.1 M NH_3-solution (∼43 mℓ) (elution of residual As and Sb)	
	As and Sb are determined by anodic-stripping voltammetry	
	The limits of detection are 0.056 and 0.028 ppm for As and Sb, respectively	
	Before the separation, the sample (0.5 g) is dissolved in 15 M HNO_3 (2 mℓ)	

[a] See also Table 18 in the chapter on Gold, Volume III and Tables 74 and 77 in the chapter on Copper, Volume III.

REFERENCES

1. **Sandhu, S. S. and Nelson, P.**, Concentration and separation of arsenic from polluted waters by ion exchange, *Environ. Sci. Technol.*, 13(4), 476, 1979.
2. **Sandhu, S. S.**, Arsenic determination by the silver diethyldithiocarbamate method and elimination of metal ion interference, Report EPA/600 4-78/038, U.S. Environmental Protection Agency, Off. Res. Dev., 1978.
3. **Ficklin, W. H.**, Separation of arsenic(III) and arsenic(V) in ground waters by ion exchange, *Talanta*, 30, 371, 1983.
4. **Henry, F. T. and Thorpe, T. M.**, Determination of arsenic(III), arsenic(V), monomethylarsonate, and dimethylarsinate by differential pulse-polarography after separation by ion exchange chromatography, *Anal. Chem.*, 52, 80, 1980.
5. **Austenfeld, F. A. and Berghoff, R.**, Improved method for selective determination of trace quantities of arsenite and arsenate in plant material, *Plant Soil*, 64(2), 267, 1982.
6. **Aggett, J. and Kadwani, R.**, Anion-exchange method for speciation of arsenic and its application to some environmental analyses, *Analyst (London).*, 108, 1495, 1983.
7. **Nelson, F. and Kraus, K. A.**, Anion exchange studies. XVIII. Germanium and arsenic in HCl solutions, *J. Am. Chem. Soc.*, 77, 4508, 1955.
8. **Maher, W. A.**, Determination of inorganic and methylated arsenic species in marine organisms and sediments, *Anal. Chim. Acta*, 126, 157, 1981.
9. **Wise, W. M. and Williams, J. P.**, Coulometric titration of total arsenic and arsenic(III) in glass, *Anal. Chem.*, 36, 19, 1964.
10. **Pyles, R. A. and Woolson, E. A.**, Quantitation and characterisation of arsenic compounds in vegetables grown in arsenic-acid-treated soil, *J. Agric. Food Chem.*, 30(5), 866, 1982.
11. **Grabinski, A. A.**, Determination of arsenic(III), arsenic(V), methanearsonate and dimethylarsinate by ion-exchange chromatography with flameless atomic absorption spectrometric detection, *Anal. Chem.*, 53, 966, 1981.
12. **Woolson, E. A. and Aharonson, N.**, Separation and detection of arsenical-pesticide residues and some of their metabolites by high pressure liquid chromatography-graphite-furnace atomic absorption spectrometry, *J. Assoc. Off. Anal. Chem.*, 63(3), 523, 1980.
13. **Malvano, R., Grosso, P., and Zanardi, M.**, Single-step radiochemical separations by column procedures in activation analysis, *Anal. Chim. Acta*, 41, 251, 1968.
14. **Kalinin, A. I., Kuznetsov, R. A., and Osipchuk, I. L.**, Ion-exchange behavior of arsenic in hydrofluoric acid media, *Vestn. Leningr. Gos. Univ. Fiz. Khim.*, 3(16), 155, 1968.
15. **Faris, J. P.**, Adsorption of the elements from hydrofluoric acid by anion exchange, *Anal. Chem.*, 32, 520, 1960.
16. **Faix, W. G., Caletka, R., and Krivan, V.**, Element distribution coefficients for hydrofluoric acid/nitric acid solutions and the anion exchange resin Dowex 1-X8, *Anal. Chem.*, 53, 1719, 1981.
17. **Chakravarty, T. N.**, Ion-exchange studies of arsenic separation from antimony, *Sci. Cult.*, 28(8), 392, 1962.
18. **De, A. K. and Chakrabarty, T.**, Ion-exchange studies of arsenic, antimony and bismuth and their separations, *Indian J. Chem.*, 7(2), 180, 1969.
19. **Klement, R. and Kühn, A.**, Separation of arsenic, antimony and tin by anion-exchange, *Fresenius' Z. Anal. Chem.*, 152, 146, 1956.
20. **Everest, D. A. and Popiel, W. J.**, Ion-exchange studies of solutions of arsenites, *J. Chem. Soc.*, No. 474, 2433, 1957.
21. **McCarthy, J. P., Caruso, J. A., and Fricke, F. L.**, Speciation of arsenic and selenium via anion exchange h.p.l.c. with sequential plasma emission detection, *J. Chromatogr. Sci.*, 21(9), 389, 1983.
22. **Seymour, M. D. and Fritz, J. S.**, Determination of metals in mixed hydrochloric and perchloric acids by forced-flow anion exchange chromatography, *Anal. Chem.*, 45, 1394, 1973.
23. **Koch, H. and Koch, B.**, Determination of the purity of selenium by activation analysis, *Kerntechnik*, 5(6), 248, 1963.
24. **Buldini, P. L., Ferri, D., and Zini, Q.**, Differential pulse polarographic determination of inorganic and organic arsenic in natural waters, *Mikrochim. Acta*, 1 (1-2), 71, 1980.
25. **March, S. F., Alarid, J. E., Hammond, C. F., McLeod, M. J., Roensch, F. R., and Rein, J. E.**, Anion-exchange of 58 elements in hydrobromic acid and hydroiodic acid, Report LA-7084, *Los Alamos Scientific Laboratory, Los Alamos, N.M.*, February 1978.
26. **de Corte, F., van den Winkel, P., Speecke, A., and Hoste, J.**, Distribution coefficients for twelve elements in oxalic acid medium on a strong anion-exchange resin, *Anal. Chim. Acta*, 42, 67, 1968.
27. **van den Winkel, P., de Corte, F., Speecke, A., and Hoste, J.**, Adsorption of some elements in acetic acid medium on Dowex 1-X8, *Anal. Chim. Acta*, 42, 340, 1968.

28. **Fish, R. H., Brinkman, F. E., and Jewett, K. L.**, Finger-printing of inorganic and organoarsenic compounds in *in situ* oil shale retort and process waters using a liquid chromatograph coupled with an atomic absorption spectrometer as detector, *Environ. Sci. Technol.*, 16(3), 174, 1982.
29. **Holak, W.**, Determination of arsenic and selenium in foods by electroanalytical techniques, *J. Assoc. Off. Anal. Chem.*, 59(3), 650, 1976.
30. **Kurosawa, S., Yasuda, K., Taguchi, M., Yamazaki, S., Toda, S., Morita, M., Uehiro, T., and Fuwa, K.**, Identification of arsenobetaine, a water soluble organo-arsenic compound in muscle and liver of a shark, prionace glaucus, *Agric. Biol. Chem.*, 44(8), 1993, 1980.
31. **Zvereva, M. N. and Dimitrieva, E. A.**, Separation of arsenic, antimony and tin by means of anionites, *Uch. Zap. Leningrad. Gos. Univ.*, No. 297, 41, 1960.
32. **Bruno, M. and Belluco, U.**, Separation and determination of micro quantities of arsenic and phosphorus, *Ric. Sci.*, 26(11), 3337, 1956.
33. **Lemmo, N. V. and Faust, S. D.**, Assessment of the chemical and biological significance of arsenical compounds in a heavily contaminated watershed. I. The fate and speciation of arsenical compounds in aquatic environments — a literature review, *J. Environ. Sci. Health*, A18(3), 335, 1983.
34. **Kling, R. and Lindeman, J.**, Rapid method for the determination of the total arsenic in commercial calcium arsenate by means of an ion exchanger, *Chem. Anal. (Warsaw)*, 2(4), 331, 1957,
35. **Iverson, D. G., Anderson, M. A., Holm, T. R., and Stanforth, R. S.**, Evaluation of column chromatography and flameless atomic-absorption spectrophotometry for arsenic speciation as applied to aquatic systems, *Environ. Sci. Technol.*, 13(12), 1491, 1979.
36. **Faust, S. D., Winka, A., Belton, T., and Tucker, R.**, Assessment of the chemical and biological significance of arsenical compounds in heavily contaminated watershed. II. Analysis and distribution of several arsenical species, *J. Environ. Sci. Health Part A*, 18(3), 389, 1983.
37. **Persson, J. A. and Irgum, K.**, Determination of dimethylarsinic acid in seawater in the sub-ppb range by electrothermal atomic-absorption spectrometry after pre-concentration on an ion exchange column, *Anal. Chim. Acta*, 138, 111, 1982.
38. **Tagawa, S. and Kojima, Y.**, Fractionation of arsenic(III), arsenic(V), methanearsonate and dimethylarsinic acid by using strongly acidic cation-exchange resin, *Japan Analyst*, 29(3), 216, 1980.
39. **Tagawa, S.**, Confirmation of arsenate, arsenite, methylarsonate and dimethylarsinate in an aqueous extract from a brown seaweed, *Bull. Jpn. Soc. Sci. Fish*, 46(10), 1257, 1980.
40. **Haywood, M. G. and Riley, J. P.**, Spectrophotometric determination of arsenic in seawater, potable water and effluents, *Anal. Chim. Acta*, 85, 219, 1976.
41. **Duval, G. R., Ironside, R., and Russell, D. S.**, Analysis of arsenic-iron alloys, *Anal. Chim. Acta*, 25, 51, 1961.
42. **Akimov, V. K., Efremova, L. V., and Rudzit, G. P.**, Reactions between arsenic and some pyrazolinethione derivatives, *Zh. Anal. Khim.*, 33(5), 934, 1978.
43. **Krishnan, S. S. and Erickson, N. E.**, Estimation of arsenic in biological materials by neutron-activation analysis, *J. Forensic Sci.*, 11(1), 89, 1966.
44. **Tam, K. H., Charbonneau, S. M., Bryce, F., and Lacroix, G.**, Separation of arsenic metabolites in dog plasma and urine following intravenous injection of arsenic-74, *Anal. Biochem.*, 86(2), 505, 1978.
45. **Buratti, M., Calzaferri, G., Caravelli, G., Colombi, A., Maroni, M., and Foa, V.**, Significance of arsenic metabolic forms in urine. I. Chemical speciation, *Int. J. Environ. Anal. Chem.*, 17(1), 25, 1984.
46. **Norin, H. and Vahter, M.**, A rapid method for the selective analysis of total urinary metabolites of inorganic arsenic, *Scand. J. Work Environ. Health*, 7, 38, 1981.
47. **Department of the Environment & National Water Council**, G. B., Arsenic in potable and seawaters by spectrophotometry (arsenomolybdenum blue procedure) 1978, Tentative method, Her Majesty's Stationary Office, London 1980, 21.
48. **Parcey, G. E. and Ford, J. A.**, Arsenic speciation by ion-exchange separation and graphite-furnace atomic absorption spectrophotometry, *Talanta*, 28, 935, 1981.
49. **Haswell, S. J., O'Neill, P., and Bancroft, K. C. C.**, Arsenic speciation in soil-pore waters from mineralized and unmineralized areas of South-West England, *Talanta*, 32, 69, 1985.
50. **Tam, G. K. H., Charbonneau, S. M., Lacroix, G., and Bryce, F.**, Confirmation of inorganic arsenic and dimethylarsinic acid in urine and plasma of dogs by ion-exchange and thin-layer chromatography, *Bull. Environ. Contam. Toxicol.*, 21(3), 371, 1979.
51. **Gurevich, A. B. and Kalina-Zhikhareva, V. I.**, Determination of arsenic in high-arsenic alloys by means of cationites and complexometric titration, *Tr. Nauchno Tekh. Ova. Chern. Metall. Ukr. Resp. Pravl.*, 4, 127, 1956.
52. **Overby, L. R., Bocchieri, S. F., and Fredrickson, R. L.**, Chromatographic, electrophoretic and ion-exchange identification of radioactive organic and inorganic arsenicals, *J. Assoc. Off. Agric. Chem.*, 48,(1), 17, 1965.
53. **Sherma, J.**, Separation of certain cations from mixtures of various cations on ion-exchange papers-arsenic, barium, cadmium, tin, and zinc, *Anal. Chem.*, 36, 690, 1964.

54. **Naraski, H. and Ikeda, M.,** Automated determination of arsenic and selenium by atomic-absorption spectrometry with hydride generation, *Anal. Chem.*, 56, 2059, 1984.
55. **Maeda, H., and Egawa, H.,** Removal and recovery of arsenic(III) ion in aqueous solution by chelating resin, *J. Chem. Soc. Jpn.*, No. 7, 1177, 1981.
56. **Sastri, C. S. and Krivan, V.,** Determination of arsenic in refractory metals by radiochemical charged-particle activation analysis, *Anal. Chim. Acta.*, 141, 399, 1982.
57. **Slovak, Z. and Docekal, B.,** Sorption of arsenic, antimony and bismuth on glycolmethacrylate gels with bound thiol groups for direct sampling in electrothermal atomic-absorption spectrometry, *Anal. Chim. Acta.*, 117, 293, 1980.
58. **Yoshida, I., Ueno, K., and Kobayashi, H.,** Selective separation of arsenic(III) and arsenic(V) ions with the iron(III) complex of chelating ion exchange resin, *Sep. Sci. Technol.*, 13(2), 173, 1978.
59. **Hamilton, T. W., Ellis, J., and Florence, T. M.,** Determination of arsenic and antimony in electrolytic copper by anodic-stripping voltammetry at a gold-film electrode, *Anal. Chim. Acta,* 119, 225, 1980.
60. **Narasaki, H.,** Determination of traces of arsenic and selenium by hydride generation atomic absorption spectrometry, *Fresenius' Z. Anal. Chem.*, 321, 464, 1985.

ANTIMONY

To isolate Sb from solutions of complex composition, such as those obtained from natural materials and also industrial products, both anion and cation exchange procedures can be employed. For this purpose mainly strongly basic and acidic resins have been used. Chelating polymers have found but very limited application for separations involving Sb.

ANION EXCHANGE RESINS

The Sb(V) species predominating in $9 \simeq 12\ M$ hydrochloric acid (HCl) is $SbCl_6^-$ and as such Sb is retained by strongly basic anion exchange resins with distribution coefficients of ~ 1, 500, $\sim 10^4$, $\sim 10^5$, and $>10^5$ in 2, 4, 6, 8, and 10 to 12 M acid. Coadsorbed with Sb(V) from 6 M HCl media, which have been used in the procedures presented in Table 1*,[1,2] are Fe(III), UO_2(II), Mo(VI), Cu, Zn, Cd, Hg(II), Ga, In, Tl(III), Sn(II and IV), Bi, and several other elements so that the selectivity of separations is not high. When adsorbing Sb(V) from the 12 M acid (see example in Table 1),[3] the selectivity is even lower, because from the concentrated acid Ti, Zr, Hf, Fe(II), Mn(II), Co, and Ge are retained in addition to those elements which are coadsorbed from the 6 M acid. To ensure that all the Sb is present in the pentavalent state the sorption solutions contain holding oxidants such as Na bromate[1] and Br^-[3] or the Sb is oxidized with H_2O_2[2] in concentrated HCl before preparation of the solution.

Sb(V) is also strongly adsorbed on anion exchange resins from hydrofluoric acid (HF) media of all concentrations.[4]

Sb(III) shows higher adsorbability than Sb(V) on strongly basic resins at low HCl concentrations, i.e., below 4 M. In this acidity range distribution coefficients for Sb(III) of $\sim 10^3$ are observed in the 2 to 3 M acid,[1] but when increasing the concentration of HCl the adsorption decreases. Thus, in 4, 6, 8, 10, and 12 M HCl distribution coefficients for Sb(III) of ~ 500, ~ 100, ~ 80, ~ 50, and ~ 20, respectively, have been measured. In the 9 to 12 M acid the predominating Sb(III) species is $SbCl_4^-$ and it is this anionic chloro complex which is retained by anion exchangers. Adsorption of Sb(III) from 3 and 7 M HCl has been employed in connection with the determination of Sb in synthetic mixtures with other elements[5] and in meteorites,[6] respectively. From these media, as well as from 8 M HCl - 1 M HF (1:1),[7] the same elements are coadsorbed as with Sb(V) (see preceding paragraph).

From dilute HF media Sb(III) is adsorbed on basic resins to about the same extent as Al, i.e., only in very dilute acid solutions (<2 M) adsorption values in the range of 10 to 100 are reached.[4] At higher acidities the adsorption is negligible.

In the presence of organic complexing agents such as malonic and tartaric acids Sb(III) forms anionic complexes which are adsorbable on anion exchangers. For example, adsorption from 5% malonate solution of pH 4.8 can be used to isolate Sb from synthetic mixtures[8] and from U fission products.[9]

Elution of Sb that has been adsorbed from the systems mentioned above can be effected with 0.5 to 1 M NaOH,[3] 1[5] and 4 N[10] sulfuric acid, 0.5 M HCl - 10% ammonium fluoride solution,[6] and 0.3 M HCl - 1 M HF.[7] No elution of Sb is required whenever this element can be determined directly on the resin, e.g., by radiometric measurements (see Table 1).[1,2]

In addition to the applications mentioned above and presented in Table 1, anion exchange separations of Sb have also been utilized in connection with the determination of this element in glass[11] and in In and Zn metal.[10]

---

* Tables for this chapter appear at the end of the text.

CATION EXCHANGE RESINS

On resins of the sulfonic acid type Sb(V) is strongly adsorbed in the range of 9 to 12 M HCl where the distribution coefficient has a value of 2×10^3. Sb(III) is much less strongly adsorbed at these acid concentrations ($K_d <100$). From dilute perchloric acid solutions, e.g., 1 M, Sb(III) is retained as SbO$^+$ on Amberlite® IR-120 to an extent which makes it possible to perform the separation procedure described in Table 2.[12] At HCl concentrations below 2 M, Sb(III) is precipitated. In the 2 M acid Sb(III) has a distribution coefficient of ~3, so that effective elution of this element can be achieved at this acid molarity (see Table 2).

Virtually no adsorption of Sb(III and V) is observed on cationic resins from HF media. Sb(III) is also not retained on strongly acid cation exchange resins from tartaric acid solutions,[13-17] and this fact has variously been utilized for separations of the type outlined in Table 2. From this table it is seen that the elution of Sb as anionic tartrate complex allows this element to be separated from Cu, Ni, Cd, Co, Zn, Mn, Fe(III), Cr(III), Pb, Ce, Ag, Al, Tl, alkalies, and alkaline earth elements. No separation is obtained from Sn which, like Sb, forms an anionic tartrate complex (see also Table 7 in the chapter on Tin).

Procedures that can be used for the determination of Sb in rocks, minerals, and industrial products are described in the chapter on Copper, Volume III (see Tables 40, 51, 55, and 56).

CHELATING RESINS

On a polymer containing mercaptoacetoxymethyl chelating groups attached to the benzene rings of XAD-4 resin, Sb(III) can be separated from Sn(IV) and As(III) with 2 M HCl as eluent.[18] After removal of these two elements Sb is eluted with 6 M HCl. This method was applied to the determination of Sb in solutions prepared from NBS standard alloys containing Sn, Pb, Cu, and smaller amounts of other metals.

Other separations of Sb on chelating resins are outlined in Table 9 in the chapter on Arsenic; Table 18 in the chapter on Gold, Volume III; and Table 74 in the chapter on Copper, Volume III (see also Reference 19).

Table 1
DETERMINATION OF Sb IN ORGANIC[a] AND INORGANIC[b] MATERIALS AFTER ANION EXCHANGE SEPARATION

Material	Ion exchange resin, separation conditions, and remarks	Ref.
Orchard leaves and tuna samples	Dowex® 2-X8 (20-50 mesh; Cl⁻ form) Column containing 10 mℓ of the wet resin and operated at a flow rate of 3 mℓ/min a) ~6 M HCl - 0.05 M HF sample solution (~110 mℓ, containing NaBrO₃ [0.2 g, to oxidize Sb to Sb {V}] and H₃BO₃ [0.5 g, to complex F⁻] [adsorption of Sb {V}]) b) 6 M HCl (20 mℓ) (as a rinse) c) Sb is determined radiometrically directly on the resin Before the ion exchange separation, the sample is irradiated with neutrons and decomposed with HNO₃ in the presence of carriers for Sb and As; subsequently, As is separated from Sb by adsorption on a column of SnO₂ from 1 M HCl - 0.1 M HF	1
Samples used to estimate the distance from which a bullet was fired	Dowex® 2-X8 (100-200 mesh; chloride form) Column containing 1.5 g of the resin conditioned with 6 M HCl and operated at a flow rate of ~1 mℓ/min a) 6 M HCl sample solution (adsorption of Sb) b) 6 M HCl (50 mℓ) (as a rinse) c) The resin is taken from the column and ¹²²Sb is determined directly on the resin by radiometric measurement Before the separation, the sample (e.g., from clothing) is irradiated with neutrons and then treated with conc HCl + H₂O₂ (in the presence of 4 mg of Sb-carrier)	2
Metallic Pb	Dowex® 1-X8 (Cl⁻ form) Column: 5 × 2 cm operated at a flow rate of 1 mℓ/cm² min a) 12 M HCl sample solution containing a few drops of Br₂ (adsorption of Sb [V]; Pb passes into the effluent) b) 12 M HCl (50 mℓ) followed by water (10 mℓ) and 1 M HNO₃ (50 mℓ) (elution of remaining Pb and other impurities) c) 0.5—1 M NaOH (elution of Sb) Sb is determined colorimetrically with rhodamine B Before the separation, the sample (1 g) is dissolved in HNO₃ (1 + 4)	3

[a] See also Table 10 in the chapter on Copper, Volume III; Tables 12 and 13 in the chapter on Zinc, Volume IV; and Table 5 in the chapter on Mercury, Volume IV.

[b] See also Tables 1, 2, and 4 in the chapter on Tin; Table 2 in the chapter on Tungsten, Volume IV; Table 3 in the chapter on Mercury, Volume IV; Tables 6, 16—19, 22, and 27 in the chapter on Copper, Volume III; Tables 8—10 in the chapter on Gold, Volume III: Table 11 in the chapter on Actinides, Volume II: Table 11 in the chapter on Lead; Table 23 in the chapter on Zinc, Volume IV; Table 88 in the chapter on Rare Earth Elements, Volume I; Table 3 in the chapter on Selenium, Tellurium, and Polonium; Table 10 in the chapter on Gallium; and Table 3 in the chapter on Bismuth.

Table 2
DETERMINATION OF Sb IN BIOLOGICAL MATERIAL AND SYNTHETIC MIXTURES WITH OTHER ELEMENTS AFTER CATION EXCHANGE SEPARATION[a]

Material	Ion exchange resin, separation conditions, and remarks	Ref.
Human hair	Amberlite IR-120 (100-140 mesh; H$^+$ form) Column: 8 × 0.2 cm conditioned with 1 M HClO$_4$ and operated at a forced flow rate of 0.4—0.5 mℓ/min a) 1 M HClO$_4$ sample solution (adsorption of Sb[III] and other heavy metals such as Hg[II], Bi, and Fe[III]) b) 2 M HCl (this eluent is used 1 min after [a] has been injected) (elution of Sb[III]) c) 4 M HCl (this eluent is used 3.5 min after sample injection) (elution of coadsorbed heavy metals) Sb is determined coulometrically Before the separation, the sample (0.3—1 g) is dissolved by wet ashing with HClO$_4$-HNO$_3$ and Sb is reduced to Sb (III) with NaHSO$_3$	12
Sb(III) in mixtures with Cu, Ni, Cd, Co, Zn, Mn, Fe(III), and Cr(III)	Zeo-Karb® 225 (NH$_4^+$ form) Column containing 10—15 g of the resin and operated at a flow rate of 20—30 drops/min a) Sample solution (containing 0.05—0.1 g of Sb[III] and any of the other elements to which Na-K-tartrate (0.25—0.5 g) or tartaric acid (0.5—1 g) has been added (adsorption of the other elements; Sb passes into the effluent as anionic tartrate complex) b) Water (elution of residual Sb) c) 2—6 M HCl (elution of adsorbed metals) The method is accurate to ±1%	13
Sb(III) in binary mixtures with Cu(II), Fe(III), Bi, and Pb	Dowex® 50-X8 (50-100 mesh; H$^+$ form) Column of 1 cm^2 containing 8 mℓ of the resin a) 0.5 M HClO$_4$ - 0.4 M tartaric acid (50 mℓ) (adsorption of Pb, Cu, and Fe; Sb passes into the effluent as anionic tartrate complex) b) 0.5 M HClO$_4$ (100 mℓ) (elution of residual Sb) c) 6 M HCl (25—40 mℓ) (elution of Pb, Cu, Bi, and Fe) The method can be used for the separation of milligram amounts of the elements	14
Sb(III) or Sn(IV) in mixtures with Rb, Cs, Mg, Ba, Ce, Ag, Al, Tl, and Pb	Amberlite® IR-120 (NH$_4^+$ form) After adsorption of the elements on a column of the resin the following eluents are used a) 0.33 M tartaric acid (elution of Sb or Sn) b) 2 M HCl or NaNO$_3$ (elution of the other metals)	15

[a] See also Table 7 in the chapter on Bismuth.

REFERENCES

1. **Rengan, K., Haushalter, J. P., and Jones, J. D.**, Simultaneous determination of arsenic and antimony in environmental samples by radiochemical neutron-activation analysis, *J. Radioanal. Chem.*, 54(1-2), 347, 1979.
2. **Baumgärtner, F., Stärk, H., and Schöntag, A.**, Determination of antimony in millimicrogram amounts by activation analysis, in order to estimate the distance from which a bullet is fired, *Fresenius' Z. Anal. Chem.*, 197, 424, 1963.
3. **Yoshino, Y. and Kojima, M.**, Analytical studies on microgram quantities of antimony. II. Ion-exchange separation of antimony in high-purity metallic lead, *Jpn. Analyst*, 6(3), 160, 1957.
4. **Faris, J. P.**, Adsorption of the elements from hydrofluoric acid by anion exchange, *Anal. Chem.*, 32, 520, 1960.
5. **Kawabuchi, K.**, Separation of antimony(III), tin(IV) and molybdenum(VI) by anion exchange, *J. Chem. Soc. Jpn. Pure Chem. Sect.*, 87(3), 262, 1966.
6. **Kiesl, W.**, Determination of trace elements in meteorites by activation analysis. I. Selenium, arsenic, antimony, tin, and mercury, *Fresenius' Z. Anal. Chem.*, 227, 13, 1967.
7. **Ivanova, M. N., Ogloblina, I. P., Genel', S. A., Mitina, V. V., Kalinin, A. I., and Lambrev. V. G.**, Group radio-chromatographic separation of trace impurities in neutron-activation analysis of special-purity silica *Zh. Anal. Khim.*, 32(6), 1066, 1977.
8. **Akki, S. B. and Khopkar, S. M.**, Anion exchange behaviour of antimony(III) in malonate solution, *Fresenius' Z. Anal. Chem.*, 255, 130, 1971.
9. **Allam, S., Zakaria, N., and Abdel-Rassoul, A. A.**, Anion-exchange separations of antimony and tin isotopes isolated from fission products, Report 127, Atomic Energy Research Establishment, Cairo, 1971.
10. **Katsman, B. Kh.**, Separation of small amounts of antimony from indium or zinc by ion-exchange chromatography, *Zh. Anal. Khim.*, 23(8), 1234, 1968.
11. **Wise, W. M. and Williams, J. P.**, Coulometric titration of total antimony and antimony(III) in glass, *Anal. Chem.*, 36, 1863, 1964.
12. **Taylor, L. R. and Johnson, D. C.**, Determination of antimony using forced-flow liquid chromatography with a coulometric detector, *Anal. Chem.*, 46, 262, 1974.
13. **Khorasini, S. S. M. A. and Khundkar, M. H.**, Separation of antimony(III) by ion-exchange, *Anal. Chim. Acta*, 25, 292, 1961.
14. **Šulcek, Z., Boseova, M., and Dolezal, J.**, Separation of antimony and bismuth by ion-exchange, *Collect. Czech. Chem. Commun.*, 34(3), 787, 1969.
15. **Sherma, J. and Bryden, A. K.**, Separation of antimony or tin by ion exchange, *Chemist Analyst*, 52(3), 69, 1963.
16. **Khorasani, S. S. M. A. and Khundkar, M. H.**, Separation of antimony(V) from iron(III), copper (II), cobalt(II), and cadmium(II) by ion exchange, *Anal. Chim. Acta*, 21, 406, 1959.
17. **Sherma, J. and Cline, C. W.**, Separation of certain cations from mixtures of various cations on ion-exchange papers. II. Antimony and arsenic, *Talanta*, 10, 787, 1963.
18. **Phillips, R. J. and Fritz, J. S.**, Chromatography of metal ions with a "thioglycolate" chelating resin, *Anal. Chem.*, 50, 1504, 1978.
19. **Khmylev, I. V., Semushin, A. M., Belousov, E. A., and Yakovlev, V. A.**, Sorption of antimony(III) and antimony(V) on ion exchangers with different functional groups from strong sulfuric acid media, *Zh. Prikl. Khim. (Leningrad)*, 58(6), 1241, 1985.

BISMUTH

Adsorption of anionic complexes of Bi with chloride, bromide, and nitrate on anion exchange resins is the basis of most methods so far reported for the isolation of Bi from natural and industrial materials. These techniques are of much higher selectivity than are procedures in which cation exchange resins or chelating polymers are used for the separations.

ANION EXCHANGE RESINS

General

In 0.5 to 2 and 2 to 12 M hydrochloric acid (HCl) solutions the predominating Bi(III) species are $BiCl_5^{2-}$ and $BiCl_6^{3-}$ respectively, and it is the pentachloro complex which is retained by strong base anion exchange resins. Distribution coefficients for Bi of $>10^4$, $\sim 10^4$, $>10^3$, $<10^3$, ~ 200, ~ 100, ~ 60, and ~ 30 have been measured in 0.5, 1, 2, 4, 6, 8, 10, and 12 M HCl, respectively. Similar adsorption behavior is shown by Bi in dilute HCl media containing organic solvents such as aliphatic alcohols[1] and acetone.[2] In the case of acetone the distribution coefficients for Bi are very high, even at 90% acetone. This is probably due to the formation of a complex, i.e., $Cl_3Bi-(O=C[CH_3]_2)$ which is extracted in the water-rich phase inside the resin where anionic chloro complexes can be formed and fixed. This is different from the behavior of other metals such as Zn, Mo, Ag, Au, etc. which show decreasing adsorption functions with increasing acetone concentrations. Usually this increase has the same effect as an increase of HCl molarity in pure aqueous systems. Bi is also strongly retained from HCl solutions by weakly basic resins. This adsorption has been utilized in connection with the separation of Bi from As(III), Te(IV), and Cd on Amberlite® CG 4B using 4.5, 1, and 0.1 M HCl, respectively, for the elution of these elements before Bi is desorbed with 3 M sulfuric acid.[3]

The strong adsorption of Bi from HCl media has been employed for many separations involving this element (see examples in Tables 1 to 3 and 5*). For this purpose Bi was adsorbed from 0.1,[4] 0.25,[5-7] 0.5,[8-10] 1,[11,12] 2,[13,14] 2.5,[15] 4,[16] 6,[17] and 8 M[18] HCl solutions. The most selective separations of Bi from accompanying elements are achieved when this adsorption is performed from the very dilute acid, e.g., 0.1 to 2 M HCl. Coadsorbed under these conditions are only a very limited number of metals which includes Zn, Cd, Hg(II), Cu(I), In, Tl(III), Pb, Sn, Sb(III), Te(IV), Au(III), Ag, Re, and Pt metals. At higher acid concentrations Fe(III), UO_2(II), Ga, and other elements are also retained by anionic resins, but can be separated from the Bi by selective elution with dilute HCl media, as, for example, with the 0.01,[13] 0.02,[16] and 0.5 M[17] acid. Coadsorbed organic matter can be removed by washing the resin with an acetone-HCl mixture (see Table 2).[4]

Comparison of the adsorption behavior of Bi on strongly basic resins in the hydrohalic acids shows that the sorption order is HI > HB > HCl. Consequently, very strong adsorption of Bi is also observed in hydrobromic (HBr) and hydroiodic acid media, especially at low acid concentrations both in the absence or presence of organic solvents.[19] If this adsorption of Bi from dilute HBr solutions is performed from media also containing dilute nitric acid, e.g., from 0.03 M HBr - 2 M HNO_3 (see Table 5)[20] and 0.05 M HBr - 0.5 M HNO_3 (see Table 4),[21] the selectivity of separations is higher than when using HCl systems. This is because Pb, Zn, Cd, and other elements which are retained from the HCl media (see above) are not adsorbed from the mixed HBr-nitric acid solutions.

From very dilute sulfuric acid medium containing K iodide, the Bi is adsorbed as a yellow anionic complex, i.e., BiI_4^-. This fact has been utilized in connection with the determination of Bi in water using ion exchanger colorimetry (see Table 1).[22]

---

* Tables for this chapter appear at the end of the text.

While Bi is only negligibly adsorbed on strongly basic resins from very dilute sulfuric acid solutions (0.1 to 0.5 N) and not at all adsorbed from 1 to 24 M hydrofluoric acid media, it is moderately retained from nitric acid solutions. In the latter distribution coefficients for Bi of ~10, ~15, ~20, ~15, ~10, ~5, and ~1 have been measured in the 1, 2, 4, 6, 8, 10, and 12 M acid, respectively. This adsorption of Bi is considerably increased in the presence of high concentrations of organic solvents such as methanol or methyl glycol. Consequently, the retention of Bi on basic resins from, e.g., 90% methanol - 10% 5 M nitric acid (K_{dBi} = >10³)[23] or from 90% methyl glycol - 10% 5 M nitric acid (K_{dBi} = 290)[24] can be used for selective separations of the type illustrated in Table 4.[24] Coadsorbed with Bi from these as well as from pure aqueous nitric acid solutions are only a few elements which include Th, Pb, and rare earth metals. After adsorption from the organic solvent systems mentioned above the coadsorbed elements can be removed by elution with 90% methanol - 10% 6 M HCl.[23]

Adsorption of Bi on anionic resins has also been observed in media containing organic complexing agents such as in 2% maleic acid solution,[25] 0.5% succinic acid solution,[25] and malonic acid solution of pH 4.5.[26]

Bi that has been adsorbed on basic resins from the systems mentioned in the preceding paragraphs is usually eluted with 1 M sulfuric acid[5-7,16,17,25,27] or 1 M nitric acid.[10,13,21,23,24] Other eluents which were used for this elution include: 2 M sulfuric acid,[11] 0.5 M[18] and concentrated nitric acid,[9] a mixture of sulfuric and nitric acids,[14] 5% thiourea solution,[12] dilute sulfuric acid containing thiourea,[15] 0.5 M NaOH followed by water and 1 M perchloric acid,[8] 0.05 M ethylenediaminetetraacetic acid (EDTA) - 0.1 M nitric acid,[17] 0.01 M diethylenetriaminepentaacetic acid (DTPA) - 0.1 M nitric acid (see Table 4),[21] and 0.05 M DTPA - 0.1 M ammonium nitrate (pH 4.5) (see Table 5).[20]

Applications

Adsorption of Bi on anion exchange resins has been utilized rather frequently for separations used in connection with the determination of this element in geological materials (see Table 1), as, for example, in meteorites,[17] ores and concentrates,[13] natural waters (including seawater),[10,22] and marine sediments.[23] Similar separation techniques have been employed to isolate Bi from biological matrices (see Table 2)[4-6,8,27] and from solutions of industrial products (see Tables 3 and 4) which include steel[14,15,23,28] and various alloys.[18,25] To separate Bi from synthetic mixtures with other elements the procedures outlined in Table 5 can be used.

CATION EXCHANGE RESINS

General

Bi is essentially not retained by strongly acidic cation exchange resins from dilute or concentrated solutions of HCl or HBr (both in the absence or presence of organic solvents), because in these media stable anionic complexes are formed (see preceding section) which prevent the adsorption on the cation exchangers. Thus, for example, in 0.5 M HCl or at higher acidities, the distribution coefficient of Bi has a value of ≃1. In comparison, distribution coefficients for Bi of ~10³, 235, >10², and ~80 have been measured in 0.5 M solutions of perchloric-, sulfuric-, hydrofluoric-, and nitric acids, respectively. Adsorption of Bi from these dilute acids has been employed for the separations illustrated in Tables 6 to 8.[29-32] Coadsorbed with the Bi under these conditions are virtually all cationic constituents of the sorption solutions which include Al, Fe, alkaline earth elements, and alkalies, so that selective separations of Bi are not obtained.

Elution of the adsorbed Bi can be effected, for example, with the following eluents: 0.5[33,34] 1,[32] and 3 M[30] HCl with 3[31] or 4 M[32] nitric acid, or with 4 M sulfuric acid.[32]

Procedures described in Tables 6 to 8[29,35-37] show that from EDTA solutions Bi is not adsorbed on cation exchange resins, so that it can be separated form a number of metals which do not form anionic EDTA complexes under the conditions selected.

Applications

From Tables 6 and 7 it is seen that cation exchange separations of Bi have been employed in connection with the determination of this metal in industrial products of the type listed. Other applications involve the analysis of Pb-based bearing metal,[38] Al-, Pb-, Cu-, and Sn-based materials,[39] and Pt-Rh alloys.[34] To separate Bi from synthetic mixtures with other elements numerous procedures can be used.[32,33,36-38,40,41] Examples are shown in Table 8.

CHELATING RESINS AND RESINS USED FOR MICROCHEMICAL DETECTION

Separation procedures based on the retention of Bi on chelating polymers are illustrated in Table 9,[42,43] while in Table 10 two methods are presented which can be used for the microchemical detection of trace amounts of Bi employing resin spot tests.[44,45]

Table 1
DETERMINATION OF Bi IN GEOLOGICAL MATERIALS AFTER ANION EXCHANGE SEPARATION[a]

Material	Ion exchange resin, separation conditions, and remarks	Ref.
Meteorites (chondrites)	Dowex® AG1-X8 (100-200 mesh; chloride form) Column containing 36 mℓ of the resin a) 6 M HCl sample solution (20 mℓ) (adsorption of Bi) b) 6 M HCl (20 mℓ) (elution of residual accompanying elements) c) 0.5 M HCl (75 mℓ) (elution of coadsorbed elements, e.g., Fe[III]) d) Water (50 mℓ) (removal of HCl) e) 1 M H$_2$SO$_4$ (125 mℓ) (elution of Bi) After additional radiochemical purification steps based on precipitations, Bi is determined radiometrically Before the ion exchange separation, the sample (0.7—1.4 g) is irradiated with neutrons and then dissolved in the presence of Bi-carrier (100 mg); subsequently, interfering radionuclides and other elements are removed by extraction (diethyl ether), distillation (HBr-H$_2$SO$_4$), and precipitation of Bi as the sulfide which finally is dissolved in HCl	17
Ores and concentrates	Anionite EDE-10P (Cl$^-$ form) Column containing the resin to a height of 7—10 cm and operated at a flow rate of 5—6 mℓ/min a) $\simeq$2 M HCl sample solution ($\sim$15 mℓ) (adsorption of Bi; Fe passes into the effluent) b) $\simeq$2 M HCl (few milliliters) (as a rinse) c) HCl (1:1200) (until negative reaction for Fe with NH$_4$SCN) d) Hot (90°C) HNO$_3$ (1:20) ($\simeq$200 mℓ) (elution of Bi) Bi is determined polarographically Before the separation, the sample (0.2—2 g) is dissolved in HCl	13
Fresh water and industrial effluents	Dowex® 1-X2 (200-400 mesh; sulfate form) To the water sample (1 ℓ, containing 0.022—1.3 μmol of Bi), conc H$_2$SO$_4$ (25 mℓ), KI (2.5 g), 0.1 M Na$_2$S$_2$O$_3$ (5 mℓ), and the resin (0.5 g) are added and the mixture is stirred for 30 min to adsorb the yellow anionic Bi-iodide complex After filtration, the resin is subjected to ion exchanger colorimetry The detection limit is 1.3 ppb of Bi; large amounts of Cu(II), Ag, and Pb cause interference, as does chloride at the level present in seawater	22

[a] See also Table 1 in the chapter on Cadmium, Volume IV; Table 3 in the chapter on Zinc, Volume IV; Table 5 in the chapter on Lead; Table 36 in the chapter on Antinides, Volume II; and Table 86 in the chapter on Rare Earth Elements, Volume I.

Table 2
DETERMINATION OF Bi IN BIOLOGICAL MATERIALS AFTER ANION EXCHANGE SEPARATION

Material	Ion exchange resin, separation conditions, and remarks	Ref.
Marine organisms (e.g., seaweed, fish muscle, and skeleton)	Bio-Rad® AG1-X8 (100-200 mesh; chloride form) Column: 6 × 0.5 cm conditioned with 0.6 M HCl (10 mℓ) and operated at a flow rate of 0.7 mℓ/min a) ∼0.5 M HCl sample solution (∼20 mℓ) (adsorption of Bi; matrix elements pass into the effluent) b) 0.6 M HCl (10 mℓ) (elution of remaining matrix constituents) c) Water (5 mℓ) followed by 0.5 M NaOH (15 mℓ), water (5 mℓ), and 1 M HClO$_4$ (10 mℓ) (elution of Bi) Bi is determined by anodic stripping voltammetry Before the separation, the sample is dry ashed at 450°C and 1 g of the ash is dissolved in HCl and HNO$_3$	8
Blood	Amberlite® CG-400 (100-200 mesh; Cl$^-$ form) Column containing 2 mℓ of the resin a) ≃0.25 M HCl sample solution (100 mℓ, in which 10 mℓ of the blood sample containing heparin or citrate have been dissolved) (adsorption of Bi) b) 1 M H$_2$SO$_4$ (20 mℓ) plus water (10 mℓ) (elution of Bi) Bi is determined spectrophotometrically with dithizone; down to 50 μg of Bi per liter of blood can be determined and at a level of 500 μg/ℓ the precision is 10%	5
Blood and urine	Amberlite® IRA-400 (Cl$^-$ form) Column: 3 × 1 cm a) 0.25 M HCl sample solution (100 mℓ) (adsorption of Bi) b) 0.25 M HCl followed by water and acetone and again water (as rinses) c) 1 M H$_2$SO$_4$ (elution of Bi) Bi is determined by anodic stripping voltammetry; detection limits are 7 ppb in blood and 3 ppb in urine Before the separation, the sample of blood (5 mℓ) or urine (10 mℓ) is diluted to 100 mℓ with 0.25 M HCl to prepare solution (a)	6
Urine	Dowex® 1 or Amberlite® IRA-400 (or Zerolit® FF) (Cl$^-$ form) Column containing the resin to a height of 6 cm a) Urine sample (containing >40 μg of Bi and acidified with HCl to pH 1) (adsorption of Bi and organic matter) b) 2:1 Mixture of acetone and HCl (100 mℓ) (elution of organic matter) c) Water (as a rinse) d) 2 N H$_2$SO$_4$ (25 mℓ) (elution of Bi) Bi is determined polarographically	4
	Amberlite® IRA-400 (Cl$^-$ form) a) Urine sample containing 40—50 μg of Bi and made ≃0.1 M in HCl (adsorption of Bi) b) 6 M HCl (as a rinse) c) Water (until washings are neutral) d) 2 N H$_2$SO$_4$ (elution of Bi) Bi is determined polarographically	27

Table 3
DETERMINATION OF Bi IN INDUSTRIAL PRODUCTS AFTER ANION EXCHANGE SEPARATION IN HCl MEDIA[a]

Material	Ion exchange resin, separation conditions, and remarks	Ref.
Solder alloys	Diaion® SA-100 (Cl⁻ form) Column: 10 × 1.1 cm operated at a flow rate of 1 mℓ/min a) 8 M HCl sample solution (20 mℓ) containing a few drops of Br-water (adsorption of Bi, Sb, and Sn; Pb passes into the effluent) b) 8 M HCl (40 mℓ) (elution of remaining Pb) c) 2 M HNO₃ (50 mℓ) (elution of Sn) d) 0.5 M HNO₃ (200 mℓ) (elution of Bi) e) 1 M NaOH (elution of Sb) Bi is determined spectrophotometrically; before the separation, the sample (0.1 g) is dissolved in HNO₃-HCl-Br-water	18
Steel	Dowex® 2-X8 (200-400 mesh; Cl⁻ form) Column of 2 cm length containing 1—2 g of the resin a) 2.5 M HCl sample solution (adsorption of Bi; Fe and other elements pass into the effluent) b) Dilute H₂SO₄ containing thiourea (elution of Bi) Bi is determined spectrophotometrically	15
High-alloyed steel	Anionite EDE-10P (grain size: 0.25—0.5 mm; Cl⁻ form) Column: 25 × 1 cm conditioned with 2 M HCl and operated at a flow rate of 1 mℓ/min a) 2 M HCl sample solution (50 mℓ containing 1.5 g of the sample) (adsorption of Bi, Sn, Sb, Pb, Zn, and Cd; Fe passes into the effluent) b) 2 M HCl (as a rinse) c) The resin is taken from the column and Bi and the other elements are batch-eluted with H₂SO₄-HNO₃ Bi and the other impurity metals (1—100 ppm) are determined spectrographically	14
High-purity Cu	Dowex® 1-X8 (100-200 mesh; Cl⁻ form) Column: 20 × 1.5 cm a) 1 M HCl sample solution (adsorption of Bi; Cu passes into the effluent) b) 2 M H₂SO₄ (elution of Bi) Bi is determined by anodic-stripping voltammetry Alternatively, Cu can be adsorbed on a column (25 × 3 cm) of Dowex® 50-X8 (H⁺ form) from solution (a); under these conditions Bi passes into the effluent	11
Proton-irradiated Pb targets	Dowex® 1-X8 (100-200 mesh; chloride form) Column: 2 × 0.4 cm a) 0.5 M HCl sample solution (adsorption of Bi; elution of Pb) b) 0.5 M HCl (elution of remaining Pb) c) Conc HNO₃ (elution of ²⁰⁵Bi and ²⁰⁶Bi) This method can be used for the preparation of the Bi-radionuclides with 90—95% yield Before the separation, the irradiated sample is dissolved in 3 M HNO₃ and the bulk of Pb is removed by precipitation with cold HCl	9

[a] See also Table 3 in the chapter on Silver, Volume III; Table 6 in the chapter on Iron, Volume V; Table 10 in the chapter on Lead; Table 11 in the chapter on Cadmium, Volume IV; Table 16 in the chapter on Copper, Volume III; and Tables 25 and 26 in the chapter on Zinc, Volume IV.

Table 4
DETERMINATION OF Bi IN INDUSTRIAL MATERIALS AFTER ANION EXCHANGE SEPARATION IN HYDROBROMIC AND NITRIC ACID MEDIA[a]

Material	Ion exchange resin, separation conditions, and remarks	Ref.
Cyclotron-bombarded Pb targets	Bio-Rad® AG1-X4 (200-400 mesh; nitrate-form) Column of 0.7 cm I.D. containing 1.5 mℓ (0.5 g) of the resin conditioned with 0.5 M HNO$_3$ - 0.05 M HBr and operated at a flow rate of 1.4 ± 0.3 mℓ/min a) Sample solution (150 mℓ) which is 0.5 M in HNO$_3$ and 0.05 M in HBr (adsorption of Bi; Pb passes into the effluent) b) 0.5 M HNO$_3$ - 0.05 M HBr (10 × 5 mℓ) (as a rinse) c) 2 M HNO$_3$ - 0.03 M HBr (5 × 10 mℓ) (elution of remaining Pb) d) 1 M HNO$_3$ (10 mℓ) followed by 0.1 M HNO$_3$ - 0.01 M DTPA (40 mℓ) or 0.2 M HNO$_3$ (40 mℓ) (elution of Bi) The ^{206}Bi activity is measured radiometrically; less than 0.2 μg of Pb is found in the Bi eluate Before the separation, the sample (up to 10 g) is dissolved in HNO$_3$	21
Cu-base alloys	Dowex® 1-X8 (100-200 mesh; nitrate form) Column: 10 × 0.6 cm conditioned with eluent (b) (50 mℓ) and operated at a flow rate of 0.5 mℓ/min a) Sample solution (20 mℓ) consisting of 5 M HNO$_3$ (2 mℓ, containing 400 mg of the sample) and methyl glycol (18 mℓ) (adsorption of Bi, Pb, Th, and La; into the effluent pass Cu, Ni, Zn, etc.) b) 90% Methyl glycol - 10% 5 M HNO$_3$ (150 mℓ) (elution of remaining Cu and other nonadsorbed elements) c) 1 M HNO$_3$ (100 mℓ) (elution of Bi) Bi is determined by EDTA titration Before the separation, the sample (2 g) is dissolved in HNO$_3$	24

[a] See also Table 27 in the chapter on Copper, Volume III.

Table 5
ANION EXCHANGE SEPARATION OF Bi FROM SYNTHETIC MIXTURES WITH OTHER ELEMENTS

Elements separated	Ion exchange resin, separation conditions, and remarks	Ref.
Bi from binary mixtures with Cd, Pb, Zn, In, Ga, Fe(III), Al, Mn(II), Ni, Cu, U(VI), and numerous other elements	Bio-Rad® AG1-X4 (200-400 mesh; nitrate form) Column: ≃7 × 1 cm containing 1 g (2.9 mℓ) of the resin conditioned with eluent (b) and operated at a flow rate of 1.6 ± 0.3 mℓ a) 2 M HNO$_3$ - 0.03 M HBr (50 mℓ) (adsorption of Bi) b) 2 M HNO$_3$ - 0.03 M HBr (120 mℓ) (elution of Cd, Pb, and the other elements) c) 1 M HNO$_3$ (10 mℓ) followed by 0.05 M DTPA- 0.1 M NH$_4$NO$_3$ (pH 4.5) (30—40 mℓ) (elution of Bi) Bi is determined by atomic absorption spectroscopy The method can be used for the separation of trace and milligram amounts of Bi from up to several grams of the other metals; only Hg(II), Au(III), and some Pt metals accompany Bi	20
Bi from Cu, Fe, Pb, Co, Mn, Ni, and Cr	Anionite TM (Cl$^-$ form) Column: 50 × 0.7 cm After adsorption of the elements the following eluents are used a) 4 M HCl (elution of Cu, Co, Mn, Ni, and Cr) b) 0.02 M HCl (elution of Fe, Pb, and residual amounts of the other elements, except Bi) c) 2 N H$_2$SO$_4$ (elution of Bi) This method can be used to determine Bi, Pb, Fe, and Cu in Sn, for which purpose the sample solution (0.5 M in HCl) is passed through the column to adsorb Bi and Pb (Cu and Fe pass into the effluent); these two elements are then separated using eluents (b) and (c) To separate Cu from Fe, a pyrophosphate solution of pH 10—11 of the two metals is passed through a column of the cation exchange resin Wofatit® R to adsorb Cu (Fe passes into the effluent as anionic pyrophosphate complex) For the elution of Cu, HCl (1:4) is employed	16

Table 6
DETERMINATION OF Bi IN INDUSTRIAL PRODUCTS AFTER CATION EXCHANGE SEPARATION

Material	Ion exchange resin, separation conditions, and remarks	Ref.
Galena (PbS) and pure Pb	Dowex® 50W-X8 (50-100 mesh; H⁺ form) Column of 1.2 cm I.D. containing 10 mℓ of the resin and operated at a flow rate of 100 mℓ/hr a) Sample solution (80—100 mℓ) acidified with 5 M HClO$_4$ (7—10 mℓ) and containing 0.5 g ascorbic acid (adsorption of Bi and Pb; Fe[II] and other elements pass into the effluent) b) 0.3—0.5 M HClO$_4$ (50—100 mℓ) (elution of residual non-adsorbed elements) c) 0.3—0.5 M HClO$_4$ - 0.008 M EDTA (100 mℓ) (elution of Bi; coadsorbed Pb is not eluted) Bi is determined colorimetrically with thiourea or by means of atomic absorption spectroscopy; down to 2 ppm of Bi in Pb and 30 ppm of Bi in galena may be determined by use of the colorimetric procedure; interference due to Sb may be overcome by adding tartaric acid to solution (a) before adsorption Before the separation, the sample is dissolved in HCl (for ores; 0.1—0.6 g) or HNO$_3$ (for metallic Pb, oxide, or salts: 0.5 g) and evaporated with HClO$_4$	29
Sb(III)-oxide	Dowex® 50-X2 Column: 15 × 1 cm a) Sorption solution (~100 mℓ) (adsorption of Bi; into the effluent passes Sb as anionic tartrate complex) b) 0.5 M HClO$_4$ (150 mℓ) (elution of remaining Sb) c) 3 M HCl (25 mℓ) followed by water (70 mℓ) (elution of Bi) This separation procedure is repeated and then Bi is determined by anodic-stripping voltammetry Before the ion exchange separation, the sample (0.1 g) is dissolved in 1% tartaric acid solution (100 mℓ) and after addition of conc HClO$_4$ (4 mℓ) the solution (a) is obtained	30

Table 7
DETERMINATION OF Bi AND OTHER ELEMENTS IN INDUSTRIAL PRODUCTS AFTER CATION EXCHANGE SEPARATION

Material	Ion exchange resin, separation conditions, and remarks	Ref.
High-purity Al chloride	Cationite KU-2-X8 (H$^+$ form) Column containing 15 g of the resin and operated at a flow rate of 0.64 mℓ/min a) 10 mM Na$_2$EDTA sample solution ($\sim$10 mℓ containing 1 g of the Al sample) (adsorption of Al, Cr, and Co; into the effluent pass the anionic EDTA complexes of Bi, Pb, Sb, Cd, Mo, In, Sn, Ni, V, Ga, Ag, and Ti) b) Water (15 mℓ) (elution of residual Bi and of the other non-adsorbed metals) Bi and the other impurity elements ($\simeq$10 ppm each) are determined by spectrographic analysis To concentrate Co and Cr as well as the other impurity metals, the sample (1 g) is dissolved in 10 mM NH$_4$SCN (10 mℓ) and the solution is stirred with EDE-10P and VP-1APX14 anion exchange resins (250 mg of each) at 60—80°C for 1 hr to adsorb the anionic thiocyanate complexes of the elements; then the resin is mixed with C powder, ashed at 550°C, and the elements are determined spectrographically	35
Al antimonide	Dowex® 50W-X8 a) 0.1 M HF—0.1 N H$_2$SO$_4$ sample solution (adsorption of Bi, Ga, Fe, In, Cd, Ca, Co, Mg, Mn, Cu, Ni, Pb, Tl, Cr, and Zn; into the effluent pass Al and Sb as anionic fluoride complexes) b) 3 M HNO$_3$ (elution of Bi and other adsorbed elements) Bi and the other impurity elements are determined spectrographically	31

[a] See also Tables 48, 52, and 54 in the chapter on Copper, Volume III.

Table 8
CATION EXCHANGE SEPARATION OF Bi FROM SYNTHETIC MIXTURES WITH OTHER ELEMENTS

Elements separated	Ion exchange resin, separation conditions, and remarks	Ref.

Bi from mixtures with Cu, Cd, Sn, Al, Ni, Mg, V(V), Mn(II), Hg, Fe(III), Zn, Ag, Co, Ba, Ca, Zr, Th, and Ti	Dowex® 50W-X8
	After adsorption of the elements on a column of the resin the following eluents are used
	A) H_2SO_4 system
	  a) 1 M H_2SO_4 (elution of Cu, Cd, Sn, Al, Ni, Mg, V, Mn, Hg, and Fe)
	  b) 4 M H_2SO_4 (elution of Bi)
	B) HNO_3 system (gradient elution)
	  a) 1 M HNO_3 (elution of Zn, Ag, Co, Ba, and Ca)
	  b) 4 M HNO_3 (elution of Bi)
	C) HCl system
	  a) 1 M HCl (elution of Bi)
	  b) 8 M HCl (elution of Zr, Th, and Ti)
		32

Bi from binary mixtures with Ni, Ca, Cu, Co, Mg, Mn, Zn, and Fe	Zerolit® 225 (H^+ form)
	Column: 15 × 2.5 cm conditioned with 0.2 M HCl and operated at a flow rate of 2 mℓ/min
	  a) 0.2 M HCl sample solution (20 mℓ) (adsorption of the elements)
	  b) 0.2 M HCl (few milliliters) (as a rinse)
	  c) 0.5 M HCl (120 mℓ) (elution of Bi)
	  d) 2 M HCl (300—400 mℓ) (elution of Ni, Cu, Fe, and Ca)
	  e) 4 M HCl (300—400 mℓ) (elution of Co, Mg, Mn, and Zn)
		33

Bi from Zn, Cd, and Pb	Amberlite® IR - 112 (NH_4^+ form)
	  a) Solution of pH 1 containing ≃0.01 M each of the metals and a small excess of EDTA with respect to Bi (adsorption of Zn, Cd, and Pb; Bi passes into the effluent as anionic EDTA complex)
	  b) Dilute HNO_3 of pH 1 (elution of residual Bi)
	  c) 3 M HCl (100 mℓ) (elution of Zn and Cd)
	  d) 2% Ammonium acetate solution (100 mℓ) (elution of Pb)
		36,37

Table 9
DETERMINATION OF Bi AFTER SEPARATION ON CHELATING RESINS[a]

Material	Ion exchange resin, separation conditions, and remarks	Ref.
Rocks	Chelex® 100 (100-200 mesh; H⁺ form) The sample (1—20 g containing ≮ 0.2 ppm of Bi) is decomposed with HF and HNO$_3$ and fumed several times with HNO$_3$; the residue is dissolved in 2 M HNO$_3$ and any residue is filtered off; the combined filtrate and washings (0.01 M HNO$_3$) are diluted two to three times with water and the pH is adjusted to 2 using NaHCO$_3$ or Na acetate; then an aliquot of ascorbic acid (three to five times in excess of the Fe present) and the resin (180 mg) are added; to adsorb the Bi the mixture is equilibrated for 24 hr on a sample rotator, and after filtration Bi is determined directly on the pelletized resin using X-ray fluorescence spectroscopy Coadsorbed with the Bi are Cu and Pb, but not Fe(II)	42
Bi in mixtures with Fe and other elements	Resin® Sel-K4 (condensation product of m- and p-cresols with formaldehyde containing nicotine as functional groups) After adsorption of BiI$_4^-$ on a 30-cm column of the resin the following eluents are used a) 1 M HNO$_3$ (175 mℓ) (elution of Fe) b) 2 M NH$_3$ solution ($\simeq$25 mℓ) (as a rinse) c) 2 N H$_2$SO$_4$ ($\sim$250 mℓ) (elution of Bi)	43

[a] See also Table 9 in the chapter on Arsenic and Reference 46.

Table 10
MICROCHEMICAL DETECTION OF Bi BY RESIN SPOT TESTS[a]

Ion exchange resin	Experimental conditions and remarks	Ref.
Cation exchange resin, e.g., Dowex® 50W-X2.5	On a white spot plate a few beads of the resin are mixed with 1 drop of the sample solution, and after ≮10 min 1 drop of a saturated solution of thiourea is added; the orange-yellow color given by the Bi-thiourea complex is examined with a lens ($\times$ 20) The limit of detection is 0.1 µg in a limiting concentration of 1 in 4 $\times$ 10^5 The concentration of neutral salts in the sample should not exceed 0.5 M; V(V), Cr(VI), and U(VI) have a marked effect only when present in large quantity	44
Dowex® 50W-X1 and Dowex® 1-X1	Catechol violet is adsorbed either by the cation or anion exchange resin and thus becomes more sensitive to Bi at pH 2—4 (limit of detection $\simeq$0.05 µg) The orange color of the anion exchanger changes to red and then to violet; in the latter case increase of ionic strength (e.g., addition of KNO$_3$) decreases the sensitivity (0.04 µg)	45

[a] See also Table 80 in the chapter on Copper, Volume III.

REFERENCES

1. **Fritz, J. S. and Pietrzyk, D. J.**, Non-aqueous solvents in anion exchange separations, *Talanta*, 8, 143, 1961.
2. **Peters, J. M. and del Fiore, G.**, Distribution coefficients for 52 elements in hydrochloric acid water-acetone mixture on Dowex 1-X8, *Radiochem. Radioanal. Lett.*, 21, (1-2), 11, 1975.
3. **Kuroda, R., Ishida, K., and Kiriyama, T.**, Adsorption behavior of a number of metals in hydrochloric acid on weakly basic anion exchange resin, *Anal. Chem.*, 40, 1502, 1968.
4. **Lanza, P., Cancialini, V., and Benati, A.**, Determination of bismuth content of urine: polarographic method, *J. Electroanal. Chem.*, 17(3-4), 395, 1968.
5. **Palliere, M. and Gernez, G.**, Determination of traces of bismuth in blood, *Ann. Pharm. Fr.*, 38(2), 123, 1980.
6. **Kauffmann, J. M., Patriarche, G. J., and Christian, G. D.**, Rapid determination of trace amounts of bismuth in urine and blood using differential pulse anodic-stripping voltammetry at the hanging-mercury electrode, *Anal. Lett. Part B.*, 14(15), 1209, 1981.
7. **Mottola, H. A.**, Removal of interferences in absorptiometric determination of bismuth with thoron I, *Anal. Chim. Acta*, 29, 261, 1963.
8. **Florence, T. M.**, Determination of bismuth in marine samples by anodic stripping voltammetry, *Electroanal. Chem. Interfacial Electrochem.*, 49, 255, 1974.
9. **Birattari, C., Bonardi, M., and Gilardi, M. C.**, Production of bismuth, radio-tracers in the Milan AVF cyclotron for environmental toxicology and proton-activation analysis, *Radiochem. Radioanal. Lett.*, 49(1), 25, 1981.
10. **Portmann, J. E. and Riley, J. P.**, Determination of bismuth in sea and natural waters, *Anal. Chim. Acta*, 34, 201, 1966.
11. **van Dyck, G. and Verbeek, F.**, Determination of traces of bismuth in copper by anodic-stripping voltammetry, *Fresenius' Z. Anal. Chem.*, 249, 89, 1970.
12. **Darbinyan, M. V. and Kapantsyan, E. E.**, Separation of bismuth and tellurium by ion-exchange chromatography, *Arm. Khim. Zh.*, 21(2), 103, 1968.
13. **Martỹnova, L. T. and Sochevanov, V. G.**, Polarographic determination of bismuth in ores and concentrates after separation on an anionite, *Byull. Nauchno. Tekh. Inf. Gos. Geol. Kom. SSSR Otd. Nauchno Tekh. Inf. VIEMSa*, 3(53), 86, 1964.
14. **Komarovskii, A. G., Vasil'eva, L. A., and Tikhomirova, N. N.**, Use of ion-exchange chromatography for determining deleterious impurities in highly alloyed steel by a spectrographic method, *Zavod. Lab.*, 37(2), 179, 1971.
15. **Léontovitch, N.**, Determination of small amounts of bismuth in ferrous materials, *Chim. Anal. (Paris)*, 47(9), 458, 1965,
16. **Stepin, V., Ponosov, V. I., and Silaeva, E. V.**, Separation of small amounts of bismuth, lead, cobalt, nickel, phosphorus, iron, and copper with ion exchange resins, in *Khromatografiya ee Theoriya i Primenenie*, Akad. Nauk SSSR, Moscow, 1960, 230.
17. **Laul, J. C., Case, D. R., Schmidt-Bleck, F., and Lipschutz, M. E.**, Bismuth contents of chondrites, *Geochim. Cosmochim. Acta*, 34, 89, 1970.
18. **Onuki, S.**, Spectrophotometric determination of bismuth in solder alloys after separation by anion exchange, *Jpn. Analyst*, 12(9), 844, 1963.
19. **Klakl, E. and Korkisch, J.**, Anion-exchange behavior of several elements in hydrobromic acid-organic solvent media, *Talanta*, 16, 1177, 1969.
20. **Strelow, F. W. E. and Van der Walt, T. N.**, Quantitative separation of bismuth from lead, cadmium and other elements by anion-exchange chromatography with hydrobromic acid-nitric acid elution, *Anal. Chem.*, 53, 1637, 1981.
21. **Van der Walt, T. N., Strelow, F. W. E., and Haasbroek, F. J.**, Separation of bismuth-206 from cyclotron bombarded lead targets using anion exchange chromatography in nitric hydrobromic acid mixtures, *Int. J. Appl. Radiat. Isot.*, 33, 301, 1982.
22. **Toshimitsu, Y., Yoshimura, K., and Ohashi, S.**, Ion-exchanger colorimetry. IV. Micro-determination of bismuth in water, *Talanta*, 26, 273, 1979.
23. **Ahluwalia, S. S. and Korkisch, J.**, Anion-exchange separation of bismuth from lead, *Fresenius' Z. Anal. Chem.*, 208, 414, 1965.
24. **Feik, F. and Korkisch, J.**, Separation of bismuth from lead, copper and other elements by anion exchange, *Talanta*, 11, 1585, 1964.
25. **Sobhana, K., Madhavankutty, P., and Savariar, C. P.**, Separation of bismuth(III) by anion-exchange chromatography in maleate and succinate solutions, *J. Indian Chem. Soc.*, 55(5), 458, 1978.
26. **Chakravorty, M. and Khopkar, S. M.**, Anion-exchange chromatographic separation of bismuth and antimony from arsenic, tin and other elements in malonic acid solution, *Chromatographia*, 12(7), 459, 1979.

27. **Lanza, P. and Concialini, V.,** Polarographic determination of bismuth in urine, *Ric. Sci.*, 36(5), 372, 1966.
28. **Capalla, T. and Sheybal, I.,** Spectrophotometric determination of traces of bismuth and tin in transformer steel, *Hutnik*, 47(8-9), 349, 1980.
29. **Šulcek, Z., Povondra, P., and Kratochvil, V.,** Rapid analytical methods for the investigation of metals and inorganic raw materials. XVIII. Determination of traces of bismuth in galena and pure lead, *Collect. Czech. Chem. Commun.*, 34,(12), 3711, 1969.
30. **Petak, P. and Koubova, V.,** Determination of small amounts of bismuth in antimony(III) oxide by using anodic-stripping voltammetry, *Analyst (London)*, 103, 179, 1978.
31. **Chuchalina, L. S., Yudelevich, I. G., Buyanova, L. M., Starshinova, N. P., and Vall, G. A.,** Spectrochemical and spectrographic methods for analysis of aluminum antimonide, *Zh. Anal. Khim.*, 24(6), 905, 1969.
32. **Akki, S. B. and Khopkar, S. M.,** Separation of bismuth(III) from other ions by cation exchange chromatography, *Sep. Sci.*, 5(6), 707, 1970.
33. **Burriel-Marti, F. and Alvarez Herrero, C.,** Separation of Bi-Ni, Bi-Ca, Bi-Cu, Bi-Co, Bi-Mg, Bi-Mn, Bi-Zn, and Bi-Fe by means of ion exchange. XIII, *Rev. Port. Quim.*, 7(3), 130, 1965.
34. **Widtmann, V.,** Determination of bismuth, lead and antimony in a platinum-rhodium alloy, *Hutn. Listy*, 25(10), 733, 1970.
35. **Semenenko, V. K., Shchetinina, T. V., Otmakhova, Z. I., Chashchina, O. V., and Fomkina, L. N.,** Use of ion exchange in the analysis of high-purity aluminum chloride, *Zh. Anal. Khim.*, 36(1), 73, 1981.
36. **Taketatsu, T.,** Chemical analysis with the aid of EDTA complexes. II. Relation between the dissociation of complex anions and the adsorbability on ion-exchange resin at various pH values, *J. Chem. Soc. Jpn. Pure Chem. Sect.*, 78(1), 148, 1957.
37. **Taketatsu, T.,** Chemical analysis with the aid of EDTA complexes. III. Separation of bismuth from zinc, cadmium and lead, *J. Chem. Soc. Jpn. Pure Chem. Sect.*, 78(1), 151, 1957.
38. **Willis, R. B. and Fritz, J. S.,** Determination of bismuth by forced-flow liquid chromatography, *Talanta*, 21, 347, 1974.
39. **D'Amore, G. and Corigliano, F.,** Spectrophotometric determination of bismuth as thiocyanate in aluminum-lead-copper- and tin-based materials, *Atti Soc. Peloritana Sci. Fis. Mat. Nat.*, 11(3) 239, 1965.
40. **Qureshi, M. and Husain, W.,** Cation-exchange behavior of several elements in formic acid solution, *Talanta*, 18, 399, 1971.
41. **Fritz, J. S. and Garralda, B. B.,** Cation-exchange separation of metal ions with hydrobromic acid, *Anal. Chem.*, 34, 102, 1962.
42. **Blount, C. W., Leyden, D. E., Thomas, T. L., and Guill, S. M.,** Application of chelating ion exchange resins for trace element analysis of geological samples using X-ray fluorescence, *Anal. Chem.*, 45, 1045, 1973.
43. **Lewandowski, A. and Szczepaniak, W.,** Ion exchanger specific for bismuth, *Chem. Anal. (Warsaw)*, 7(3), 593, 1962.
44. **Fujimoto, M.,** Micro-analysis with the aid of ion-exchange resins. IX. Detection of small quantities of bismuth with thiourea, *Bull. Chem. Soc. Jpn.*, 30(1), 83, 1957.
45. **Murase, T.,** Micro-detection by the use of ion-exchanger granules. Detection of bismuth with catechol violet, *J. Chem. Soc. Jpn. Pure Chem. Sect.*, 78(8), 983, 1958.
46. **Brajter, K. and Slonawska, K.,** Application of ion exchanger Cellex P for the selective separation of bismuth from other elements in the presence of tetrene as complexing reagent, *Fresenius' Z. Anal. Chem.*, 320, 142, 1985.

SULFUR

The majority of the ion exchange methods so far employed for the separation of S when present as sulfate are based on its nonadsorption on cation exchange resins of the strongly acidic type. By means of this technique S can be isolated rapidly from virtually all natural and industrial materials of which the cationic constituents are retained from weak acid or neutral solutions. Essentially the same selectivity of separations is achieved by adsorbing sulfate and other species on anion exchange resins from which cationic elements are not retained. The latter technique also allows the fractionation of the different S species, but the separations need more time than the cation exchange procedures which do not require a subsequent elution step before S can be determined.

CATION EXCHANGE RESINS

General

From acid, neutral, and alkaline solutions sulfate is not retained by strongly acidic cation exchange resins of the sulfonic acid type. This makes it possible to separate sulfate from most cationic constituents of sample solutions, provided that the adsorption of the latter is performed from media allowing their complete retention. Usually, systems of this type are dilute acid solutions adjusted to pH values of not less than[11-13] and neutral media[14-33] which include natural waters[34-68] and industrial effluents.[19,69-71] This adsorption of accompanying cations, for which numerous examples are shown in Tables 1 to 10*, can also be performed by equilibrating the cation exchangers with aqueous suspensions of the finely ground samples using the batch technique (see Tables 1 and 10).[72-76] Other systems that have been used to separate sulfate from cationic constituents include 0.3 M hydrochloric acid (HCl),[1,77,78] 0.5 M to 1 M nitric acid,[79,80] 80% ethanol,[81] 50% isopropanol,[82] 95% ethanol containing 2% NaOH,[83] and other alkaline solutions containing alkali hydroxides and carbonates or ammonia.[28,84-93] Coeluted with sulfate, i.e., not retained by the cation exchange resins from the media mentioned above, are other anions as, for example, chloride, phosphate, nitrate, and other anionic species (e.g., thiocyanate),[94] as well as anionic complexes of metals such as those of Au and Pt metals.

In most cases, the adsorption of the metal ions is carried out on columns containing the cation exchange resins in the hydrogen forms, so that during the adsorption process, carbonates are decomposed forming CO_2 which may interfere with the performance of the ion exchange process by bubble formation in the resin bed (if sulfide is present H sulfide will form).[90] This can be avoided by preceding acidification of the sample solution, e.g., natural water to a pH value of 1 to 2 or by using the batch technique,[35,37,40,47,57,63,65,76,82,84-87,95-99] which in order to achieve complete adsorption of the metal cations can be combined with a subsequent column operation (slurry column method; see Table 10).[75,100] The coelution of chloride with sulfate can be prevented by the use of cation exchangers in the Ag form (see Table 4).[40,41] On such resins AgCl is precipitated and simultaneously the cations are adsorbed according to the reaction: $R_sSO_3^- Ag^+ + Na^+Cl^- \rightarrow R_sSO_3^- Na^+ + AgCl \downarrow$ (R_s = resin matrix). On resins in the Ag form phosphate is also precipitated as Ag_3PO_4.[101] Another example for this type of reactive ion exchange is presented in Table 11, which shows that sulfate can be precipitated as Sr sulfate on a cationic resin in the Sr form.[102] Similarly, sulfate can also be precipitated as Ba sulfate on a resin in the Ba form.[54] Adsorption of S on a mixture of cation and anion exchange resins following formation of methylene blue has been employed for the determination of sulfide using ion exchanger colorimetry (see Table 3).[95]

---

* Tables for this chapter appear at the end of the text.

Retention of accompanying elements from the systems mentioned above is usually effected on the H or Na forms of strongly acidic resins of the sulfonic acid type. However, on a resin in the H form, which has been conditioned and then stored in the column for a relatively long time before passage of the sorption solution, a very slight decomposition of the organic matrix of the polymer may occur.[103] This entails the formation of trace quantities of sulfate (from the fixed sulfonic acid groups) so that on subsequent use of the resin, without preliminary reconditioning with acid followed by water, the sulfate (from the resin) will accompany the sulfate contained in the sample solution. Therefore, the pretreatment of a resin in the H form should be done just before the start of the sorption process. On cationic resins in the Na form formation of sulfate was not observed.[103] In case that very small amounts of sulfate (e.g., micro- or submicrogram amounts) have to be separated from relatively large quantities of cations such as Fe(III) or other ions which may cause some oxidation of the cation exchanger (forming sulfate), it is advisable either to use a holding reductant such as hydroxylamine hydrochloride or to perform the separation on an anion exchange resin which on oxidation does not produce sulfate ions.

After adsorption of the cationic constituents residual sulfate remaining in the resin bed can be eluted by the passage of water[1-4,8,10,13-17,19-25,27,28,46,53,71-73,79,83,85,86,87,91,100,104-109] or dilute acid solutions such as 0.004,[6] 0.1,[1] and 0.2 M[77] HCl or 0.1 M nitric acid.[80] In most cases, however, this additional washing step is not required, because sulfate can be determined in a suitable fraction of the effluent while the sorption is still going on. In other words, after the first milliliters of the effluent (which should correspond to more than the void volume of the resin column) have been rejected, a suitable volume of the subsequent percolate is taken for analysis. Therefore, cation exchange separations of sulfate are much more rapid than anion exchange procedures which necessitate elution of the adsorbed sulfate before the determination can be carried out.

Applications

Separations on cation exchange resins have very frequently been employed in connection with the determination of S in geological materials which include coal (see Tables 1 and 10),[5,14-16,72,76,110,111] shales (see Table 1),[73] ores (see Tables 1 and 10),[1] such as pyrites,[86,91,107,112-114] soils (see Table 1),[7,12,115,116] air (see Table 2),[83,117] atmospheric particulates,[92,118] and natural waters (see Tables 2 to 4).[9,11,32,34-44,46-68,88,95,96,109,110,113,119-132] Similar separation procedures have been used for the analysis of biological materials,[2,8,17,18,98,110,133-135] as, for example, urine (see Table 5)[2,8,17,136] and plants (see Table 6),[18,99,137,138] and a great variety of industrial products which include fly ash and other industrial emissions (see Table 7),[6,45,82] industrial effluents (waste waters and leach liquors) (see Table 7),[13,19,25,58,69-71,129] pulp and paper mill process streams,[139] synthetic organic compounds (including pharmaceuticals) (see Table 8),[20-22,84,87,97,105,140-144] organometallic compounds,[98] cellulose nitrate,[145] Pu (see Table 9),[79,80,146] U (see Table 9),[94,104,147] Th nitrate,[108] Pb sulfate (see Table 10),[100] cements (see Table 10),[23,74,85] rubber articles (vulcanisates) (see Table 10),[85,148] gypsum (see Table 10),[75] boiler deposits (see Table 10),[76] precipitator dusts (see Table 10),[76] metals (Al, Cu, Mg, Fe, and Ni) (see Table 11),[148-153] steel,[152] ferrovanadium,[77] Cu sulfate (see Table 11),[154] chrome alum,[121] metal sulfides,[106,141,155,156] ferrous sulfamate,[157] elementary S,[28,159] glasses,[29,90] rare earth concentrates,[10] high-purity ammonia solutions,[159] mixtures of alkali metal and ammonium sulfates and H sulfates,[160] corrosion products,[161] pickling bath solutions,[26] color anodization baths,[162] Cu plating electrolytes,[3] and electrolytic capacitor electrolytes.[81]

To separate sulfate from other S species the procedure described in Table 11 or other methods can be employed.[93,163]

Furthermore, cation exchange separation methods have variously been employed to separate sulfate from synthetic mixtures with cations which interfere with the determination of sulfate.[30,117,164-175]

ANION EXCHANGE RESINS

General

On anion exchange resins sulfate is more strongly adsorbed than chloride, so that usually resins in the chloride form are used for separations involving this S species. The adsorption of sulfate is best performed from very dilute acid or neutral media, as, for example, from 0.004 to 0.1 M HCl solutions,[6,176-180] dilute acetic acid solutions of pH 4,[181,182] or samples of natural waters.[183] Coadsorbed with sulfate from virtually all these systems are other anions such as phosphate, nitrate, and other anionic S species as well as anionic chloro complexes of a few metals, e.g., Au and Pt metals. Not adsorbed together with sulfate from the dilute HCl media are selenite[180] and tellurite[179] as well as cationic constituents of the sample solutions, so that selective separations can be achieved. Removal of coadsorbed orthophosphate can be effected by its elution with 0.1 M HCl (see Table 12)[6,177,184] or 0.1 M HCl - 0.19 M KCl.[185]

For the elution of the adsorbed sulfate, acid and alkaline media as well as salt solutions can be used. Acid eluents include 0.5,[176] 1,[180] and 2 M[6] HCl (see Table 13), while alkaline elutions can be effected with the following media: 0.2[186] and 0.5 M[176] NaOH, 0.5[177] and 1 M[181,182] ammonia solutions, and 6% ammonium carbonate solution.[109] Salt solutions that can be employed for the elution of sulfate are 0.25 M,[187] 3.5%,[178] and 10%[179] NaCl solutions, 2 M KCl,[185] and 10% ammonium acetate.[188]

Resins in the hydroxide forms can be used for the adsorption of H sulfide (see Table 12) (reaction: $2R_sOH + H_2S \rightarrow [R_s]_2S + H_2O$; R_s = resin matrix),[189,190] sulfide anions, e.g., S^{2-} and CS_3^{2-}[191,192] (sulfide and thiocarbonate), as well as SO_2.[190] The latter may be determined by using a procedure based on reactive ion exchange (see Table 14).[193] An indirect determination of sulfate is also possible by the use of a resin in the thiocyanate form, which when contacted with the sulfate solution releases thiocyanate ions which are determined spectrophotometrically.[194]

Elution of the adsorbed sulfide and sulfite can be effected with 4 M NaOH[189] or 5% NaCl solution[195] and 2 M NaOH,[196] respectively. Other eluents for these as well as other S species are listed in Tables 14 and 15 in which procedures are presented that can be used for the fractionation of mixtures containing sulfide, sulfite, sulfate, thiosulfate, thiocyanate, and polythionates. The adsorbability of the latter species on anion exchange resins, e.g., Dowex® 1-X2, was found to increase with increasing number of S atoms in the oxyanion, i.e., pentathionate ($S_5O_6^{2-}$) > tetrathionate ($S_4O_6^{2-}$) > trithionate ($S_3O_6^{2-}$) > dithionate ($S_2O_6^{2-}$).[197]

Applications

Anion exchange resins have been used in connection with the determination of sulfate and sulfide in geological and biological materials which include air (see Table 12),[189] water (see Table 12),[177,183,189] plants (see Table 12),[178,198] and body fluids.[181,182,188,199] Similar separation methods were utilized for the analysis of various S species in industrial products, e.g., Se (see Table 13),[180] Fe alloys (see Table 13),[176] Cu,[200] fuel oils,[201] common salt,[186] neutron-irradiated alkali chlorides (see Table 14),[184,202,203] viscose,[191,192] photographic gelatin (see Table 14),[204,205] and black liquors.[206]

For mutual separation of oxyanions of S the methods outlined in Tables 14 and 15 can be employed.[202,203] Similar procedures have been described in the literature.[207,208] In one of the methods the eluent for sulfide, sulfite, and thiosulfate is 0.3 M Na nitrate in 10% ethanol at pH 9.7.[208]

To separate sulfate and also sulfite from binary mixtures with other anions the procedures illustrated in Table 16 can be used. A method for the separation of sulfate from oxalate has also been described.[209]

Ion chromatographic separations of S species on anion exchange resins are outlined in the chapter "Special Analytical Techniques Using Ion Exchange Resins", Volume I.

RESINS USED FOR MICROCHEMICAL DETECTION

The resin spot tests described in Table 17 can be employed for the microchemical detection of sulfide and ions of other elements.

Table 1
DETERMINATION OF SULFATE IN GEOLOGICAL MATERIALS AFTER CATION EXCHANGE SEPARATION[a]

Material	Ion exchange resin, separation conditions, and remarks	Ref.
Coal	Dowex® 50-X8 (50-100 mesh; H⁺ form) Column: 12 × 1.2 cm operated at a flow rate of 6 mℓ/min a) Aqueous sample solution (~50 mℓ, e.g., containing 1—10 mg of sulfate) (adsorption of cationic constituents; sulfate passes into the effluent) b) Water (50 mℓ) (elution of residual sulfate) c) HCl (1:4) (30 mℓ) (elution of adsorbed cations) Sulfate is determined via titration of unconsumed Ba with 0.01 N H_2SO_4 in the presence of arsenazo III; before the separation, the sample (50—150 mg, passing a 60-mesh sieve) is ashed by using the oxygen-flask method or the Eschka procedure	14—16
Coal ash, slag, and cinders	Cation exchange resin KPS 200 (H⁺ form) Column: ~17 × 3 cm operated at a flow rate of 2—6 drops/sec The sample (0.1—0.2 g; ground to pass through a 0.16-mm sieve) is mixed with water (100 mℓ) at 50°C and transferred to the column; the effluent contains the sulfate which is also present in subsequent water washes until 800 mℓ of percolate has been collected The free acid (H_2SO_4 and HCl) is titrated with 0.1 M NaOH and the chloride is determined either potentionmetrically or argentimetrically; the sulfate content is obtained by difference	72
Combustible shales	Cationite (H⁺ form) Column containing 25 g of the resin and operated at a flow rate of 10—15 mℓ/min The sample ash (0.9—1 g) is mixed with water and shaken for 15 min with the resin (10 g) to adsorb cationic constituents (if the solution is not acid to methyl red a further 5 g of the resin is added and the mixture is again agitated for 15 min) Subsequently, the resin is filtered off and the filtrate is passed through the column to adsorb remaining cations; in the effluent and water washings, sulfate is determined titrimetrically	73
Ores	Cation exchange resin, e.g., cationite KU-1 (H⁺ form) in a column operated at a flow rate of 80—120 drops/min a) Sample solution (>50 mℓ) which is ≯0.3 M in HCl (adsorption of heavy metals and other cationic constituents; SO_4^{2-} passes into the effluent) b) 0.1 M HCl (75 mℓ) followed by water (50 mℓ) (elution of remaining SO_4^{2-}) Following precipitation as $BaSO_4$ the determination is performed according to the Lunge method; an accuracy within 0.2% is attained Before the separation, the sample is decomposed in the presence of an oxidant to convert all S into SO_4^{2-}	1
Soil	Zeo-Karb® 225 (50-60 mesh; H⁺ form) The air-dried sample is shaken with Morgan reagent (3% acetic acid solution which is 10% in Na acetate), filtered, and the filtrate is passed through a 20-cm column of the resin to adsorb the cationic constituents of the extract; the first 15 mℓ of the effluent is discarded and in the next 5 mℓ sulfate is determined colorimetrically The recovery of up to 100 ppm of SO_4^{2-} is 96—105% in the range of 5—80 ppm	12

[a] See also Table 11 in the chapter on Iron, Volume V.

Table 2
TITRIMETRIC DETERMINATION OF SULFATE IN NATURAL WATERS AND AIR AFTER CATION EXCHANGE SEPARATION[a]

Material	Ion exchange resin, separation conditions, and remarks	Ref.
River- and mineral waters	Dowex® 50W-X8 (H⁺ form) Column: 20 × 1.9 cm The sample is passed through the resin column to adsorb cations and in 50—75 mℓ of the effluent (the first 150 mℓ are discarded) sulfate is determined by indirect spectrophotometric EDTA titration of excess of Ba ions From 20—1000 ppm of sulfate can be determined with a coefficient of variation of 2.2% (34 results)	34
Mineral waters	Dowex® 50-X8 (20-50 mesh; H⁺ form) A portion (10 mℓ) of the sample (20—90 mg/ℓ of sulfate) is mixed with the resin (10 g) and in 5 mℓ of the supernate, sulfate is determined titrimetrically There is no interference from phosphate, chloride, nitrate, or fluoride in concentrations < 200 mg/ℓ	35
Rain, potable, and surface waters	Dowex® 50W-X8 (20-50 mesh; H⁺ form) Column: 20 × 0.5 cm The sample (15 mℓ) is passed through the resin bed to adsorb cationic constituents and in the effluent sulfate is determined titrimetrically	36
Drinking water	Amberlite® IR-120 (H⁺ form) To ~100 mℓ of the sample, 2 g of the resin is added and the mixture is shaken for 30 min to adsorb cationic constituents; after filtration, sulfate is determined titrimetrically in a suitable aliquot of the filtrate (EDTA titration of excess of Ba added to the sample)	37
Air	Dowex® 50W-X10 (≯50 mesh; H⁺ form) Column of 2.5 cm OD filled with the resin to a height of 10 cm and operated at a flow rate of 2.5 mℓ/min a) Sample solution (adsorption of Na and other cationic interferences; H_2SO_4 passes into the effluent) b) Distilled water (5 × 5 mℓ) (elution of residual H_2SO_4) Sulfate is determined titrimetrically; before the separation, the H_2SO_4 collected on an air filter is absorbed by 0.15 mℓ of 2% NaOH in 95% ethanol	83

[a] See also Table 3 in the chapter on Silicon.

Table 3
SPECTROPHOTOMETRIC DETERMINATION OF SULFATE AND SULFIDE IN NATURAL WATERS AFTER CATION EXCHANGE SEPARATION

Material	Ion exchange resin, separation conditions, and remarks	Ref.
Hot-spring water	Aqueous suspensions of the macroreticular-type ion exchange resins Amberlyst® 15 (H$^+$ form) and Amberlyst® A-27 (chloride form)	95
	To 25 mℓ of 1% Zn acetate solution, the sample (containing < 10.3 nmol of S^{2-}) is added and the solution is diluted to 50 mℓ with the Zn acetate solution; then 5 mℓ of 0.1% N,N-dimethyl-p-phenylenediammoniumsulfate in 5 M H$_2$SO$_4$ and 1 mℓ of 0.1 M FeCl$_3$ in 6 M HCl are added and the mixture is shaken vigorously for 1 min to form methylene blue	
	After 60 min 4- and 2-mℓ portions, respectively, of the anion and cation exchange resin suspensions are added and the mixture is shaken for 10 min to adsorb the dye onto the coagulated resins; the latter are subjected to ion exchanger colorimetry; nitrite interfers seriously	
Rain water	The sample (containing 0.6—3.8 mg/ℓ of sulfate) is drawn through a column (22 × 0.32 cm) of Dowex® 50-X8 (20-50 mesh; H$^+$ form) to remove potentially interfering cations, then injected manually at 150-sec intervals from a 1.1-mℓ loop into water flowing at 2 mℓ/min into a stream of reagents, allowing the determination of sulfate by flow-through spectrophotometry; the precision was 4.1%	38
Riverwater	Amberlite® IR 120B	9
	Column: 10 cm × 1.5 cm^2	
	The sample is passed through a membrane filter (0.45 μm) and to the filtrate (60—100 mℓ), Ag$_2$O (20 mg) is added to remove Cl$^-$; after 2 hr of shaking the AgCl precipitate is removed by centrifugation and the supernate is passed through the resin bed to adsorb cations, and in the effluent (pH 3—4) sulfate is determined spectrophotometrically	
	Between 2 and 10 ppm of sulfate could be determined with a recovery of 94—101%	
Water samples	Dowex® 50W-X8 (20-50 mℓ; H$^+$ form)	39
	Column: 20 × 0.5 cm	
	After passage of the sample through the resin bed, sulfate is determined spectrophotometrically in a 1-mℓ aliquot of the percolate	

Table 4
DETERMINATION OF SULFATE IN NATURAL WATERS AFTER CATION EXCHANGE SEPARATION USING TWO COLUMN PROCEDURES

Material	Ion exchange resin, separation conditions, and remarks	Ref.
Mineral waters and seawater	Strongly acidic cation exchanger Merck I (Ag^+ and H^+ forms) Columns: ~15 × 0.5 cm	40
	A) Column separation (two columns connected in series, the first containing the resin the Ag^+ form, and the second in the H^+ form) (flow rate: 1—3 mℓ/min)	
	After passage of the sample (50—100 mℓ, containing 1—100 mg of Cl^-) through both columns the effluent containing the sulfate is free from chloride (removed on the Ag-form resin) and HCO_3^- (decomposed on the H^+-form resin)	
	B) Batch separation	
	To the sample (50—100 mℓ) 0.5—1 g of the resin in the Ag^+ form is added and the mixture is stirred magnetically for 20—30 min; then the mixture is filtered and to the filtrate the resin in the H^+ form (0.5—1.0 g) is added; after stirring for ~20—30 min the resin is removed by filtration obtaining a filtrate containing the sulfate (free from Cl^- and HCO_3^-)	
	After separation (A) or (B) sulfate is determined titrimetrically using the solid-state Pb-selective electrode; the error is generally < 5%; down to 10 ppm of sulfate can be determined in the presence of up to a tenfold excess of chloride; phosphate interfers	
Fresh water	Rexyn® 101 or equivalent sulfonic acid-type resin (16-50 mesh; H^+ and Ag^+ forms)	41
	Two columns of 0.8 cm OD containing each ~2 mℓ of the resin to a height of ~10 cm (contained in an automated continuous-flow analytical system)	
	A) Separation on H^+-form resin	
	On this resin cations, e.g., Ca and Mg, are exchanged for H^+ ions yielding an acidic solution of H_2SO_4 and HCl which are measured by electric conductance	
	B) Separation on Ag^+-form resin	
	Cations are exchanged for Ag^+ ion with Cl^- precipitating as AgCl and remaining in the resin bed; the effluent from this column contains an H^+-ion concentration which is equivalent to sulfate (measurement by electric contuctance)	
	The difference in H^+-ion concentration between the effluents from columns (A) and (B) is equivalent to the chloride content	
	The same separation procedure can also be performed manually[42-44]	

Table 5
DETERMINATION OF SULFATE IN BIOLOGICAL MATERIALS AFTER CATION EXCHANGE SEPARATION

Material	Ion exchange resin, separation conditions, and remarks	Ref.
Urine	Dowex® 50-X8 a) Digested urine sample adjusted to pH 2—3 with NH_3 solution (adsorption of cationic constituents such as ^{40}K; SO_4^{2-} and PO_4^{3-} pass into the effluent) b) Water (elution of remaining SO_4^{2-} and PO_4^{3-}) After precipitation as $BaSO_4$ and $BiPO_4$, respectively ^{35}S and ^{32}P are determined by liquid scintillation counting; the corresponding recoveries are 95 and 70%	8
	Zeo-Karb® 225 (52-100 mesh; H^+ form) Column: 10 × 1 cm a) Acid-hydrolyzed urine sample (for total sulfate) or untreated sample (for inorganic sulfate) (<3 mℓ in each case) (adsorption of cationic constituents; SO_4^{2-} passes into the effluent) b) Water (15—20 mℓ) (elution of remaining SO_4^{2-}) Sulfate is determined titrimetrically	2
Urine and feces	Dowex® 50W-X4 (H^+ form) a) Aqueous sample solution (5 mℓ) (adsorption of cationic constituents; SO_4^{2-} and phosphates pass into the effluent) b) Water (20 mℓ) (elution of remaining SO_4^{2-} and phosphates) Sulfate is determined titrimetrically Before the ion exchange separation, the sample (containing 2 — 8 μeq of S) is subjected to Schöniger combustion	17

Table 6
DETERMINATION OF SULFITE AND S OXIDES IN BIOLOGICAL MATERIALS AFTER CATION EXCHANGE SEPARATION

Material	Ion exchange resin, separation conditions, and remarks	Ref.
Biological materials (e.g., liver extract)	Dowex® 50W Column: 6 × 0.7 cm a) Aqueous sample solution (0.5 mℓ) (adsorption of cationic constituents; S-sulfo-L-cystein passes into the effluent) b) Water (6 × 0.5 mℓ) (elution of residual S-sulfo-L-cystein which is then determined with ninhydrin) The recovery is 96.8 ± 0.3%; before the separation, the sample is deproteinized with anhydrous acetic acid in ethanol and the supernate is treated with S-([2-amino-2-carboxyethyl] sulphonyl)-L-cystein which reacts with sulfite (which is to be determined); after 30 min the mixture is centrifuged, the supernatant solution is evaporated, and the residue (containing the reaction product, i.e., S-sulfo-L-cystein) is dissolved in water, thus obtaining solution (a)	134
Plants (e.g., spruce needles)	Dowex® 50W (H^+ form) Column: 10 × 1.5 cm The sample solution (5 mℓ of aqueous H_2O_2 in which the oxides of S were absorbed) is passed through the resin bed to adsorb cationic constituents and in the effluent H_2SO_4 is determined spectrophotometrically with dimethylsulfonazo III Before the separation, the dried sample (≈100 mg) is subjected to Schöniger combustion; the method was applied in the determination of adsorbed atmospheric pollutants	18

Table 7
DETERMINATION OF S IN INDUSTRIAL EMISSIONS AND EFFLUENTS AFTER CATION EXCHANGE SEPARATION[a]

Material	Ion exchange resin, separation conditons, and remarks	Ref.
Fly ash	A) Cation exchange separation (Dowex® 50-X8, 100-200 mesh; H$^+$ form) Column: 10 × 1 cm operated at a flow rate of 0.5 mℓ/min a) 0.00375 M HCl sample solution (5 mℓ) (adsorption of cationic impurities, e.g., ^{24}Na,^{140}La,^{153}Sm, and ^{46}Sc; ^{35}SO$_4^{2-}$ and other anions pass into the effluent) b) 0.00375 M HCl (20 mℓ) (elution of residual sulfate and other anions) B) Anion exchange separation (Dowex® 1-X8, 100-200 mesh, Cl$^-$ form) Column: 10 × 1 cm operated at a flow rate of 1 mℓ min a) 0.00375 M HCl sample solution (5 mℓ, obtained after evaporation of eluates [a] and [b] following separation [A]) (adsorption of sulfate and impurity anions) b) 0.1 M HCl (100 mℓ) (elution of ^{82}Br$^-$ and H$_2$^{32}PO$_4^-$) c) 2 M HCl (30 mℓ) (elution of sulfate) S is determined radiometrically; the mean chemical yield for the ^{35}S separation was 98%; before the cation exchange separation, the sample is irradiated with neutrons and decomposed with HCl-HNO$_3$-HF	6
Diesel-engine exhaust particulates	Amberlite® CG-120 (100-200 mesh; H$^+$ form) The filter sample is extracted by ultrasonic vibration with 50% isopropanol (100 mℓ) for 1 hr at ambient temperature and to the extract 5 mℓ of the resin is added and digestion is carried out for 3 hr at 60°C to adsorb metal cations; after removal of the resin by filtration, sulfate is determined titrimetrically using the Ba(ClO$_4$)$_2$-thoron method Before the ion exchange separation the particulates are collected on a Si-fiber filter	82
Stack gases	Dowex® 50-X16 (50-100 mesh; H$^+$ form) Column: 10 × 3 cm After adsorption of total S oxides in a 1:1 mixture (50 mℓ) of 3% H$_2$O$_2$ solution and 0.2 M NaOH, the solution is boiled to destroy peroxide, acidified with HCl, boiled to remove CO$_2$, and diluted to 500 mℓ; an aliquot (75 mℓ) of this solution is passed through the resin bed to adsorb Na and the H$_2$SO$_4$ passing into the effluent is determined titrimetrically	45
Mining effluents	Dowex® 50-X8 (50-100 mesh; H$^+$ form) Column: 20 × 1 cm a) Aqueous sample solution (adsorption of metal ions; sulfate passes into the effluent) b) Distilled water ($\simeq$60 mℓ) (elution of residual sulfate) From the eluate, sulfate is precipitated by adding excess BaCl$_2$ and the unconsumed Ba is titrated with EDTA (o-cresolphthalein complexan as indicator) The complete method consists in determining (1) the total sulfate obtained from the initial sulfate plus that produced by oxidizing SO$_3^{2-}$ with iodine; (2) the total sulfate obtained by oxidizing the polythionate ions including the S$_4$O$_6^{2-}$ produced by oxidizing S$_2$O$_3^{2-}$ with I as in (1), with acidic H$_2$O$_2$; and (3) the total sulfate obtained by oxidizing all the thio-salts, including S$_2$O$_6^{2-}$ with H$_2$O$_2$, KClO$_4$, and conc HNO$_3$; in each instance metal ions are removed after the oxidizing step on the cation exchange resin column	19

Table 7 (continued)
DETERMINATION OF S IN INDUSTRIAL EMISSIONS AND EFFLUENTS AFTER CATION EXCHANGE SEPARATION[a]

Material	Ion exchange resin, separation conditons, and remarks	Ref.
Effluent from paper manufacture	Dowex® 50-X8 (H⁺ form) Column: 30 × 1.1 cm containing 10 mℓ of the resin The filtered water sample (50—100 mℓ) is passed through the resin column to adsorb cations and in the effluent (the first 30 of which are discarded) sulfate is determined titrimetrically	69
Industrial waters	Dowex® 50W-X2 (50-100 mesh; Na⁺ form) Column: 5 × 1.5 cm operated at a flow rate of 10—25 mℓ/min The sample (100 mℓ) is passed through the resin (to adsorb cations) and the first 50 mℓ of effluent is discarded; in the residual effluent sulfate is determined spectrophotometrically.	70
Lixivation water from blast-furnace slags	Dowex®50 (H⁺ form) Column: 15 × 2 cm containing 30 mℓ of the resin and operated at a flow rate of 5—6 mℓ/min a) Sample (50—100 mℓ) acidified with 20% acetic acid (10 mℓ) and to which 0.1 N I solution has been added until it assumes a permanent yellow color (adsorption of Ca; sulfate passes into the effluent) b) Distilled water (20—30-mℓ portions until the effluent is neutral) (elution of residual sulfate) Sulfate is determined gravimetrically as BaSO₄	13

[a] See also Table 8 in the chapter on Nitrogen; Table 6 in the chapter on Boron; and Table 11 in the chapter on Sodium, Volume V.

Table 8
DETERMINATION OF S IN SYNTHETIC ORGANIC COMPOUNDS AFTER CATION EXCHANGE SEPARATION

Material	Ion exchange resin, separation conditions, and remarks	Ref.
Organic compounds such as sulphanilic acid and phenoxymethyl penicilin	Amberlite® IR-120 (30-60 mesh; H⁺ form) Column: 8 × 0.7 cm operated at a rate of 1 mℓ/min a) Aqueous sample solution (5 mℓ) (adsorption of cations; sulfate passes into the effluent) b) Water (25 mℓ) (elution of residual sulfate) Sulfate is determined titrimetrically Before the separation, the sample (2—5 mg) is wet ashed with fuming HNO$_3$ (Carius tube), the residue is treated with 0.02 M NH$_3$ solution (5 mℓ), and after evaporation to dryness the final residue is taken up in water to prepare solution (a)	20
Organic compounds	Amberlite® IR-120 (A.R.) (H⁺ form) The sample (3—10 mg) is heated with Na$_2$O$_2$ (0.8 g), the melt is dissolved in water, and the alkaline solution (50 mℓ) is stirred for 5 min with 27 g of the resin; in an aliquot of the supernate sulfate is determined titrimetrically with Ba (thoron indicator)	84
Pharmaceutical products (antibiotic and alkaloidal sulfates)	Dowex® 50-X4 or -X8 (200-400 mesh; H⁺ form) Column: 20 × 1.5 cm a) Aqueous solution (10 mℓ) of the sample (200 mg) (adsorption of cationic constituents; H$_2$SO$_4$ passes into the effluent) b) Water (≃50 mℓ) (elution of remaining H$_2$SO$_4$) Sulfate is determined titrimetrically	21
Carbohydrates (e.g., monosaccharide sulfate and sulfated polysaccharides)	Bio-Rad® AG 50W-X8 (200-400 mesh; H⁺ form) Column containing 1 mℓ of the resin a) Aqueous sample solution (250 µℓ containing ≃3 mg of the sample) (adsorption of cationic constituents; SO$_4^{2-}$ passes into the effluent) b) Water (4.75 mℓ) (elution of remaining SO$_4^{2-}$) Sulfate is determined by flame photometry	22

Table 9
DETERMINATION OF S IN Pu AND U AFTER CATION EXCHANGE SEPARATION

Material	Ion exchange resin, separation conditions, and remarks	Ref.
Pu-sulfide and Pu-U-sulfide ceramic fuel	Dowex® 50-X8 (100-200 mesh; H⁺ form) Column: 10 × 0.7 cm a) ≃0.5 M HNO$_3$ sample solution (adsorption of Pu; sulfate passes into the effluent) b) Water (elution of remaining sulfate) In the combined effluent (≃40 mℓ) from (a) and (b) sulfate is titrated amperometrically; the recovery of S is ≃99.7% Before the separation, the sample (0.3—0.5 g) is dissolved in Br-saturated conc HNO$_3$ at ≃ −120°C	79
Pu solutions	Dowex® 50-X8 (50-100 mesh; H⁺ form) Column: 10 × 0.4 cm conditioned with 0.5 M HNO$_3$ a) 0.5—1 M HNO$_3$ sample solution (adsorption of Pu; sulfate passes into the effluent) b) 0.1 M HNO$_3$ (10 mℓ) (elution of residual sulfate) Sulfate is determined by direct titration with Pb nitrate (PAN as indicator)	80
Uranyl salts	Dowex® 50-X12 (200-400 mesh; H⁺ form) Column of 2 cm ID containing ~16 mℓ of the resin and operated at a flow rate of 8 mℓ/min a) Sample solution (35 mℓ, containing 1 g of UO$_3$ dissolved in a measured volume of standard acid of the anion under consideration) (adsorption of UO$_2$[II]; the anions pass into the effluent) b) Water (65 mℓ) (elution of remaining anions) The acid effluent is titrated with NaOH to determine one of the following anions: SO$_4^{2-}$, NO$_3^-$, Cl$^-$, ClO$_4^-$ and Cr$_2$O$_7^{2-}$	104

Table 10
DETERMINATION OF S IN VARIOUS INDUSTRIAL MATERIALS AFTER CATION EXCHANGE SEPARATION USING BATCH ADSORPTION PROCEDURES

Material	Ion exchange resin, separation conditions, and remarks	Ref.
Pb sulfate (Pb storage battery plates)	Amberlite® IR-120 (particle size: 0.15—0.3 mm; H⁺ form) Column of 1 cm ID containing the resin to a height of 2—3 cm a) The Na_2CO_3 sample solution is boiled with a batch of the resin for ≃20 min (adsorption of bulk of Na and removal of CO_2) and then the slurry is transferred to the column (H_2SO_4 passes into the effluent) b) Distilled water (50 mℓ) (elution of residual H_2SO_4) H_2SO_4 is determined by titration with 0.1 M NaOH; before the separation, the plate sample (~6 g) is boiled for 1 hr in a solution containing 0.5 g of Na_2CO_3 ($PbSO_4$ + Na_2CO_3 → $PbCO_3$ + Na_2SO_4); then the solution is saturated with CO_2 and the precipitated $PbCO_3$ is filtered off; this filtrate is the sample solution (a)	100
Cements, pigments, rubber articles, and pyrites	Amberlite® IR-120 (H⁺ form) The sample (0.5—0.7 g of cement) is fused with Na_2O_2 and Na_2CO_3 (1 + 1) (8 g), then extracted with water and made up to 200 mℓ; a 50-mℓ portion is treated with a batch of the resin to remove most of the CO_2 and the slurry is transferred to a column (18 × 1.8 cm) containing the same resin. In the effluent and washings (100 mℓ of water) H_2SO_4 is determined titrimetrically In the analysis of pyrite, the sample (0.3—0.7 g) is fused with Na_2O_2 (5 g) and the melt is extracted with water and the solution boiled to decompose H_2O_2; the resulting solution is then passed through a column of the resin to remove Fe and other elements	85, 86
Portland cement and clinker	Zeo-Karb® 225 (14-52 mesh; H⁺ form) To a suspension of the resin (1 g) in water (5 mℓ), the sample (5—50 mg containing ≃2 mg of SO_3) is added and the mixture is stirred slowly with a magnetic stirrer for 3 hr at room temperature; after filtration, sulfate in the filtrate is determined conductimetrically; fluoride if present is determined spectrophotometrically	74
Mixtures of gypsum, anhydrite, CaO, and CaS	Dowex® 50 (H⁺ form) An aqueous suspension of the sample (≃0.3 g) and the resin (25 g) is stirred for ≃15 min at 90°C and then applied to a short column of the same resin to complete the adsorption of Ca; H_2SO_4 in the effluent is determined by titration with 0.1 M NaOH	75
Coal ash, boiler deposits, and precipitator dusts	Zeo-Karb® 225 (H⁺ and Na⁺ forms for P_2O_5 contents of ≤1 and >1%, respectively) The sample (20 mg; −200 mesh) is mixed with 1.5 g of the resin and water (20 mℓ) and the mixture is shaken mechanically for 1 hr to adsorb the cationic constituents; then, in an aliquot of the supernate, sulfate is determined spectrophotometrically with Ba chloranilate	76

Table 11
DETERMINATION OF S IN INDUSTRIAL PRODUCTS AFTER CATION AND ANION EXCHANGE SEPARATIONS

Material	Ion exchange resin, separation conditions and remarks	Ref.
High-purity metals (Al, Cu, Mg, Fe, and Ni)	A) First column operation (Dowex® 50-X8, 200-400 mesh; H⁺ form) Column: 25 × 0.8 cm From ≃12 M HCl ^{46}Sc is retained by the cation exchange resin while the radionuclides of S, P, and other elements pass into the effluent B) Second column operation (Dowex® 1-X8; similar column as in [A]) On passing the 12 M HCl effluent from (A) through the resin bed ^{122}Sb, ^{233}Pa, ^{59}Fe, ^{60}Co, and ^{65}Zn are adsorbed; ^{35}SO$_4^{2-}$ and ^{32}PO$_4^{3-}$ pass into the effluent in which they are determined radiometrically Before the first ion exchange separation step the sample is irradiated with neutrons	149—151
Mixtures containing S$_2$O$_3^{2-}$, SO$_3^{2-}$, and SO$_4^{2-}$	A) Separation on cation exchange resin Column: 15 × 1 cm containing the resin in the Sr^{2+} form a) Aqueous sample solution (retention of sulfite and sulfate; thiosulfate passes into the effluent) b) Water (elution of residual thiosulfate) B) Separation on anion exchange resin Column containing the resin in the Cl$^-$ form a) Aqueous sample solution (retention of the S-anions) b) 0.05 M Sr-nitrate or -chloride solution (elution of thiosulfate)	102
CuSO$_4$	Coupled columns containing Mykion® PS (1.5 mℓ) (grain size: 0.1—0.2 mm; Na⁺ form) (upper column) and Mykion® PA (grain size: 0.1—0.2 mm; Cl$^-$ form) (lower column) and operated at a flow rate of ~2 mℓ/min; the diameter of the columns is 0.7 cm a) Sample solution (4—5 mℓ containing ~8—16 mg of CuSO$_4$) (adsorption of Cu[II] on upper column; Na$_2$SO$_4$ passes into the lower resin column where SO$_4^{2-}$ is adsorbed, liberating an equivalent amount of Cl$^-$ which as NaCl passes into the effluent from both columns) b) Distilled water (15 mℓ) (elution of remaining NaCl) Sulfate is determined indirectly via argentiometric titration of chloride	154

Table 12
DETERMINATION OF SULFIDE AND SULFATE IN ENVIRONMENTAL MATERIAL AFTER ANION EXCHANGE SEPARATION[a]

Material	Ion exchange resin, separation conditions, and remarks	Ref.
Air and water	Amberlite® IRA-400 (20-50 mesh; OH⁻ form) Column: 1 × 0.8 cm containing 0.5 g of the resin and which is attached to a flow meter and vacuum pump a) Air sample (retention of H_2S) (flow rate: 1 mℓ/min for H_2S concentrations of 1—20 ppm, and 30 mℓ/min for 0.07 ppm until at least 1 µg of H_2S is adsorbed) b) 4 M NaOH (10 mℓ) (elution of sulfide at a rate of 1 mℓ/min) Sulfide is determined spectrophotometrically; the method is applicable to the determination of 1—20 µg of H_2S in air or water at concentrations of 0.07—20 ppm and down to 0.1 part per 10^9; the ion exchange column can retain the H_2S for up to 10 days without loss; in the analysis of waters, the sample is passed through the column at 30—100 mℓ/min until 1—20 µg of H_2S is adsorbed	189
Underground water	Amberlite® IR-143 (weakly basic anion exchange resin; 50-100 mesh; chloride form) Column: 2—2.5 × 1.5 cm operated at a flow rate of 5 mℓ/min, except with eluent (c) which is passed at a rate of 1 mℓ/min a) Sample (500 mℓ, acidified with 1—2 mℓ of 6 M HCl and containing <100 mg of sulfate per liter) (adsorption of sulfate; cationic constituents pass into the effluent) b) 0.1 M HCl (20 mℓ) (elution of phosphate and remaining cations) c) 0.5 M ammonia solution (30 mℓ) (elution of sulfate) Sulfate is determined gravimetrically as $BaSO_4$ Eluate (c) also contains chloride and nitrate	177
Water	Anionite AV-17 Column: 40 × 1 cm a) Water sample containing 50—100 µg SO_4^{2-} per liter (adsorption of sulfate; cationic constituents pass into the effluent) b) 6% ammonium carbonate solution (50 mℓ) (elution of SO_4^{2-}) Sulfate is determined titrimetrically	183
Plants (e.g., hay)	Amberlite® IRA-400 (Cl⁻ form) a) 0.01 M HCl sample solution (adsorption of SO_4^{2-}; cationic constituents pass into the effluent) b) 3.5% NaCl solution (elution of SO_4^{2-}) Sulfate is determined spectrophotometrically; with this method from 0.1—2.4% of SO_4^{2-} in the sample can be determined	178

[a] See also Table 1 in the chapter on Nitrogen and Table 25 in the chapter on Actinides, Volume II.

Table 13
DETERMINATION OF SULFATE IN INDUSTRIAL PRODUCTS AFTER ANION EXCHANGE SEPARATION[a]

Material	Ion exchange resin, separation conditons, and remarks	Ref.
High-purity Se	Dowex® 1-X8 (60-100 mesh; chloride form) Column: 7 × 0.8 cm a) Aqueous sample solution (≤0.1 M in HCl) (~15 mℓ) (adsorption of sulfate; SeO_3^{2-} passes into the effluent) b) Water (20 mℓ) (elution of Se) c) 0.1 M HCl (20 mℓ) (elution of remaining Se) d) Water (5 mℓ) (as a rinse) e) 1 M HCl (30 mℓ) (elution of sulfate) Eluent (e) is treated with a Ba-chromate suspension (which precipitates sulfate as $BaSO_4$) and the liberated amount of chromate is determined spectrophotometrically using diphenylcarbazide Before the separation, the sample (2 g) is decomposed with HNO_3 + Br (conversion of S [>0.5 ppm] to sulfate)	180
Fe alloys	Amberlite® IRA-410 (480-490 mesh; Cl^- form) Column operated at a flow rate of 20—30 drops/min a) Slightly acidified (HCl) sample solution (300—500 mℓ) (adsorption of sulfate and phosphate; Fe and other elements pass into the effluent) b) Water (250—300 mℓ) (elution of residual nonadsorbed elements) c) 0.5 M NaOH (100 mℓ) followed by water (until neutral) and 0.5 M HCl (100 mℓ) and again water (elution of SO_4^{2-} and PO_4^{3-}; the alkaline and acid elutions are repeated twice) Sulfate and phosphate are determined gravimetrically; before the separation, the sample (3—5 g) is dissolved in HNO_3 and nitrates are removed by repeated evaporations with HCl	176

[a] See also Table 2 in the chapter on Tungsten, Volume IV.

Table 14
DETERMINATION OF SULFITE AND OTHER S SPECIES IN INDUSTRIAL PRODUCTS AFTER ANION EXCHANGE SEPARATION

Material	Ion exchange resin, separation conditions, and remarks	Ref.
Aqueous solutions of SO_2	Anion exchange resin Amberlite® IRA-461 (Cl^- form containing Tl[III] adsorbed as $TlCl_4^-$ and which also contains radioactive ^{204}Tl [0.5 mCi]) Column: 10 × 1 cm containing 1 g of the resin On passing the sample solution (1 mℓ, containing 2—12 ppm of SO_2) through the resin bed an equivalent amount of Tl(III) is reduced to nonadsorbed Tl(I) which passes into the effluent where it is determined radiometrically (SO_2 is the most effective eluent for Tl adsorbed as Tl[III]; see in the chapter on Thallium) The amount of radioactive ^{204}Tl contained in this effluent is proportional to the SO_2 concentration of the sample The column should be washed with 0.2 M HCl between determinations and should be protected from sunlight (to prevent reduction of Tl[III] by the resin)	193
Neutron-irradiated alkali chlorides, S-containing salts, and ionic crystals doped with ^{35}S	Rexyn® 201 anion exchange resin (100-200 mesh; Cl^- form) Column operated at a flow rate of 12—15 mℓ/min a) 0.04 M NaCl at pH 10.6 (elution of sulfide) b) 0.04 M NH_4NO_3 containing 30% of acetone and maintained at pH 11 (elution of sulfite) c) 0.2 M NaCl (elution of sulfate) d) 1 M NaCl (elution of thiosulfate) e) 1 M $NaClO_4$ (elution of thiocyanate)	203
Alkali halides	Amberlite® IRA-410 (chloride form) Column: 8 × 1 cm operated at a flow rate of ~20 drops/min a) 0.3 M NaOH sample solution (adsorption of SO_3^{2-}, S^{2-}, SO_4^{2-}, and $S_2O_3^{2-}$) b) Water (as a wash) c) 0.2 M NH_4NO_3 - 30% acetone (adjusted to pH ~10) (elution of SO_3^{2-} followed by S^{2-}) d) 0.1 M $NaNO_3$ (elution of SO_4^{2-} followed by $S_2O_3^{2-}$) The ^{35}S activity in each fraction is determined radiometrically Before the separation, the sample (≃500 mg of NaCl, KCl, or RbCl) is irradiated with neutrons ($^{35}Cl[n,p]^{35}S$) and then dissolved in de-aerated 0.3 M NaOH in the presence of carriers for the various S species that may be present	202
Photographic gelatin	Amberlite® IRA-400 (50-100 mesh; Cl^- form) a) 7% gelatin solution (1 mℓ) (adsorption of $S_2O_3^{2-}$, polythionates, and sulfites) b) Water at 37°C (elution of gelatin) c) 20% NaCl solution (elution of $S_2O_3^{2-}$ and sulfites; polythionates are not eluted) Thiosulfate is determined titrimetrically	205

Table 15
MUTUAL SEPARATION OF OXYANIONS OF S BY ANION EXCHANGE

Elements separated	Ion exchange resin, separation conditions, and remarks	Ref.
Sulfite, thiosulfate, trithionate, tetrathionate, and pentathionate	Deacidite® FF-X2 (52-100 mesh BSS; chloride form) Column: 20 cm × 0.75 cm^2 After adsorption of the anionic S species the following eluents (250 mℓ each) are used for their sequential desorption a) 2 M K-H-phthalate (elution of SO_3^{2-} and $S_2O_3^{2-}$) c) 3 M HCl (elution of $S_3O_6^{2-}$) d) 6 M HCl (elution of $S_4O_6^{2-}$) e) 9 M HCl (elution of $S_5O_6^{2-}$) A similar scheme has been used for the qualitative separation of tri-, tetra-, penta-, and hexa-thionates[211]	210
Polythionates	Dowex® 1-X2 (50-100 mesh; chloride form) Column: 11 × ~0.9 cm containing 2.16 g of the resin After adsorption of the polythionate species the following eluents are used for their fractionation a) 1 M HCl (60 mℓ) (elution of $S_2O_6^{2-}$) b) 3 M HCl (60 mℓ) (elution of $S_3O_6^{2-}$) c) 6 M HCl (50 mℓ) (elution of $S_4O_6^{2-}$) d) 9 M HCl (50 mℓ) (elution of $S_5O_6^{2-}$) Other oxy-acids (SO_4^{2-}, SO_3^{2-}, and $S_2O_3^{2-}$) are eluted with eluent (a) before $S_2O_6^{2-}$ appears in the eluate	197
Sulfate, sulfite, thiosulfate, and sulfide	Diaion® SA 100 (100-200 mesh; nitrate form) Column of 0.7 cm ID containing 4.9 g of the resin and operated at a flow rate of 0.1 mℓ/min After adsorption of the anionic species the following eluents are used for their fractionation a) 0.1 M NH$_4$NO$_3$ (80 mℓ) (adjusted to pH 9.7 with NH$_3$ and which is 30% in acetone) (elution of SO_3^{2-} followed by S^{2-}) b) 0.1 M NaNO$_3$ (65 mℓ) (elution of SO_4^{2-}) c) 1 M NaNO$_3$ (20 mℓ) (elution of $S_2O_3^{2-}$) The recovery is >97% for each ion; for a mixture of sulfate, sulfite, and thiosulfate the column is eluted first with 0.1 M NaNO$_3$ (100 mℓ) adjusted to pH 9.7 which removes sulfite and sulfate in that order, and then with eluent (c) to desorb the thiosulfate	212
S^{2-}, SO_3^{2-}, SO_4^{2-}, and $S_2O_3^{2-}$	Dowex® 1-X8 (100-200 mesh; Cl$^-$ form) Column: 18.5 × 0.85 cm After adsorption of the S-anions the following eluents are used a) 0—0.1 M KCl (gradient elution of sulfide, sulfite, and sulfate) b) 1 M KCl (elution of thiosulfate)	213

Table 16
ANION EXCHANGE SEPARATION OF SULFATE AND SULFITE FROM OXYANIONS OF OTHER ELEMENTS

Elements separated	Ion exchange resin, separation conditions, and remarks	Ref.
SO_4^{2-} from TeO_4^{2-}	Amberlite® IRA-400 (80-120 mesh; chloride form) Column: 5 × 1 cm operated at a flow rate of 2 mℓ/min a) Dilute HCl (below 0.1 M) to neutral sample solution containing <10 mg each of the two anions (adsorption of sulfate; tellurate passes into the effluent) b) Water (as a rinse) c) 10% NaCl solution (30 mℓ) (elution of sulfate) This method is applicable to the radiochemical analysis of fission products, especially for the detection of neutron-induced ^{35}S	179
Sulfate from phosphate	Diaion® SA-100 (100-200 mesh; Cl$^-$ form) Column: 14.5 × 1.3 cm operated at a flow rate of 1.8—2 mℓ/min After adsorption of the two anions (milligram amounts) the following eluents are used a) 0.1 M HCl - 0.19 M KCl (60 mℓ) (elution of phosphate) b) 2 M KCl (35 mℓ) (elution of sulfate)	185
SO_4^{2-} from CrO_4^{2-}	Anionite AV-17 (or AV-28) (OH$^-$ form) Column: 30 × 0.3 cm containing 1 g of the resin and operated at a flow rate of 0.5—0.6 mℓ/min a) Sample solution containing ≃1 mg of S and Cr per milliliter (adsorption of SO_4^{2-} and CrO_4^{2-}) b) 0.25 M NaCl (elution of SO_4^{2-}) c) 1 M NaOH (elution of CrO_4^{2-})	187
Sulfite from selenite and tellurite	Dowex® 1-X8 (100-200 mesh; OH$^-$ form) Column: 13 × 1 cm containing 5 g of the resin and operated at a flow rate of 0.1 mℓ/min After adsorption of the anionic species the following eluents are used for their separation a) 0.5 M NaOH - 3 M ammonia solution (220 mℓ) (elution of tellurite and selenite which are contained in the 100—130- and 140—210-mℓ fractions, respectively) b) 2 M NaOH (40 mℓ) (elution of sulfite)	214

Table 17
MICROCHEMICAL DETECTION OF SULFIDE BY RESIN SPOT TESTS[a]

Ion exchange resin	Experimental conditions and remarks	Ref.
Amberlite® IRA-400 (OH⁻ and nitroprusside forms) and Dowex® 50-X8 (20-50 mesh; Ag⁺ form)	The test solution (5 or 6 drops; aqueous solution of fusion with Na of an organic compound containing S, Cl, I, and N) is added to the anion exchange resin in the nitroprusside form (three to four beads on a spot plate) which turns violet and later on becomes black if sulfide is present; subsequently, fresh beads of the same resin are added until all of the S has been removed from the solution; the resulting solution which is free from S is divided into two parts; to one part, four to five beads of Dowex® 50 are added, and if Cl⁻ is present a white precipitate (AgCl) develops on the resin surface which, after 2 or 3 min, turns black; to the second part four to five beads of the anion exchanger in the OH⁻ form are added followed by 30% H_2O_2 (1 drop) and Cl-water (a few drops) to oxidize iodide to I which is strongly retained by the resin beads; these are then separated from the solution and mixed with 1% starch solution in 10% acetic acid (1 drop); the beads turn deep violet if the mixture contains I; the solution phase (from which the beads were removed) is evaporated and to the residue the following reagents are added in this order: water (1 drop), four to five resin beads in the OH⁻ form, 0.01% $CuSO_4$ (1drop), 5% NH_4-molybdate solution (few drops), and conc HCl until a blue color appears on the resin surface which shows the presence of N To prepare the nitroprusside form of the resin the exchanger in the OH⁻ form is soaked in 20% alcoholic solution of $Na_2Fe(CN)_5NO$ in 50% alcohol	215
Dowex® 50-X8 (50-100 mesh; H⁺ form	A small quantity of solid *p*-aminodimethylaniline or *p*-phenylenediamine is placed on a spot plate with 1 drop of 0.1 *M* $FeCl_3$ in 12 *M* HCl and set aside for ~5 minutes; then a few beads of the resin are added, followed after a few minutes by 1 drop of the sample solution; in the presence of sulfide a green or blue-green color appears after a few minutes on the edges of the resin, being readily observable with a hand lens With *p*-aminodimethylaniline the detection limit is 0.05 µg of sulfide and the limiting concentration is 1 in 0.8×10^6	216

[a] See also Table 80 in the chapter on Copper, Volume III.

REFERENCES

1. **Ionescu, M., Demetrescu, A., and Mitran, E.,** Application of ion-exchange resin to the Lunge method for the determination of sulphur in ores, *Rev. Minelor (Bucharest)*, 9, 473, 1958.
2. **Morrison, A. R.,** Titrimetric method for determination of inorganic and total sulfates in urine, *Lab. Pract.*, 21(10), 726, 1972.
3. **Kotik, F. I. and Kutneva, E. R.,** Rapid volumetric determination with nitchromazo as indicator of sulfate in copper-plating electrolytes containing ethylenediamine, *Zavod. Lab.*, 34(1), 33, 1968.
4. **Kšir, O., Jankovský, J., and Follpracht, K.,** Determination of sulphur with the use of ion-exchange, *Rudy*, 6(2, Add.1), 1, 1958.
5. **Houzim, V. and Zeman, P.,** Rapid determination of total sulfur in fuel, *Chem. Prum.*, 16(3), 167, 1966.
6. **Li, M. and Filby, R. H.,** Separation of sulphur-35 from multiradionuclide solutions by ion-exchange acid alumina adsorption, *Radiochem. Radioanal. Lett.*, 59(1), 35, 1983.
7. **Grindel, N. M. and Dzysyuk, N. V.,** Titrimetric determinatin of sulphur in soils with chlorophosphonazo III as indicator, *Pochvovedenie*, No. 9, 138, 1980.
8. **Kramer, G. H.,** Separation of sulphur-35 and phosphorus-32 from urine and their subsequent estimation, Report AECL-7162, Atomic Energy of Canada Ltd., 1981.
9. **Koita, T., Miyata, H., and Toei, K.,** Determination of sulfate ion in river water with the barium salt of 1-anthraquinoneazo R acid, *Jpn. Analyst*, 29(3), 176, 1980.
10. **Dedkova, V. P., Akimova, T. G., Nikitina, I. E., Paskhina, S. I., and Savvin, S. B.,** Determination of sulfate in solutions of high mineral content containing lanthanoids with use of orthanilic K, *Zh. Anal. Khim.*, 38(1), 166, 1983.
11. **Procházková, L.,** Use of the chloranilic acid method for the colorimetric determination of sulfate in water, *Fresenius' Z. Anal. Chem.*, 182, 103, 1961.
12. **Magar, W. Y. and Pollard, A. G.,** Colorimetric determination of sulfate in Morgan extracts of soil, *Chem. Inc.*, No. 16, 505, 1961.
13. **Blasius, E., Wagner, H., and Ziegler, K.,** Determination of sulfur compounds in lixivation water from blast-furnace slags, *Arch. Eisenhuettenwes.*, 42(7), 473, 1971.
14. **Ahmed, S. M. and Whalley, B. J. P.,** Analysis for total sulphur in Canadian coals by a modified oxygen-flask method using arsenazo III, *Fuel (London)*, 51(3), 190, 1972.
15. **Ahmed, S. M. and Whalley, B. J. P.,** Volumetric finish for Eschka analyses using arsenazo III, *Fuel (London)*, 51(4), 334, 1972.
16. **Ahmed, S. M. and Whalley, B. J. P.,** Use of arsenazo III in the determination of total sulphur in coal by oxygen-flask method, *Fuel (London)*, 48, 217, 1969.
17. **Camien, M. N.,** Titrimetric micro-determination of sulfur in biological materials, *Anal. Biochem.*, 15(1), 127, 1966.
18. **Bartels, U. and Thi, T. P.** Spectrophotometric determination of sulphur in plants using Schöniger combustion and dimethylsulphonazo III, *Fresenius' Z. Anal. Chem.*, 310, 13, 1982.
19. **Makhija, R. and Hitchen, A.,** Titrimetric determination of sulfate, thiosulfate and polythionates in mining effluents, *Anal. Chim. Acta* 105, 375, 1979.
20. **White, D. C.,** Titrimetric micro-determination of sulfate using lead nitrate as titrant and dithizone as indicator, *Mikrochim. Acta*, No. 2, 254, 1959.
21. **Roets, E. and Vanderhaeghe, H.,** Titrimetric determination of sulfate in pharmaceutical products, *J. Pharm. Pharmacol.*, 24(10), 795, 1972.
22. **Barker, S. A., Kennedy, J. F., Somers, P. J., and Stacey, M.,** Method for the micro-determination of sulfate content of carbohydrates, *Carbohydr. Res.*, 7, 361, 1968.
23. **Donskaya, E. V. and Volkova, M. G.,** Determination of sulphuric anhydride in Portland cement by means of a cationite, *Zh. Prikl. Khim.*, 29(10), 1598, 1956.
24. **Zabiyako, V. I. and Sharapova, G. N.,** Determination of sulfate ion in chrome alum, *Tr. Ural. Nauchno Issled. Khim. Inst.*, No. 11, 49, 1964.
25. **Soffer, N.,** Determination of dithionate, sulphite and sulfate in manganese leach liquors, *Analyst (London)*, 86, 843, 1961.
26. **Nesh, F. and Haas, E. C.,** Analysis of pickling bath solution using ion exchange, *Anal. Chem.*, 28, 2034, 1956.
27. **Johnson, A. R. and McVicker, G. B.,** Determination of sulphur in inorganic sulfides by using oxygen-flask combustion, *Anal. Chem.*, 38, 913, 1966.
28. **Légrádi, L.,** Determination of elementary sulphur by means of ion-exchange resins, *Magy. Kem. Lapia*, 20(3), 167, 1965.
29. **Kuznetsov, V. V. and Mezhlumyan, P. G.,** Determination of total sulphur and chromium in glasses by spectrophotometric titration, *Zavod. Lab.*, 44(5), 524, 1978.
30. **Lambert, J. L., Yasuda, S. K., and Grotheer, M. P.,** Colorimetric determination of sulfate ion, *Anal. Chem.*, 27, 800, 1955.

31. **Fritz, J. S. and Yamamura, S. S.**, Rapid micro-titration of sulfate, *Anal. Chem.*, 27, 1461, 1955.
32. **Basu, D. K.**, Sulphur-35: its radiochemical estimation in seawater, *J. Indian Chem. Soc.*, 42(4), 263, 1965.
33. **Healy, C. and Atkins, D. H. F.**, Determination of atmospheric sulphur dioxide after collection on impregnated filter-paper, Report AERE-R7956, U.K. Atomic Energy Authority 1975.
34. **Mackellar, W. J., Wiederanders, R. E., and Tallman, D. E.**, Indirect determination of sulfate ion by spectrophotometric titration of excess of barium(II) ions with ethylenediaminetetraacetate, *Anal. Chem.*, 50, 160, 1978.
35. **Pagenkopf, G. K., Brady, W., Clampert, J., and Purcell, M. A.**, Titrimetric determination of sulfate in mineral waters, *Anal. Chim. Acta*, 98, 177, 1978.
36. **Reijnders, H. F. R., van Staden, J. J., and Griepink, B.**, Titrimetry in a continuous flow system. II. Application of the sulfate determination to environmental samples, *Fresenius' Z. Anal. Chem.*, 293, 413, 1978.
37. **Hoevers, J. W.**, Complexometric determination of sulfate in the analysis of drinking water, *Chem. Weekbl.*, 40, 546, 1964.
38. **Madsen, B. C. and Murphy, R. J.**, Flow-injection and photometric determination of sulfate in rain-water with methylthymol blue, *Anal. Chem.*, 53, 1924, 1981.
39. **Reijnders, H. F. R., van Staden, J. J., and Griepink, B.**, Batchwise photometric determination of sulfate in water samples, *Fresenius' Z. Anal. Chem.*, 298, 156, 1979.
40. **Mascini, M.**, Titration of sulfate in mineral waters and seawater by using the solid-state lead electrode, *Analyst (London)*, 98, 325, 1973.
41. **Stainton, M. P.**, An automated method for determination of chloride and sulfate in fresh water using cation exchange and mesaurement of electrical conductance, *Limnol. Oceanogr.*, 19(4), 707, 1974.
42. **Mackereth, F. J. H.**, Ion exchange procedures for the estimation of (I) total ionic concentration (II) chlorides, and (III) sulfates in natural waters, *Mitt. Int. Ver. Theor. Angew.*, 4, 1, 1955.
43. **Mackereth, F. J. H.**, Rapid microestimation of the major anions in freshwater, *Proc. Soc. Water Treat. Exam.*, 4, 27, 1955.
44. **Mackereth, F. J. H.**, Some methods of water analysis for limnologists, *Freshwater Biol. Assoc. Sci. Publ.*, 21, 68, 1963.
45. **Seidman, E. B.**, Determination of sulphur oxides in stack gases, *Annl. Chem.*, 30(10), 1680, 1958.
46. **Beisova, M. P. and Kryukov, P. A.**, Conductimetric titration of sulfates in natural waters, *Gidrokhim. Mater.*, 26, 190, 1957.
47. **Ákos, Szabó, Zs., and Inczédy, J.**, Rapid determination of sulfate in natural water by radio-frequency titration after ion-exchange, *Magy. Kem. Lapia*, 23(9), 528, 1968.
48. **Geyer, R. and Doerffel, K.**, Volumetric determination of sulfate in waters, *Fresenius' Z. Anal. Chem.*, 158, 418, 1957.
49. **Nogina, A. A. and Kobyak, G. G.**, Volmetric semi-micro determination of sulfate in natural water by titration without a buret, *Uch. Zap. Permsk. Univ.*, 25(2), 73, 1963.
50. **Petrova, T. V., Khakimkhodzhaev, N., and Savvin, S. B.**, New organic reagents for the photometric determination of barium, strontium, calcium and sulfate, *Izv. Akad. Nauk SSSR Ser. Khim.*, No. 2, 259, 1970.
51. **Wagner, A.**, Rapid potentiometric determination of sulfate, nitrate and chloride ions in water and in aqueous solutions, *Mitt. Ver. Grosskesselbesitzer*, No. 52, 50, 1958.
52. **Ceauşescu, D.**, Rapid determination of sulfate, chloride and nitrate in water in a single sample, *Fresenius' Z. Anal. Chem.*, 165, 424, 1959.
53. **Matveev, A. A. and Nechiporenko, G. N.**, Simultaneous determination of trace amounts of sulfates and chloride in a single water sample, *Gidrokhim. Mater.*, 33, 134, 1961.
54. **Samchenko, Z. A., Nekryach, E. F., and Gordnovskii, I. T.**, Determination of sulfate and chloride ions in natural water by ion-exchange precipitation chromatography, *Khim. Tekhnol. Vody*, 4(4), 319, 1982.
55. **Nechiporenko, G. N.**, Determination of sulfate ions by direct titration with lead nitrate using diphenylcarbazone indicator, *Gidrokhim. Mater.*, 29, 214, 1959.
56. **Kondo, O., Miyata, H., and Toei, K.**, Determination of sulfate in river-water by flow-injection analysis, *Anal. Chim. Acta*, 134, 353, 1982.
57. **Lewis, W. M.**, Determination of sulfate and chloride in water by direct titration with diphenylcarbazone as indicator, *Proc. Soc. Water Treat. Exam.*, 16, 287, 1967.
58. **Adamski, J. M. and Villard, S. P.**, Applications of the methyl-thymol blue-sulfate method to water and waste-water analysis, *Anal. Chem.*, 47, 1191, 1975.
59. **Merks, A. G. A. and Sinke, J. J.**, Application of automated method for dissolved-sulfate analysis to marine and brackish waters, *Mar. Chem.*, 10(2), 103, 1981.
60. **Merks, A. G. A. and Sinke, J. J.**, Appplication of automated method for dissolved-sulfate analysis to marine and brackish waters, *Mar. Chem.*, 10(2), 103, 1981.

61. **Basargin, N. N. and Nogina, A. A.**, Determination of sulfate in the presence of phosphates in natural and boiler-freed water by direct titration with barium salts, with nitchromazo as indicator, *Zh. Anal. Khim.*, 22(3), 394, 1967.
62. **Wagner, A.**, Rapid potentiometric determination of sulfates, chlorides and nitrates in water, *Bull. Cent. Belge Etud. Docum. Eaux.*, No. 37, 164, 1957.
63. **Budesinsky, B. W.**, Determination of sulfate in waters, *Microchem. J.*, 20(3), 360, 1975.
64. **Johannesson, J. K.**, Note on the determination of low concentrations of sulfate in rain and other waters, *N. Z. J. Sci.*, 1(3), 423, 1958.
65. **Hulanicki, A., Lewandowski, R., and Lewenstam, A.**, Elimination of ionic interference in the determination of sulfates in water using the lead-sensitive ion-selective electrode, *Analyst (London)*, 101, 939, 1976.
66. **White, D. C.**, Micro-determination of sulfate with lead nitrate as titrant and dithizone as indicator, *Mikrochim. Acta*, No. 2, 282, 1960.
67. **Isagai, K.**, Determination of sulfate ion in natural water by use of ion-exchange resin and disodium ethylenediamine-tetra-acetate, *J. Chem. Soc. Jpn.*, 75(6), 613, 1954.
68. **Isagai, K.**, Determination of sulfate ion in natural water containing large amounts of ferric and ferrous ions, *Jpn. Analyst*, 4(3), 171, 1955.
69. **Aldrich, L. C.**, Volumetric method for the determination of sulfates, *TAPPI*, 57(7), 122, 1974.
70. **Blasius, E. and Ziegler, K.**, Photometric and tritrimetric determination of different sulfur compounds dissolved in the same aqueous solution, *Arch. Eisenhuettenwes.*, 44(9), 669, 1973.
71. **Kotik, F. I.**, Determination of sulfate in effluents from electroplating works, *Zavod. Lab.*, 37(5), 541, 1971.
72. **Nicolai, H. and Ellermann, G.**, Titrimetric determination of sulfate in coal ash, slag and cinders by using a strongly acid cation exchanger, *Chem. Tech. (Berlin)*, 18(10), 634, 1966.
73. **Vil'bok, Kh. O.**, Ion-exchange-chromatographic volumetric determination of sulfate in the ash of combustible shales, *Tr. Tallin. Politekh. Inst. Ser. A*, No. 210, 83, 1964.
74. **Bowley, M. J.**, Micro-determination of sulfate and fluoride in lime-silica-sulfate-fluoride phase systems, *Analyst (London)*, 94, 787, 1969.
75. **Robbins, L. A. and Wheelock, T. D.**, Rapid titrimetric determination of sulfate in mixtures of gypsum, anhydrite, calcium oxide and calcium sulfide by using ion exchange, *Anal. Chem.*, 36, 429, 1964.
76. **Schafer, H. N. S.**, Improved spectrophotometric determination of sulfate with barium chloronilate as applied to coal ash and related materials, *Anal. Chem.*, 39, 1719, 1967.
77. **Verbitskaya, V. A., Stepin, V. V., and Onorina, I. A.**, Determination of sulfur and phosphorus in ferrovanadium by ion exchange chromatography, *Tr. Vses. Nauchno Issled. Inst. Stand. Obraztsov*, 4, 136, 1968.
78. **Zinov'eva, L. D., Gladysheva, K. F., and Zelenina, T. P.**, Determination of sulphur by using ion-exchange chromatography, *Sb. Nauchn. Tr. Vses. Nauchno Issled. Gornometall. Inst. Tsvetn. Met.*, No. 9, 118, 1965.
79. **Evans, H. B. and Mori, S.**, Amperometric determination of sulfur in plutonium sulfide and plutonium-uranium sulfide ceramic fuels, *Anal. Chem.*, 40, 217, 1968.
80. **Boase, D. G., Eisenzimmer, M., and Thomson, R. G.**, Determination of sulfate in nitric acid-sulfuric acid mixtures, Report AECL-3676, Atomic Energy of Canada Ltd., 1971.
81. **Priscott, B. H., Hand, T. G., and Young, E. J.**, Analysis of electrolytic-capacitor electrolyte-determination of chloride and sulfate in the parts per million range, *Analyst (London)*, 91, 48, 1966.
82. **Schuetzle, D., Skewes, L. M., Fisher, G. E., Levine, S. P., and Gorse, R. A.**, Determination of sulfates in diesel particulates, *Anal. Chem.*, 53, 837, 1981.
83. **Dubois, L., Baker, C. J., Teichman, T., Zdrojewski, A., and Monkman, J. L.**, Determination of sulfuric acid in air, a specific method, *Mikrochim. Acta*, No. 2, 269, 1969.
84. **Inglis, A. S.**, The micro-determination of sulphur in organic compounds, *Mikrochim. Acta*, No. 12, 1834, 1956.
85. **Hirano, S., Kurobe, M., and Ito, F.**, Determination of total sulphur in cements, pigments and rubber articles by an ion exchange method followed by alkalimetric titration, *Jpn. Analyst*, 4(9), 565, 1955.
86. **Hirano, S. and Kurobe, M.**, Determination of total sulphur in pyrites by anion-exchange method followed by alkalimetric titration, *Jpn. Analyst*, 4(9), 552, 1955.
87. **Mikailov, V. V. and Tarasenko, T. I.**, Determination of sulfur in organophosphorus compounds with the indicator nitchromazo, *Zavod. Lab.*, 33(11), 1380, 1967.
88. **Tovbin, M. V. and Dyatlovitskaya, F. G.**, Application of ion-exchange resins to the analysis of water. II, *Ukr. Khim. Zh.*, 20(4), 434, 1954.
89. **Strukova, M. P. and Lapshova, A. A.**, Determination of sulphur in organometallic compounds and other materials, *Zh. Anal. Khim.*, 24(10), 1577, 1969.
90. **Oehme, F.**, High-frequency titrimetric determination of hydrogen sulfide and thiols in technical gases after absorption in alkaline solution, *Erdoel Kohle*, 13(6), 394, 1960.

91. Řezáč, Z. and Straka, K., Determination of sulphur in pyrite concentrates and the possibility of determining sulphur in volatile organic and inorganic substances, *Fresenius' Z. Anal. Chem.*, 166, 161, 1959.
92. Johnson, D. A. and Atkins, D. H. F., Air-borne system for sampling and analysis of sulphur dioxide and atmospheric aerosols, *Atmos. Environ.*, 9(9), 825, 1975.
93. Polčin, J., Rapid determination of total sulphur in sulfides, sulfites and thiosulfates by means of ion exchange and conductimetric titration, *Chem. Zvesti*, 11(8), 494, 1957.
94. Danse, A., Cook, E. B. T., Stoch, H., and Steele, T. W., The determination of thiocyanate in uranium solutions, Report NIM-167 National Institute of Metallurgy, Johannesburg, South Africa, April 1967.
95. Matsuhisa, K., Ohzeki, K., and Kambara, T., Coagulated ion-exchanger colorimetry for the determination of trace amounts of sulfide as methylene blue, *Bull. Chem. Soc. Jpn.*, 56, 3847, 1983.
96. Zocchi, A., Volumetric determination of sulfates: simple modification for routine application of the method, *Rass. Chim.*, 32(2), 61, 1980.
97. Smith, J. and Syme, A. C., A direct micro-alkalimetric titration method for determining sulphur in organic compounds, *Analyst (London)*, 81, 302, 1956.
98. Stoffyn, P. and Keane, W., Spectrophotometric micro- and sub-micro determination of sulfur in organic substances with barium chloranilate, *Anal. Chem.*, 36, 397, 1964.
99. Likussar, W., Raber, H., Huber, H., and Grill, D., Spectrophotometric determination of sulfate, chloride and fluoride in plant materials, *Anal. Chim. Acta.*, 87, 247, 1976.
100. Gabrielson, G., Determination of lead sulfate by means of cation exchangers, *Anal. Chim. Acta.*, 15, 426, 1956.
101. Haiduc, I., Cormos, D. C., and Nandrea, M., Potentiometric determination of sulfate by titration with lead(II) nitrate with use of a lead selenide synthetic singe-crystal electrode, *Stud. Univ. Babes Bolyai Ser. Chem.*, 21(1), 56, 1976.
102. Vladescu, L. and Ghinea, S., Separation and concentration of thiosulfate with ion exchange resins, *Rev. Roum. Chim.*, 29(6), 507, 1984.
103. Watanuki, K., Dissolution of sulfuric acid from an acid-form cation-exchange resin. The effect on the determination of sulfate ions, *Jpn. Analyst*, 7(6), 385, 1958.
104. Day, H. O., Gill, J. S., Jones, E. V., and Marshal, W. L., Use of a cation-exchange resin for total anion analysis of aqueous solutions containing uranyl ion, *Anal. Chem.*, 26, 611, 1954.
105. Kartseva, V. D., Lokshin, G. B., Libinson, G. S., and Kruzhkova, N. G., Determination of sulfate ions in antibiotic sulfates by complexometric titration, *Antibiotiki (Moscow)*, 25(12), 909, 1980.
106. Kriege, O. H. and Theodore, M. L., Analysis of rare earth sulphides, selenides and tellurides, *Talanta*, 13, 265, 1966.
107. Whiteker, R. A. and Swift, E. H., Volumetric determination of sulfate and analysis of pyrites. Application of cation-exchange resins, *Anal. Chem.*, 26, 1602, 1954.
108. Clinch, J., The determination of sulfate in thorium nitrate, *Analyst (London)*, 81, 358, 1956.
109. Enaki, I. G., and Nabivanets, B. I., Determination of sulfate in natural waters having various contents of minerals, *Gidrobiol. Zh.*, 8(5), 124, 1972.
110. Karas, F., Eliášek, J., and Palatý, V., Rapid method for the determination of sulfates, *Voda*, 37(10), 316, 1958.
111. Edgcombe, L. J., Simultaneous determination of sulphur and chlorine in coal, *Fuel (London)*, 34(4), 429, 1955.
112. Aleskovski, V. B. and Kheifets, Z. I., The use of cation-exchange resins for the determination of sulphur in pyrites, *Tr. Leningr. Tekhnol. Inst.* No. 27, 121, 1953.
113. Liteanu, C. and Manoliu, C., Indirect determination of sulfur in pyrites by flame-photometric titration, *Rev. Roum. Chim.*, 16(3), 411, 1971.
114. Efros, S. M., A comparison of various methods of removing tervalent ions in the determination of sulfur in pyrites, *Tr. Leningr. Tekhnol. Inst. im. Lensoveta*, No. 35, 91, 1956.
115. Kao, C. W., Graham, E. R., and Blanchar, R. W., Determination of sulfate in soils as the barium-133 sulfate precipitate, *Soil Sci.*, 112(4), 221, 1971.
116. Khramov, V. P. and Kolosova, V. S., The determination of sulfates in aqueous extracts of soils by cation exchange, *Uch. Zap. Sarat. Univ.*, 42, 93, 1955.
117. Fernandez, T., Garcia Luis, A., and Garcia Montelongo, F., Spectrophotometric determination of sulfate for measuring sulfur dioxide in air, *Analyst (London)*, 105, 317, 1980.
118. Hoffer, E. M., Kothny, E. L., and Appel, B. R., Simple method for microgram amounts of sulfate in atmospheric particultes, *Atmos. Environ.*, 13(2), 303, 1979.
119. Savvin, S. B., Akimova, T. G., Dedkova, V. P., and Varshal, G. M., Determination of sulfate in natural water and in atmospheric precipitation, *Zh. Anal. Khim.*, 24(12), 1868, 1969.
120. Dedkova, V. P., Akimova, T. G., and Savvin, S. B., Spectrophotometric determination of sulfate with orthanilic K, *Zh. Anal. Khim.*, 36(7), 1358, 1981.
121. LePeintre, M. and Richard, J., Accurate determination of sulfates in waters by direct titration, *Chim. Anal.*, 39(9), 331, 1957.

122. **Sen Gupta, R. and Ananthanarayanan, S.,** Volumetric estimation of sulfate in sea-water using benzidine hydrochloride, and determinations of sulfate to chlorinity ratio, *Indian J. Chem.*, 1(9), 403, 1963.
123. **Reijnders, H. F. R., van Staden, J. J., and Griepink, B.,** Photometry in a continuous-flow system. Determination of sulfate in a continuous-flow system in environmental samples with dimethylsulphonazo III, *Fresenius' Z. Anal. Chem.*, 295, 122, 1979.
124. **Ceausescu, D.,** Simple and rapid determination of sulfate, chloride and nitrate in a single sample of natural water, *Revta Chim.*, 19(11), 676, 1968.
125. **Lukin, A. M. and Chernŷsheva, T. V.,** Volumetric determination of sulfates with cholorophosphonazo III, *Zavod. Lab.*, 34(9), 1054, 1968.
126. **Gales, M. E., Kaylor, W. H., and Longbottom, J. E.,** Determination of sulfate by automated colorimetric analysis, *Analyst (London)*, 93, 97, 1968.
127. **Utsumi, S., Oinuma, Y., and Isozaki, A.,** Spectrophotometric determination of micro-amounts of sulfate with dimethylsulfonazo III, *Jpn. Analyst*, 27(5), 278, 1978.
128. **Ohlweiler, O. A. and de Oliveira Meditsch, J.,** Absorptiometric determination of sulfate in water, *Anal. Chim. Acta*, 25, 233, 1961.
129. **Mayer, J., Hlucháů, E., and Abel, E.,** Polarographic micro-determination of sulfate, *Anal. Chem.*, 39, 1460, 1967.
130. **Pavlik, M. and Mach, M.,** Catex method for the determination of sulfates in water and the comparison with commonly used method, *Voda*, 36(5), 123, 1957.
131. **Nasu, T.,** Fluorimetric determination of micro-amounts of sulfate using thorium-morin complex, *J. Hokkaido Univ. Educ.*, 23(1), 35, 1972.
132. **Lambert, J. L. and Ramasamy, J.,** Colorimetric determination of sulfate with barium violurate, *Anal. Chim. Acta*, 75, 460, 1975.
133. **Khan, S. U., Morris, G. F., and Hidiroglou, M.,** Micro-determination of total sulfur and inorganic sulfate in biological materials, *Microchem. J.*, 24(3), 291, 1979.
134. **Ubuka, T., Kinuta, M., Akagi, R., Kiguchi, S., and Azumi, M.,** Reaction of S-[(2-amino-2-carboxyethyl)sulphonyl]-L-cystein with sulfite: synthesis of S-sulpho-L-cysteine and L-alanine-3-sulphinic acid and application to the determination of sulfite, *Anal. Biochem.*, 126(2), 273, 1982.
135. **Kim, G., Bird, E. W., and Loup, R. J.,** Determination of total sulfur in phosphoproteins, with special reference to casein, *Anal. Biochem.*, 43(1), 199, 1971.
136. **Wainer, A. and Koch, A. L.,** Determination of inorganic sulfate in urine using barium chloranilate, *Anal. Biochem.*, 3(6), 457, 1962.
137. **Schwager, H. and Keller, T.,** Micro-determination of total sulphur in plant tissue, *Int. J. Environ. Anal. Chem.*, 4(4), 275, 1976.
138. **Raber, H., Likussar, W., and Grill, D.,** Rapid spectrophotometric determination of sulfur in plant materials, *Int. J. Environ. Anal. Chem.*, 4(4), 251, 1976.
139. **Boczkowski, R. J.,** Potentiometric determination of total sulfur in pulp- and paper-mill process streams, *Tappi*, 60(1), 134, 1977.
140. **Bishara, S. W., Attia, M. E., and Hassan, H. N. A.,** Micro-determination of organic sulfur or inorganic sulfate, *Rev. Roum. Chim.*, 19(6), 1099, 1974.
141. **Ackermann, G. and Pitzler, G.,** Distillation method for the micro-determination of sulphur, *Microchim. Acta*, No. 4, 636, 1963.
142. **Srinivasan, S. R., Radhakrishnamurthy, B., Dalferes, E. R., and Berenson, G. S.,** Determination of sulfate in glycosaminoglycans by gas chromatography, *Anal. Biochem.*, 35(2), 398, 1970.
143. **Dixon, J. P.,** Methods for the micro-determination of sulphur in organic compounds, *Analyst (London)*, 86, 597, 1961.
144. **McGillivray, R. and Woodger, S. C.,** Application of the oxygen-flask combustion technique to the determination of trace amounts of chlorine and sulfur in organic compounds, *Analyst (London)*, 91, 611, 1966.
145. **Dawoud, A. F. and Gadalla, A. A.,** Rapid determination of small amounts of sulfate in cellulose nitrate by a wet-digestion procedure, *Analyst (London)*, 95, 823, 1970.
146. **Warren, H. D. and Brunstad, A.,** Turbidimetric micro-determination of sulfate in plutonium solutions, Report HW-55349, U.S. Atomic Energy Commission, March 1958.
147. **Day, H. O., Gill, J. S., Jones, E. V., and Marshall, W. L.,** A search for reliable sulfate analysis of uranyl sulfate: the ion exchange method, Report CF-52-2-50, Oak Ridge National Laboratory, Oak Ridge, Tenn., February 7, 1952.
148. **Yamaji, I.,** Determination of sulfur content in rubber vulcanisates by high-frequency titration, *Rep. Tokyo Chem. Ind. Res. Inst.*, 50(6), 203, 1955.
149. **Albert, P., Blouri, J., Cleyrerque, C., Deschamps, N. and Le Héricy, J.,** Neutron-activation determination of traces of sulfur and phosphorus in very high-purity metals. New source of error in the determination of sulfur after irradiation with slow neutrons. I. Reactions used, and determination of sulfur and phosphorus in aluminum and magnesium, *J. Radioanal. Chem.*, 1, 297, 1968.

150. **Albert, P., Blouri, J., Cleyrerque, C., Deschamps, N., and Le Héricy, J.**, Neutron-activation determination of traces of sulfur and phosphorus in very high-purity metals. New source of error in the determination of sulfur after irradiation with slow neutrons. II. Determination of sulfur and phosphorus in copper, *J. Radioanal. Chem.*, 1, 389, 1968.
151. **Albert, P., Blouri, J., Cleyrerque, C., Deschamps, N., and Le Héricy, J.**, Neutron-activation determination of traces of sulfur and phosphorus in very high-purity metals; new source of error in the determination of sulfur after irradiation with slow neutrons. III. Determination of sulfur and phosphorus in iron and nickel and general conclusions, *J. Radioanal. Chem.*, 1, 431, 1968.
152. **Ohlweiler, O. A.**, Determination of sulphur in iron or steel by indirect polarography of sulfate, *Revta Quim. Ind. (Rio de Janeiro)*, 38(452), 16, 1969.
153. **Srivastava, R. D. and Gesser, H.**, Methods for the determination of trace impurities in nickel: boron, chloride, sulfide, and sulfate, *J. Prakt. Chem.*, 38(5-6), 262, 1968.
154. **Mikes, J. A. and Szantó, J.**, Simplified rapid method for determination of sulfate by means of ion-exchange, *Talanta*, 3, 105, 1959.
155. **Stepanova, A. N., Bulatov, M. I., and Aleskovskii, V. B.**, Spectrophotometric determination of basic components and of sulfate in thin films of lead sulfide, *Izv. Vyssh. Uchebn. Zaved. Khim. Khim. Tekhnol.*, 15(1), 35, 1972.
156. **Popova, E. D. and Miranov, K. E.**, Determination of sulfur in rare-earth-metal sulfides using chlorophosphonazo III as metallochromic indicator, *Zh. Anal. Khim.*, 31(10), 2050, 1976.
157. **U.K. Atomic Energy Authority**, Determination of sulfate in ferrous sulphamate feed solutions, Report PG 230(W), 1961.
158. **Fehér, F., Eckhard, S., and Sauer, K. H.**, Analytical characteristics of commercial and purified sulphur, *Fresenius' Z. Anal. Chem.*, 168, 88, 1959.
159. **Sun, S. and Tang, R.**, Determination of trace sulfate by molecular-emission-cavity analysis, *Fenxi Huaxue*, 12(4), 283, 1984.
160. **Kreshkov, A. P. and Kuznetsova, L. B.**, Potentiometric determination of sulfuric acid, hydrogen sulfates and sulfates, *Zavod. Lab.*, 34(11), 1295, 1968.
161. **Tyman, V. and Uhrova, M.**, Titrimetric determination of sulfates in corrosion products, *Chem. Prum.*, 22(1), 31, 1972.
162. **Shishova, E. A., Fridman, G. I., and Antonenko, L. V.**, Determination of oxalic, sulfuric and sulfosalicylic acids in color anodization baths, *Tekhnol. Legk. Splavov. Nauchno Tekh. Byull. VIL Sa*, No. 3, 108, 1972.
163. **Eriksen, T. E. and Engman, S. O.**, Determination of small quantities of sulfate and thiosulfate in aqueous solutions of sulphur dioxide, *Acta Chem. Scand.*, 26(8), 3333, 1972.
164. **Palatý, V.**, Indirect spectrophotometric determination of sulfate ion, *Talanta*, 10, 307, 1963.
165. **Jones, D. L., Moody, G. J., Thomas, J. D. R., and Hangos, M.**, Interferences of a barium-ion-selective electrode used for the potentiometric titration of sulfate, *Analyst (London)*, 104, 973, 1979.
166. **Bertolacini, R. J. and Barney, J. E.**, Ultra-violet spectrophotometric determination of sulfate, chloride and fluoride with chloranilic acid, *Anal. Chem.*, 30, 202, 1958.
167. **Bertolacini, R. J. and Barney, J. E.**, Colorimetric determination of sulfate with barium chloranilate, *Anal. Chem.*, 29, 281, 1957.
168. **Fritz, J. S. and Freeland, M. Q.**, Direct titrimetric determination of sulfate, *Anal. Chem.*, 26, 1593, 1954.
169. **Dimitt, R. L. and Graham, E. R.**, Determination of sulfate in microgram quantities as barium-133 sulfate, *Anal. Chem.*, 48, 604, 1976.
170. **Jaselskis, B. and Vas, S. F.**, Titrimetric determination of semi-micro amounts of sulfate in presence of phosphate, *Anal. Chem.*, 36, 1965, 1964.
171. **Kirsten, W. J., Hansson, K. A., and Nilsson, S. K.**, Determination of sulfate and sulfur in inorganic and organic materials, *Anal. Chim. Acta*, 28, 101, 1963.
172. **Freeland, M. Q. and Fritz, J. S.**, Direct titration of sulfate, Report ISC-667, U.S. Atomic Energy Commission, 1956.
173. **Palatý, V.**, Determination of sulfate, *Chem. Ind.*, No. 7 176, 1960.
174. **McSwain, M. R., Watrous, R. J., and Douglass, J. E.**, Improved methylthymol blue procedure for automated sulfate determinations, *Anal. Chem.*, 46, 1329, 1974.
175. **Magri, A. L., Santopadre, P., and Tomassetti, M.**, Lead-EDTA spectrophotometric method for sulfate determination. *Ann. Chim. (Rome)*, 73 (5-6), 273, 1983.
176. **Venturello, G. and Gualandi, C.**, Determination of sulphur and phosphorus in iron alloys with exchange resins, *Boll. Sci. Fac. Chim. (Bologna)*, 17(1), 5, 1959.
177. **Mizutani, Y.**, Separation and concentration of underground water by use of an anion exchanger, *Jpn. Analyst*, 5(11), 620, 1956.
178. **Bringley, J. B. and Dick, A. T.**, Semi-microdetermination of inorganic sulfate in plant material, *J. Agric. Food. Chem.*, 15(3), 539, 1967.

179. **Kimura, K., Ikeda, N., Inarida, M., and Kawanishi, H.,** Separation of sulfate and tellurate by ion exchange, *Jpn. Analyst,* 7(2), 73, 1958.
180. **Miyamoto, M.,** Studies of analytical methods for trace impurities in high-purity substances. VII. Determination of sulphur in high-purity selenium, *Jpn. Analyst,* 10(3), 211, 1961.
181. **Kan, M., Kashiwagi, H., Terao, K., and Imaeda, K.,** Determination of organic sulfate in serum and dialysate, *Biochem. Med.,* 26(2), 135, 1981.
182. **Kan, M., Kashiwagi, H., and Maeda, K.,** Determination of the total amount of inorganic sulfate removed during hemodialysis treatment, *Clin. Chim. Acta.,* 114, 275, 1981.
183. **Karalova, Z. K. and Shibaeva, N. P.,** Micro-determination of sulfate in water, *Zh. Anal. Khim.,* 19(2), 258, 1964.
184. **Deshpande, R. G.,** Extraction of sulphur-35 from pile-irradiated potassium chloride, *J. Chromatogr.,* 2(1), 117, 1959.
185. **Shiraishi, N., Iba, T., Yoshikawa, S., and Morishige, T.,** Separation and determination of sulfate in a large amount of phosphate by ion exchange chromatography, *Jpn. Analyst,* 14(5), 450, 1965.
186. **Honda, M. and Tadano, H.,** A method for determining sulfate ion with the aid of ion-exchange resins, *Jpn. Analyst,* 2(5), 451, 1953.
187. **Subbotina, A. I., Arkhangel'skaya, E. A., and Petrov, A. M.,** Chromatographic separation of sulfate and chromate, *Tr. Khim. Khim. Tekhnol.,* No 1, 118, 1963.
188. **Koopman, B. J., Jansen, G., Wolthers, B. G., Beukhof, J. R., Go, J. G., and Van der Hem, G. K.,** Determination of inorganic sulfate in plasma by reversed-phase chromatography using ultra-violet detection and its application to plasma samples of patients receiving different types of haemodialysis, *J. Chromatogr. Biomed. Appl.* 38, 259, 1985.
189. **Paez, D. M. and Guagnini, O. A.,** Isolation and ultra-micro-determination of hydrogen sulfide in air or water by use of ion-exchange resin, *Mikrochim. Acta,* No. 2, 220, 1971.
190. **Krejcar, E.,** Anion-exchange resins as sorbents of sulfur dioxide, *Chem. Prum.,* 15(2), 77, 1965.
191. **Dabrowski, A. and Popis, H.,** Ion exchange in viscose analysis, *Chem. Anal. (Warsaw),* 9(5), 953, 1964.
192. **Rahman, M.,** Spectrophotometric determination of xanthate and total sulfur in viscose, *Anal. Chem.,* 43, 1614, 1971.
193. **Rao, V. R. S. and Tataiah, G.,** Estimation of sulfur dioxide by reduction of thallium(III) held on anion-exchange resin as chlorothallate, *Radiochem. Radioanal. Lett.,* 29(4), 179, 1977.
194. **Ducret, L. and Ratouis, M.,** Determinations in inorganic analysis by extraction with the aid of coloured cations. III. Determination by extraction with the aid of basic dyestuffs. Determination of traces of sulfate by methylene blue, *Anal. Chim. Acta,* 21, 91, 1959.
195. **Ono, A. and Otsuki, T.,** Ultra-violet spectrophotometric determination of micro amounts of sulphur in steel by reduction to hydrogen sulfide, *Tetsu To Hagane,* 68(2), 333, 1982.
196. **Jasinski, R. and Trachtenberg, I.,** Application of a sulfate-sensitive electrode to natural waters, *Anal. Chem.,* 45, 1277, 1973.
197. **Iguchi, A.,** Separation of polythionates with anion-exchange resins, *Bull. Chem. Soc. Jpn.,* 31(5), 597, 1958.
198. **Madrowa, M.,** Determination of sulphur in vegetable matter by using anionite paper, *Chem. Anal. (Warsaw),* 10(6), 1147, 1965.
199. **Lundqvist, P., Martensson, J., Sorbo, B., and Ohman, S.,** Adsorption of thiocyanate by anion-exchange resins and its analytical applications, *Clin. Chem.,* 29(2), 403, 1983.
200. **Cuypers, M. Y., Le Héricy, J., Cuypers, J., and Albert, P.,** Determination of sulphur in copper by neutron activation, *C. R. Hebd. Seances Acad. Sci. Paris,* 261(25), 5494, 1965.
201. **Schreiber, B. and Pella, P. A.,** Application of anion exchange resins on filters to X-ray fluorescence determination of sulfate, *Anal. Chem.,* 51, 783, 1979.
202. **Das, N. R. and Bhattacharyya, S. N.,** Ion-exchange separation of different oxidation stages of sulphur formed by the $^{35}Cl(n,p)$ ^{35}S reaction in alkali halides, *J. Radioanal. Chem.,* 57(1), 61, 1980.
203. **Owens, C. W.,** Rapid ion exchange separation of radioactive sulphur anions, *Radiochem. Radioanal. Lett.,* 13(5-6), 325, 1973.
204. **Janus, J. W. and Nellist, D. R.,** Determination of thio-sulfate in photographic gelatin, *J. Photogr. Sci.,* 15(6), 270, 1967.
205. **Bassignana, P., Tagliafico, G. B., Valbusa, L., and Pocchiari, F.,** Quantitative determination of thiosulfate in photographic gelatins, *J. Photogr. Sci.,* 9(6), 372, 1961.
206. **Olsson, J. E. and Samuelson, O.,** Determination of sulfide, thiosulfate and polysulfide in black liquors, *Sven. Papperstidn.,* 68(6), 179, 1965.
207. **Schmidt, M. and Sand, T.,** Separation of thionates by means of anion exchangers, *Z. Anorg. Chem.,* 330(3-4), 188, 1964.
208. **Fan, B., Liu, M., and Hu, Z.,** High performance ion-exchange chromatographic separation and determination of sulfide, sulfite and thiosulfate, *Fenxi Huaxue,* 10(7), 393, 1982.

209. **Halmos, P. and Inczedy, J.**, Use of outer-sphere complex formation reactions in ion-exchange chromatography. Separation of oxalate and sulfate ions, *Talanta*, 27, 557, 1980.
210. **Pollard, F. H., Nickless, G., and Glover, R. B.**, Chromatographic studies on sulphur compounds. V. Study to separate thiosulfate, sulfite and lower polythionates by anion-exchange chromatography, *J. Chromatogr.*, 15(4), 533, 1964.
211. **Dawson, W. M. and Jones, W. F.**, Studies in qualitative inorganic analysis. XLIII. Alkali-metal salts of the sulfur oxo-acids, *Mikrochim. Acta*, No. 2, 339, 1974.
212. **Iguchi, A.**, Separation of sulfate, sulfite, thiosulfate, and sulfide ions with anion-exchange resins, *Bull. Chem. Soc. Jpn.*, 31(5), 600, 1958.
213. **Aly, H. F., Abdel-Hamid, M. M., and Abdel Rassoul, A. A.**, Gradient chromatographic elution of sulphur anions, *J. Radioanal. Chem.*, 40(1-2), 65, 1977.
214. **Iguchi, A.**, The separation of tellurite, selenite and sulfite ions by anion-exchange resins, *Bull. Chem. Soc. Jpn.*, 31(6), 748, 1958.
215. **Qureshi, S. Z. and Rathi, M. S.**, Ion-exchange method for simultaneous microgram detection of sulphur, chlorine, iodine, and nitrogen in organic compounds, *Anal. Chem.*, 47, 1424, 1975.
216. **Fujimoto, M.**, Micro-analysis with ion-exchange resins. III. Detection of small quantities of sulfides with p-aminodimethylaniline or p-phenylenediamine, *Bull. Chem. Soc. Jpn.*, 29(5), 567, 1956.

SELENIUM, TELLURIUM, AND POLONIUM

Both the adsorption and nonadsorption of Se and Te on anion and cation exchange resins have been utilized for the isolation of these two elements from complex matrices such as geological and biological materials. For the same purpose chelating polymers have also found applications. Only a very limited number of ion exchange separation methods involving Po have been reported.

ANION EXCHANGE RESINS

General

Se

Tetravalent Se is virtually not adsorbed on strongly basic resins such as Dowex® 1 at low hydrochloric acid (HCl) concentrations up to about 6 M (Se [VI] is also not sorbed). At higher concentrations of this acid it is retained by the resin to some extent (from hydrobromic acid media the adsorption is higher). Moderate adsorption of Se(IV) is observed in solutions of pH 2.6 to 3.0 (see Table 2*)[1] or higher, e.g., at pH 5.[2] This adsorption of selenite, i.e., SeO_3^{2-}, can be effected on strongly basic resins in the acetate form (see Table 2)[1] or hydroxide form (see Table 7).[3] Coadsorbed with Se(IV) are selenate and many other anions such as sulfate, phosphate, and nitrate, so that in media of this type selective separations can be obtained from cationic constituents only. At pH 1, sorption of selenite is negligible, while that of selenate is complete so that Se(IV) can be separated from Se(VI) by elution of the former with 0.1 M HCl.[4]

Elution of Se(IV) that has been adsorbed as selenite can be effected with dilute or concentrated HCl or perchloric acid (see Table 2)[1] or by means of 1 M NaOH (see Table 7).[3] Desorption of the adsorbed Se(IV) can also be effected by first reducing it with concentrated hydroiodic acid to the metal which subsequently is dissolved and eluted with concentrated nitric acid.[2]

It has been shown[5] that Se(IV) is adsorbed on Amberlite® IRA 400 from 2 M HCl in the presence of the complexing ligand 4-(5-mercapto-3-methyl-4-[4-sulfophenylazo]pyrazol-1-yl) benzene sulfonate. Elution of Se is effected with 13 M nitric acid.

For most analytical separations of Se from accompanying elements its nonadsorbability from HCl media (see above) has been utilized (see Tables 1 to 4), although for the same purpose fluoride solutions of pH 1 (see Table 1)[6] and 2 M nitric acid (see Table 4)[7] may be employed. In the latter two media Se can be separated from elements forming adsorbable anionic fluoride and nitrate complexes, respectively, while in the HCl systems, e.g., 3 to 6 M, separations from elements forming anionic chloro complexes, e.g., Te(IV) (see below), Fe(III), Ga, U(VI), Zn, Cd, Hg(II), etc., can be achieved. However, no separation of Se in any of these media is possible from the alkalies, alkaline earth elements, rare earths, Ni, and other metals.

Se(IV) and also Te(IV) are not adsorbed on Dowex® 1 at any concentration of nitric or phosphoric acid.

Te

In 0.5 to 2, 4 to 6, and 9 to 12 M HCl solutions the predominating Te(IV) species are TeO^{2+}, $TeOCl_4^{2-}$, and $TeCl_6^{2-}$, respectively, and it is the hexachloro complex which is strongly retained by anion exchange resins of the quaternary ammonium type such as Dowex®1. This adsorption of Te(IV) increases with increasing acid concentration in the range from 1 to 12 M acid. Thus, in 1, 2, 3, and 4 to 12 M HCl, distribution coefficients for Te(IV) of

---

* Tables for this chapter appear at the end of the text.

~8 to 10, ~50, <10^3, and >10^3, respectively, have been measured.[8] This strong retention of Te from HCl media has been utilized for the isolation of this element from various matrices for which purpose Te(IV) was adsorbed on anion exchange resins from 2,[9] 3,[10-13] 4,[14,15] 6,[13,16] 7,[17] 10,[6] and 12 M[18] HCl media (see Tables 1 to 4). At the latter two molarities Se(IV) is weakly coadsorbed with the Te but can readily be removed by elution with HCl solutions of lower concentrations, e.g., 6[6] or 3 M HCl.[18] Other coadsorbed elements include Fe(III), UO_2(II), Mo(VI), Zn, Cd, Hg, Cu, Co, Au, and Pt metals, of which Co and Fe(III) can be removed by elution with 4 and 2 M HCl respectively (see Table 1).[6]

Te(IV) is also strongly adsorbed from 9 M Li chloride solutions, a fact which has been used to separate it from Se(IV) which is not retained from this medium (see Table 3).[19,20] Elution of the Te can be effected with 5 M Li chloride.

At pH values of ~3 or higher Te(IV) is present as the tellurite anion (TeO_3^{2-}), and as such it is adsorbed on strongly basic resins.[21] When the pH value of the solutions is decreased to 1 and below Te is only negligibly adsorbed,[22] so that for the elution of this element usually 0.1 to 1 M HCl media is employed (see Tables 1, 3, and 4).[6,8,10,11,14-18,21] Redox elution of Te has also been recommended.[2,8] In this case the adsorbed T(IV) is reduced on the column to the metal using stannous chloride[2] or SO_2[8] dissolved in 6 M HCl as the reductants, whereafter dilute HCl (e.g., 1 M) is passed to elute coadsorbed elements. Subsequently, the Te is redissolved oxidatively using 30% H_2O_2[2] or 8 M HCl plus nitric acid and then eluted with 1 M NaCl[2] or 1 M HCl respectively. T(VI) is not adsorbed from HCl media, aso that, for example, in 3 to 12 M HCl tellurite can be readily separated from tellurate (see Table 4).[23]

Strong adsorption of Te(IV) on basic resins, e.g., Dowex® 1, is observed in hydrofluoric acid media of <5 M in which distribution coefficients of >10^2 have been measured.[24] Under the same conditions Se(IV) is only negligibly retained (K_d value of <10).[25]

Po

Like Te(IV), tetravalent Po is also strongly adsorbed on basic resins from HCl solutions. It is retained as the anionic hexachlorocomplex $PoCl_6^{2-}$, which is the predominating Po species in 2 to 12 M HCl. At acid concentrations from 0.1 to 0.5 M and 0.5 to 2 M mainly the species $Po(OH)_2^{2+}$ and $Po(OH)Cl_5^{2-}$, respectively, are present. It is to be expected that the distribution coefficients of Po will be higher at all acidities than that of Te(IV) because of the more pronounced metallic properties of the former element. In Table 4 a procedure is shown in which Po is adsorbed from 8 M HCl and finally eluted with concentrated nitric acid.[26] Adsorption of Po on Dowex® 1 from 3 M HCl has been used for the radiochemical separation of ^{210}Po from ^{210}Pb and ^{210}Bi. After elution of the latter two elements with 3 and 12 M HCl, respectively, the Po is eluted with 8 M nitric acid (see Table 12 in chapter on Lead).

Applications

From Tables 1 to 3 it is seen that separations of Se and Te on anion exchange resins have been used in connection with the determination of these elements in geological materials such as rocks (see Table 1),[6,10] meteorites (see Table 1),[11] and natural waters,[2,27] as well as in biological materials (see Table 2)[1,12] and industrial products (see Table 3).[9,17,18,20]

To separate Se, Te, and Po from synthetic mixtures with other elements the procedures outlined in Table 4 can be employed.

Procedures for the ion chromatographic determination of Se are described in the chapter "Special Analytical Techniques Using Ion Exchange Resins," Volume I.

CATION EXCHANGE RESINS

General

Se

Tetravalent Se (selenite ion) is virtually not adsorbed on strong acid resins such as Dowex® 50 from hydrochloric or nitric acid solutions of any concentration. Distribution coefficients are about unity in the 0.1 M acids and lower at higher acidities. It is expected that Se(VI) (selenate ion) would show a behavior similar to that of selenite. Se(IV) is also not adsorbed on cationic resins from neutral systems,[3] ammoniacal media,[28] and dilute alkali hydroxide solutions, e.g., 0.1 M NaOH.[29] The nonadsorbability of Se from dilute acid solutions, e.g., 0.03,[30] 0.05,[31] 0.2,[32,33] and 0.3 M[15] nitric acid and 0.01,[34-36] 0.05,[37] and 0.1 M[38-40] HCl has variously been employed to separate this element from solutions obtained by the dissolution of geological materials (see Tables 5 to 7) and biological matrices (see Table 8). This technique makes it possible to separate Se from Fe, Al, Mn, Cu, Zn, Cd, Pb, alkali metals, alkaline earth elements, and all other cationic constituents of the solutions, but not from anions such as phosphate and sulfate. Sometimes Se is first preconcentrated by coprecipitation with ferric hydroxide (see Tables 5 and 7)[32,33,38,39] and then separated from the Fe by cation exchange in 0.2 M nitric acid[32,33] or 0.1 M HCl.[38,39] In this connection it has to be mentioned that in the presence of a very large excess of Fe(III) and also Sn(IV), selenite is adsorbable on Dowex® 50 (see Table 10[36] and in Table 11 in the chapter on Iron, Volume V). Also retained by this resin is trimethylselenonium ion, i.e., $(CH_3)_3Se^+$ (see Table 8).[41,42]

After adsorption of the accompanying cations residual Se remaining in the resin bed is usually eluted with water.

Te

Tetravalent Te is moderately retained by strongly acidic cation exchange resins from very dilute hydrochloric and nitric acid solutions. Thus, on Bio-Rad® AG50W-X8 distribution coefficients for Te of ~40, 17, ~5, ~1.5, and <1 have been measured in 0.1, 0.2, 0.5, 1, and 2 to 12 M HCl, respectively.[43] Adsorption of Te(IV) is also observed in the range from pH 1 to 8.5,[44-46] as well as from 0.1 M solutions of phosphoric, oxalic, and tartaric acids[47] and from 60% acetone - 0.2 M HCl (see Table 10).[48] In the latter medium Te(IV) has a distribution coefficient of 63 and in 0.1 M thiourea solutions which are 0.1, 0.2, 0.5, and 1 M in HCl distribution coefficients for Te(IV) of $>10^4$, $>10^3$, ~10^3, and 900, respectively, have been measured on Bio-Rad® AG50W-X8.[43] Following adsorption of the cationic thiourea complex of radioactive ^{132}Te, a column of this resin can act as a source (generator) for ^{132}I (daughter of ^{132}Te).[49]

Dilute HCl media from which Te(IV) has been adsorbed on strongly acid cation exchange resins include ≤0.01 M HCl,[13,50] 0.02 to 0.05 M HCl (see Tables 6 and 9),[29,37] and 0.1[51,52] and 0.12 M HCl.[28] Coadsorbed with the Te(IV) from these systems are virtually all cationic elements which include the alkalies, alkaline earth elements, rare earths, Fe, Al, Cu, Ni, Pb, Co, etc. Not adsorbed with the Te(IV) are Te(VI) (which is not adsorbed from 0.1 to 1 M HCl or at higher acidities) and Se(IV and VI), as well as metals forming very stable anionic chloro complexes as, for example, Au(III), Pt metals,[28] Hg(II), and Cd.[48] Also not retained are oxyanions such as phosphate and sulfate, as well as simple anions, e.g., fluoride.

The elution of the adsorbed Te(IV) can be effected with 0.3 M HNO$_3$,[15,53] 0.3 M HCl,[29,37] 3 N H$_2$SO$_4$,[46] 0.5 M HNO$_3$,[45] 1 M HCl,[48,51] 1 M NaOH,[29] and ammoniacal solutions (see Table 9).[28,54,55]

On a cation exchanger in the Sn(II) form Te is reduced to the metal which can then be eluted with >1 M HCl.[56]

Po

This element shows an adsorption behavior similar to that of Te(IV), although it is more strongly adsorbed at comparable acidities. An example for a cation exchange separation involving Po is outlined in Table 10.[57]

Applications

From Tables 5 to 7 it is seen that cation exchange separation procedures have variously been used in connection with the determination of Se in silicates,[32] sulfides ores (e.g., pyrite),[15,30] ambient particulates,[31] atmospheric aerosol samples,[37] and natural waters.[3,33,38,39,58-60] From some of these matrices Te was also isolated and then determined quantitatively (see Table 6).[15,37] Similar procedures have been used for the separation of Se from biological matrices (see Table 8) which include blood,[34,35,61,62] human dental enamel,[40] human hair,[61] urine,[41,42] organs,[62] liver,[34,35,63] tuna fish,[34,35] viscera,[34,35] and plant material.[61,63,64]

Cation exchange procedures of the type shown in Table 9 have been employed in connection with the determination of Te in matrices predominantly consisting of Ag, U, and Cu.

To separate Se, Te, and Po from synthetic mixtures with other elements, the procedures outlined in Table 10[29,36,48,57] or similar methods can be employed.[65-67]

CHELATING RESINS

Chelex®100 and other chelating polymers have been utilized for the separation procedures illustrated in Tables 11 and 12 which have been used in connection with the determination of Se and Te in Cu matrices[68,69] and of Se in natural waters[70,71] and urine.[72,73]

Table 1
DETERMINATION OF Se AND Te IN ROCKS AND METEORITES AFTER ANION EXCHANGE SEPARATION[a]

Material	Anion exchange resin, separation conditions, and remarks	Ref.
Rocks	Dowex® 1-X8 (100-200 mesh; Cl⁻ form)	6
	A) First column operation (column operated at a flow rate of $\simeq$0.4—0.5 mℓ/cm² min)	
	a) Sample solution (~6 mℓ) of pH 1 which is $\simeq$0.1—0.2 M in F⁻ (adsorption of elements forming stable anionic fluoride complexes; Se, Te, and other elements pass into the effluent)	
	b) Water (elution of remaining Se, Te, and other nonadsorbed elements)	
	B) Second column operation (column of the resin preequilibrated with 10 M HCl and operated at a flow rate of ~0.5 mℓ/cm² min)	
	a) 10 M HCl solution of the residue obtained after evaporation of eluates (a) and (b) (adsorption of Se, Te, Sc, Co, Fe, and other elements)	
	b) 8 M HCl (elution of Sc)	
	c) 6 M HCl (elution of Se)	
	d) 4 M HCl (elution of Co)	
	e) 2 M HCl (elution of Fe)	
	f) 1 M HCl (elution of Te)	
	⁷⁵Se and ¹²³ᵐTe are determined radiometrically, with limits of detection of 4 and 90 ng and chemical yields of 87 ± 4 and 96 ± 3%, respectively	
	Before the first ion exchange separation step, the sample (100 mg) is irradiated with neutrons and then dissolved in HNO₃-HF (1:10) (in the presence of carriers)	
	Bio-Rad® AG1-X8 (100-200 mesh; Cl⁻ form)	10
	Column of 1 cm ID containing 5 g of the resin conditioned with 3 M HCl	
	a) 3 M HCl sample solution (~10 mℓ) (adsorption of Te)	
	b) 3 M HCl (25 mℓ) (elution of Se)	
	c) 0.3 M HCl (25 mℓ) (elution of Te)	
	After further purification Te is determined radiometrically	
	Before the ion exchange separation, the sample (100—200 mg) is irradiated with neutrons, dissolved in HF-HCl-HNO₃ and Te is preconcentrated by radiochemical precipitations	
Stony meteorites	Dowex® 1 (100-200 mesh)	11
	Column: 4 × 0.8 cm	
	a) 3 M HCl (3—5 column volumes) (adsorption of Te; elution of Se)	
	b) 0.2—0.5 M HCl (elution of Te)	
	Se and Te are determined radiometrically	
	Before the separation, the sample is irradiated with neutrons and then decomposed by fusion with Na₂O₂ in the presence of carriers; after dissolution of the melt in 6 M HCl, Se and Te are precipitated as the elements using SO₂-gas as the reductant and then further purified by radiochemical precipitations	

[a] See also Table 4 in the chapter on Platinum Metals, Volume III and Table 3 in the chapter on Mercury, Volume IV.

Table 2
DETERMINATION OF Se IN BIOLOGICAL MATERIALS AFTER ANION EXCHANGE SEPARATION[a]

Material	Ion exchange resin, separation conditions, and remarks	Ref.
Milk, blood, kale, liver, and animal muscle	Amberlite® IRA-400 (14-52 mesh; acetate form) Column: 6 × 1.5 cm conditioned with either 0.001 M HCl or acetate buffer of pH 3.0 and operated at a flow rate of 1—2 mℓ/min a) Sample solution of pH 2.6—3.0 (adsorption of Se[IV]; matrix constituents pass into the effluent) b) Distilled water (25 mℓ) (elution of residual matrix) c) Conc HCl or HClO$_4$ (0.5 mℓ) followed by water (20 mℓ) (elution of Se) Se is determined voltammetrically; Se concentrations of <1 ng/mℓ can be determined Before the ion exchange separation, the sample (0.2—1 g) is wet ashed with H$_2$SO$_4$ and HNO$_3$	1
Blood plasma and serum	Dowex® 2-X8 (200-400 mesh; chloride form) Column: 3.5 × 1 cm a) 3 M HCl sample solution (4 mℓ) (adsorption of Au, Hg, and Zn) b) 1 M HCl (6 mℓ) (elution of Se) After extraction with diethyldithiocarbamate-CCl$_4$ the ^{75}Se is determined by γ-spectrometry Before the ion exchange separation, the sample (0.1—0.5 mℓ) is irradiated with neutrons and then decomposed with H$_2$SO$_4$-HNO$_3$ (in the presence of Se-carrier)	12

[a] See also Table 5 in the chapter on Mercury, Volume IV and Table 10 in the chapter on Copper, Volume III.

Table 3
DETERMINATION OF Se AND Te IN INDUSTRIAL MATERIALS AFTER ANION EXCHANGE SEPARATION[a]

Material	Ion exchange resin, separation conditions, and remarks	Ref.
High-purity Se	Anion exchange resin L-150 (chloride form) Column of 0.5 cm ID containing ~4 mℓ of the resin and operated at a flow rate of 0.5 mℓ/min a) 7 M HCl sample solution (3 mℓ) (adsorption of Te; Se passes into the effluent) b) 7 M HCl (~15 mℓ) (elution of remaining Se) c) 1 M HCl (~20 mℓ) (elution of Te) Te is determined radiometrically A similar procedure is used to separate As from Sb on Amberlite® IRA-400; the As is eluted with 7 M HCl (20 mℓ), and after washing the column with 2 M HCl (15 mℓ) the adsorbed Sb is eluted with 2% EDTA solution (15—20 mℓ) adjusted to pH 12 with NH_3 solution Before the ion exchange separation, the sample is irradiated with neutrons and then dissolved in conc HNO_3 (in the presence of carriers); subsequently, several radiochemical separation steps based on precipitation and extraction are performed	17
Pt	Deacidite® FF (100-200 mesh; chloride form) Column: 5 × 1 cm conditioned with 12 M HCl a) 12 M HCl sample solution (5 mℓ) (adsorption of Te) b) 3 M HCl (20 mℓ) (elution of Se) c) 0.3 M HCl (20 mℓ) (elution of Te) Se and Te are determined radiometrically Before the ion exchange separation, the sample (0.1 g) is irradiated with neutrons and then dissolved in HNO_3 + HCl (in the presence of carriers); Pt is removed by precipitation as $(NH_4)_2PtCl_6$ and Se + Te are further purified by coprecipitation with ferric hydroxide	18
Te dioxide	Dowex® 1-X8 (Cl$^-$ form) a) 2 M HCl sample solution (adsorption of interfering radionuclides and Te; Se passes into the effluent) b) 2 M HCl (elution of remaining Se) Se is determined radiometrically; the chemical yield ranges from 70—90%, and the limit of detection is ≈0.1 ppm Before the separation, the sample (≈200 mg) is irradiated with neutrons and then dissolved in HNO_3-HCl (in the presence of Se-carrier)	9
Dust precipitated from the waste gas of H_2SO_4 production	Anionite AV-17 (Cl$^-$ form) a) 9 M LiCl sample solution (adsorption of Te[IV]; Se[IV] passes into the effluent) b) 5 M LiCl (elution of Te) Te is determined gravimetrically or by titration With suitable concentrations of LiCl (0.5—9 M), Te(IV) can also be separated from binary mixtures with Bi, Pb, Sb(III), Au(III), and Fe(III)[19]	20

[a] See also Table 4 in the chapter on Niobium and Tantalum, Volume IV; Tables 16, 17, 23, 25, and 26 in the chapter on Copper, Volume III; and Table 3 in the chapter on Tin.

Table 4
ANION EXCHANGE SEPARATION OF Se, Te, AND Po[a]

Elements separated	Ion exchange resin, separation conditions, remarks	Ref.
Po from Pu, U, La, and Fe	Bio-Rad® AGl-X4 (100-200 mesh; Cl⁻ form) Column: 8 × 1.3 cm a) 8 M HCl sample solution (adsorption of Po, Pu, U, and Fe; La passes into the effluent) b) 6 M HCl containing HI (reductive elution of Pu[III] and Fe[II]) c) 1 M HCl (elution of U) d) 16 M HNO$_3$ (elution of Po) ^{209}Po is determined by α spectrometry; the yield of Po was 70—90% Before the ion exchange separation, ^{209}Po tracer is coprecipitated from a sample (1 ℓ) with Fe(OH)$_3$ or La(OH)$_3$ and the precipitate is dissolved in 8 M HCl to prepare solution(a)	26
Te(IV) from Se(IV) and other elements	Anex® S-X8 (grain size: 0.15—0.3 mm; chloride form) Column: 8 cm × 0.2 cm² conditioned with 6 M HCl (5 mℓ) and operated at a flow rate of 0.2—0.5 mℓ/min a) 6 M HCl (adsorption of Te[IV]; elution of Se[IV], As, and other elements) b) 1—2 M HCl (15—20 mℓ) (elution of Te[IV]) The method is suitable for samples having ratios of Te and Se down to 1:10^4	16
Te from iodide	Amberlite® IRA-400 (80-100 mesh; chloride form) Column: 5 × 1 cm a) 4 M HCl (adsorption of iodide and Te[IV]; Te[VI] passes into the effluent) b) 0.1—1 M HCl (elution of Te[IV]) c) 10 M HCl (elution of iodide)	14
Tellurite from tellurate	Amberlite® IRA-400 (80-120 mesh; chloride form) Column: 5 × 1.2 cm From 3—12 M HCl, Te(IV) is retained by the resin while Te(VI) passes into the effluent The method can be used to separate 10-mg amounts of the two anionic Te species	23
Se from Hg	Anionite AV-17 (grain size: 0.5—0.7 mm; NO$_3^-$ form) Column: 15 × 1.8 cm a) 2 M HNO$_3$ sample solution containing, per milliliter, ≃4 μg of Se and ≃200 μg of Hg (adsorption of Hg; Se passes into the effluent) b) 2 M HNO$_3$ (7 column volumes) (elution of residual Se)	7

[a] See also Tables 6 and 12 in the chapter on Lead and Tables 25, 36, and 43 in the chapter on Actinides, Volume II.

Table 5
DETERMINATION OF Se IN GEOLOGICAL MATERIALS AFTER CATION EXCHANGE SEPARATION[a]

Material	Ion exchange resin, separation conditions, and remarks	Ref.
Silicates, seawater, and marine organisms	Zeo-Karb® 225-X8 (52-100 mesh; H$^+$ form) Column: 10 × 1.5 cm a) 0.2 M HNO$_3$ sample solution (minimum volume) (adsorption of Fe and other cations; Se passes into the effluent) b) 0.2 M HNO$_3$ (350 mℓ) (elution of remaining Se) Se is determined spectrophotometrically; the results are corrected for losses during the analysis by using ^{75}Se tracer Before the separation, the silicate sample (1—2 g) is decomposed with HNO$_3$-HF and Se is preconcentrated by coprecipitation with ferric hydroxide (60 mg of Fe) which is dissolved in conc HNO$_3$ (0.2 mℓ); organic materials may be decomposed with H$_2$SO$_4$-HClO$_4$-NH$_4$-molybdate; seawater (5 ℓ) is filtered through a 0.5-μm membrane filter and then Se is isolated using the coprecipitation step; for the elution of remaining Se, 270 mℓ of eluent (b) is required; for seawater containing ≃0.4—0.5 μg of Se per liter, the standard deviation was 0.03 μg/ℓ and for a silicate sediment and a seaweed containing ≃1.5 μg of Se per gram and 0.8 μg of Se per gram, respectively, the coefficients of variation were 8.0 and 4.7%	32
Pyrite	Dowex® 50W-X8 (H$^+$ form) Column: 8 × 1 cm operated at a flow rate of 3 mℓ/min A 10-mℓ aliquot of the sample solution is passed through the resin bed to adsorb Fe and in the effluent (the first 6 mℓ are discarded) Se is determined by atomic absorption spectrometry; the detection limit is 5 ppm Before the separation, the sample (0.5 g) is decomposed with HNO$_3$ (5 mℓ) and 60% HClO$_4$ (2 mℓ), and after evaporation the residue is dissolved in warm water plus HNO$_3$ (1 mℓ) and the solution diluted to 500 mℓ	30
Ambient particulates	Dowex® 50W-X8 (50-100 mesh; H$^+$ form) Column: 12 × 0.8 cm The sample solution (0.05 M in HNO$_3$) is passed through the resin bed to adsorb accompanying metal ions such as Fe, Mn, Cu, etc., and in the effluent Se is determined by atomic absorption spectrometry; Se recovery is 100 ± 3% Before the separation, the sample is decomposed with conc HNO$_3$ + H$_2$O$_2$	31

[a] See also Table 11 in the chapter on Iron, Volume V.

Table 6
DETERMINATION OF Se AND Te IN GEOLOGICAL MATERIALS AFTER CATION EXCHANGE SEPARATION[a]

Material	Ion exchange resin, separation conditions, and remarks	Ref.

Sulfide ores	A) First column operation (cationite KU-2, 15 g)
	  a) 0.3 M HNO$_3$ (10 mℓ) (adsorption of cations)
	  b) 0.3 M HNO$_3$ (300 mℓ) (elution of Se and Te)
	B) Second column operation (anionite AV-17, 5 g, Cl$^-$ form) (separation of Se from Te)
	  a) 4 M HCl (10 mℓ) (adsorption of Te)
	  b) 4 M HCl (40 mℓ) (elution of Se)
	  c) 1 M HCl (120 mℓ) (elution of Te)
	Se and Te are determined spectrophotometrically or polarographically
	Before the first ion exchange separation step, the sample (0.5 g) is dissolved in HNO$_3$
		15

Atmospheric aerosol samples	Amberlite® IR-120 (50-80 mesh; H$^+$ form)
	  a) 0.05 M HCl sample solution (adsorption of Te; Se passes into the effluent)
	  b) 0.05 M HCl (elution of remaining Se)
	  c) 0.3 M HCl (elution of Te)
	Se and Te are determined by atomic absorption spectroscopy; recoveries are ≃92% and the ion exchange step reduces interference from foreign ions to <10%
	Before the separation, the sample (≃0.1 g of deposit collected on a fiber-glass filter) is decomposed with HNO$_3$-HClO$_4$
		37

[a] See also Table 4 in the chapter on Platinum Metals, Volume III and Table 20 in the chapter on Cobalt, Volume V.

Table 7
DETERMINATION OF Se IN NATURAL WATERS AFTER CATION EXCHANGE SEPARATION

Material	Ion exchange resin, separation conditions, and remarks	Ref.
Seawater	Dowex® 50W-X8 (Na⁺ form) Column: 5 × 3 cm operated at a flow rate of 1.3 mℓ/min a) Sample solution (30 mℓ) adjusted to pH 1 (adsorption of Fe[III]) b) 0.1 M Na acetate solution (100 mℓ) adjusted to pH 1 with HCl (elution of Se) Se is determined fluorometrically; the recovery of Se is >90% Before the ion exchange separation, the sample (5 ℓ) is acidified with HCl (50 mℓ) and Se is preconcentrated by coprecipitation with ferric hydroxide (30 mg Fe); the precipitate is dissolved in 6 M HCl (5—6 mℓ) and 1 M Na acetate (3 mℓ) is added, and to prepare solution (a) the pH is adjusted to 1 with HCl and the volume increased to 30 mℓ by the addition of water	38
Spa and mineral waters	Dowex® 50 Column containing 3 mℓ of the resin a) Dilute HCl sample solution (10.3 mℓ) (adsorption of Fe[III]; Se passes into the effluent) b) 0.1 M HCl (15 mℓ) (elution of remaining Se) Se is determined fluorimetrically; the detection limit is 0.01 ppb of Se Before the separation, the sample (400 mℓ) is acidified with conc HCl (40 mℓ) and any Se(VI) is reduced to Se(IV) by heating in the presence of 48% HBr (20 mℓ) and saturated Br-water (0.5 mℓ) (which oxidizes Se⁰ and Se²⁻ to Se[IV]) Subsequently, Se(IV) is coprecipitated with ferric hydroxide (after addition of 1 mℓ of FeCl₃ solution; [30 mg/mℓ])	39
Effluents and waters	A) First column operation (Zeo-Karb® 225-X8, 52-100 mesh, H⁺ form) Column of 1 cm ID containing 15 mℓ of the resin a) Aqueous sample solution (adsorption of cations; Se passes into the effluent) b) Distilled water (5 × 10 mℓ) (elution of remaining Se) Eluates (a) and (b) are combined and then passed through an anion exchange resin using the following procedure B) Second column operation (Deacidite® FF, 52-100 mesh, OH⁻ form) Column of 1 cm ID containing 15 mℓ of the resin a) Combined eluates (a) and (b) obtained by (A) (adsorption of Se) b) Distilled water (5 × 10 mℓ) (removal of accompanying elements) c) 1 M NaOH (100 mℓ) (elution of Se which is contained in the fraction from 40—100 mℓ) Se is determined colorimetrically or turbidimetrically	3

Table 8
DETERMINATION OF Se IN BIOLOGICAL MATERIALS AFTER CATION EXCHANGE SEPARATION

Material	Ion exchange resin, separation conditions, and remarks	Ref.
Blood, tuna fish, liver, viscera	Amberlite® IR-120 (H$^+$ form) Column: 10 × 1 cm a) 0.01 M HCl sample solution (20 mℓ) (adsorption of cations; Se passes into the effluent) b) Distilled water (20 mℓ) (elution of residual Se) After extraction with dithizone-CCl$_4$ the Se is determined by atomic absorption spectrometry; the recovery of Se is >93% Before the separation, the sample (1 mℓ or 1 g) is dried and burnt in an O-filled flask and the Se is absorbed in 0.01 M HCl (20 mℓ)	34, 35
Human dental enamel	Dowex® 50-X8 (H$^+$ form) Column: 12 × 1.5 cm a) 0.1 M HCl sample solution (~5 mℓ) (adsorption of accompanying cations; Se passes into the effluent) b) Water (elution of remaining Se) After precipitation, ^{81}Se or ^{75}Se is determined by γ-ray spectrometry Before the ion exchange separation, the sample is irradiated with neutrons and then dissolved in conc HCl (in the presence of Se-carrier); subsequently, Se is precipitated with SO$_2$-gas and the element is dissolved in HNO$_3$ followed by evaporation and redissolution in HCl	40
Urine[a]	Dowex® 50W-X8 (NH$_4^+$ form) Column: 20 × 1 cm conditioned with 1 M NH$_3$ solution a) Urine sample (5—10 mℓ) (adsorption of trimethylselenonium ions: TMSe$^+$ = [CH$_3$]$_3$Se$^+$) b) 1 M NH$_3$ solution (50 mℓ) (elution of matrix constituents) c) Distilled water (removal of NH$_3$) d) HCl (0.05, 0.1, 0.5, 1.0, and 4 M HCl used in succession; TMSe$^+$ is eluted with the 4 M acid; the first 5 mℓ of eluate is discarded and then 40 mℓ is collected as TMSe$^+$ fraction) Se is determined by graphite-furnace atomic absorption spectrometry The detection limit is 1.0 ng/mℓ and the recovery is 82—102%	41

[a] See also Reference 74.

Table 9
DETERMINATION OF Te IN INDUSTRIAL PRODUCTS AFTER CATION EXCHANGE SEPARATION[a]

Material	Ion exchange resin, separation conditions, and remarks	Ref.
Crude AgCl and high-purity U oxide	Dowex® 50W-X8 (50-100 mesh; H$^+$ form) Column: 50-mℓ buret containing 15 mℓ of the resin and operated at a flow rate of 1 drop/sec a) ~0.3 M HNO$_3$ sample solution (100 mℓ) (adsorption of accompanying cations; Te passes into the effluent) b) Water (5 × 15 mℓ) (elution of residual Te) Te is determined spectrophotometrically Before the ion exchange separation, the sample of AgCl (⪈1 g) is fused with Na$_2$O$_2$ (5 g), the melt is leached with water (250 mℓ), and after adjustment of the pH to 8.5 the peroxides are decomposed by boiling; the precipitate containing Te is dissolved in 50% HNO$_3$ and then Te is preconcentrated by repeated coprecipitation with Fe-hydroxide; finally, the HNO$_3$ solution of the precipitate is evaporated to 2 mℓ and diluted to 100 mℓ to prepare solution (a); the U-oxide sample (5 g) is dissolved in HNO$_3$-HCl and Te is coprecipitated twice with ferric hydroxide in the presence of Na$_2$CO$_3$-NaHCO$_3$ (to complex U)	53
Cu compounds and anodic Cu	Wofatit® R (NH$_4^+$ form) From the Cu solution which is 10% in NH$_3$ (containing Na$_4$P$_2$O$_7$ or citric acid to complex Fe[III]) the Cu is adsorbed (as cationic tetramine complex) on a column (45 × 2.5 cm) of the resin using a flow rate of 6—8 mℓ/min; in the effluent and washings (10% ammonia solution) Te is determined polarographically or by titration Before the separation, the sample of metallic Cu (2 g) is dissolved in HNO$_3$ (1:1)	54, 55

[a] See also Table 49 in the chapter on Rare Earth Elements, Volume I and Table 10 in the chapter on Gallium.

Table 10
CATION EXCHANGE SEPARATION OF Se, Te, AND Po

Elements separated	Ion exchange resin, separation conditions, and remarks	Ref.
Te(IV) from binary mixtures with Au(III), In, Cd, and other elements	Bio-Rad® AG 50W-X8 (200-400 mesh; H⁺ form) Column of 1.65 cm ID containing the resin (30 mℓ) to a height of 14 cm a) 60% Acetone - 0.2 M HCl (50—100 mℓ) (adsorption of Te[IV]; Au[III], and other elements pass into the effluent) b) 60% Acetone - 0.2 M HCl (elution of residual nonadsorbed elements) c) Deionized water (100 mℓ) (removal of acetone) d) 1 M HCl (150 mℓ) (elution of Te) With this method trace amounts and up to 120 mg of Te can be separated from gram quantities of Au and other elements	48
Se from foreign cations	Dowex® 50W-X8 (100 mesh; H⁺ form) Column: 50-mℓ buret filled to the 38-mℓ mark with the resin and operated at a flow rate of 2 mℓ/min a) Sample solution (100 mℓ, adjusted to pH 2.5) containing 0.1 M Na pyrophosphate solution (4 mℓ) (adsorption of foreign cations, e.g., Ca, Mg, and Ni; Se passes into the effluent) b) HCl of pH 2.5 (4 × 5 mℓ) (elution of remaining Se) Se is determined by an indirect atomic adsorption method	36
Se from Te	Zerolit® 225 (low degree of cross-linking; ~50 mesh; H⁺ form) Column: 1.5 cm × 0.7 cm² After adsorption from 0.02—0.05 M HCl the following eluents are used a) 0.1 M NaOH (elution of Se) b) 1 M NaOH followed by 0.2 M HCl (elution of Te)	29
Po from Bi	Dowex® 50 (50-100 mesh). Column: 6.5 × 0.7 cm a) 0.1—0.3 M HCl (adsorption of Po and ^{210}Bi) b) 2 M HNO$_3$ (elution of ^{210}Bi) c) 2 M HCl (elution of Po)	57

Table 11
DETERMINATION OF Se AND Te IN Cu AFTER SEPARATION ON CHELATING RESINS[a]

Material	Ion exchange resin, separation conditions, and remarks	Ref.
Electrolytic Cu	Chelex® 100 (100-200 mesh; NH_4^+ form) Column: 12 × 1.5 cm conditioned with 0.1 M NH_3 solution a) Ammoniacal sample solution (∼5 mℓ) (adsorption of Cu as cationic amine complex; Se and Te pass into the effluent) b) 0.1 M ammonia solution (∼40 mℓ) (elution of Se and Te) Se and Te are determined by anodic stripping voltammetry; detection limits are 1.3 and 1.6 nM for Se and Te, respectively Before the separation, the sample (0.5 g) is dissolved in 15 M HNO_3 (2 mℓ), N-oxides are largely expelled by heating to ≃85°C, 0.625 mℓ of H_2O_2 solution (500 ppm) is added, and after 10 min the solution is cooled and treated with 15 M NH_3 (3 mℓ) to prepare solution (a)	68
High-purity Cu	Chelex® 100 (100-200 mesh; NH_4^+ form) Column: 12 × 1.5 cm a) Ammoniacal sample solution (10 mℓ, consisting of 4 mℓ 6 M HCl and 6 mℓ of 15 M NH_3 solution) (adsorption of Cu as cationic amine complex; Te passes into the effluent) b) 0.1 M ammonia solution (10 mℓ) (elution of residual Te) Te is determined by nondispersive atomic fluorescence spectrometry Before the separation, the sample (1 g) is dissolved in conc HNO_3 (5 mℓ)	69

[a] See also Reference 75.

Table 12
DETERMINATION OF Se IN NATURAL WATERS AND URINE AFTER SEPARATION ON CHELATING RESINS[a]

Material	Ion exchange resin, separation conditions, and remarks	Ref.
Fresh- and seawater	Chelex® 100 (100-200 mesh) Column: 5 × 1 cm conditioned with water, 1 M HCl, and formate buffer (pH 2.5) and operated at a flow rate of 2 mℓ/min a) Filtered sample (50 mℓ) buffered to pH 2.5 with formate buffer in the ratio 10:1 (adsorption of interfering metals such as Cu, Ni, Cr, Mo, Sn, V, and Te [IV]; Se passes into the effluent) b) Formate buffer of pH 2.5 (20 mℓ) (elution of remaining Se) Se is determined polarographically	70
River-, estuarine, and seawater[b]	Amberlite® IRA-400-X8 (100-200 mesh) modified with the K-salt of bismuthiol II (5-mercapto-3-phenyl-1,3,4-thiadiazoline-2-thione); for its preparation, the resin is added to an aqueous solution of bismuthiol II and the mixture is shaken at 30°C for 60 min; then the impregnated resin is washed with water and methanol, and dried Column: 5 × 1 cm operated at a flow rate of 1 mℓ/min a) Water sample (e.g., 1 ℓ) (selective adsorption of Se[IV] as a bismuthiol II complex [selenotrisulfide]; matrix constituents pass into the effluent) b) Distilled water (50 mℓ) (elution of remaining matrix) c) 0.1 M penicillamine (a dithiol) at pH 5 (elution of Se) Se is determined fluorimetrically; the adsorbed Se can also be eluted with 8—13 M HNO$_3$	71
Urine	Poly(dithiocarbamate) resin (60-80 mesh) Column: polyethylene pipette tip (Eppendorf 21375D and 21375E) containing 70 mg of the resin The filtered sample (100 mℓ) is divided into equal parts and after adjusting the two aliquots to either pH 2 or 6, the samples are passed through columns of the resin to adsorb Se(IV) and Se(VI); subsequently, the resin is dissolved by treatment with conc HNO$_3$ (1 mℓ) and a few drops of 30% H$_2$O$_2$ and Se is determined by inductively coupled plasma atomic spectroscopy; similarly, differential determination of Te(IV) and Te(VI) is also possible	72, 73

[a] See also Tables 74 and 77 in the chapter on Copper, Volume III.
[b] See also Reference 76.

REFERENCES

1. **Adeloju, S. B., Bond, A. M., Briggs, M. H., and Hughes, H. C.**, Stripping-voltammetric determination of selenium in biological materials by direct calibration, *Anal. Chem.*, 55, 2076, 1983.
2. **Bernier, W. E., and Janauer, G. E.**, In situ reduction on ion-exchange resins as a method for preconcentration of selenium and other heavy metals from aqueous solutions, in Proc. Univ. of Missouri's 10th Annu. Conf. on Trace Substances in Environmental Health, June 8 to 10, 1976, 323.
3. **Sheratt, J. G. and Conchie, E. C.**, Determination of selenium in effluents and waters, *J. Assoc. Public Anal.*, 7(4), 109, 1969.
4. **Dreipa, E. F., Pakholkov, V. S., and Luk'yanov, S. A.**, Sorption of selenium(IV) and selenium(VI) from aqueous selenous and selenic acid solutions by anionic and amphoteric ion-exchange resins, *Zh. Prikl. Khim. (Leningrad)*, 53(1), 54, 1980.
5. **Nakayama, M., Chikuma, M., Tanaka, H., and Tanaka, T.**, Selective collection of selenium(IV) on anion-exchange resin with azothiopyrinedisulphonic acid, *Talanta*, 30, 455, 1983.
6. **Mignonsin, E. P. and Roelandts, I.**, Application of radiochemical neutron-activation analysis to the determination of selenium and tellurium in geological materials, *Chem. Geol.*, 16(2), 137, 1975.
7. **Selezneva, N. A., Plotnikova, O. M., and Novikov, V. P.**, Ion-exchange separation of selenium from mercury, *Izv. Akad. Nauk Kaz. SSR Ser. Khim.*, No. 4, 76, 1969.
8. **Kleemann, E. and Herrmann, G.**, Adsorption of tellurium(IV), iodide and iodate on anion-exchange resins, *J. Chromatogr.*, 3(3), 275, 1960.
9. **Johansen, O. and Steinnes, E.**, Determination of selenium in tellurium dioxide by neutron-activation analysis, *Int. J. Appl. Radiat. Isot.*, 28(6), 599, 1977.
10. **Hughes, T. C.**, Determination of tellurium in geological materials using radiochemical neutron-activation analysis with a low-energy detector, *J. Radioanal. Chem.*, 59(1), 7, 1980.
11. **Schindewolf, U.**, Selenium and tellurium content of stony meteorites by neutron activation, *Geochim. Cosmochim. Acta*, 19(2), 134, 1960.
12. **Maziere, B., Comar, D., and Kellershohn, C.**, Determination of selenium in biological material by neutron activation, *Bull. Soc. Chim. Fr.*, No. 10, 3767, 1970.
13. **Gaibakyan, D. S. and Darbinyan, M. V.**, Ion-exchange separation of selenium and tellurium. I. Separation of selenium and tellurium in hydrochloric acid solution on cationites and anionites, *Izv. Akad. Nauk Arm. SSR Khim. Nauk*, 16(3), 211, 1963.
14. **Inareda, M.**, Separation of tellurium from iodide with anion-exchange resin, *J. Chem. Soc. Jpn. Pure Chem. Sect.*, 80(4), 399, 1959.
15. **Stepin, V. V., Ponosov, V. I., Novikova, E. V., Stashkova, N. V., Sushkova, S. G., Murashova, V. I., and Emasheva, G. N.**, Determination of selenium and tellurium in sulfide ore by using ion-exchange chromatography, *Tr. Vses. Nauchno Issled. Inst. Stand. Obraztsov*, 3, 109, 1967.
16. **Šimek, M.**, Photometric determination of tellurite after its separation from selenite and interfering ions on the strongly basic anion-exchange resin Anex S, *Chem. Listy*, 60(6), 817, 1966,
17. **Koch, H. and Koch, B.**, Determination of the purity of selenium by activation analysis, *Kerntechnik*, 5(6), 248, 1963.
18. **Morris, D. F. C. and Killik, R. A.**, Determination of traces of selenium and tellurium in samples of platinum by neutron-activation analysis, *Talanta*, 10, 279, 1963.
19. **Busev, A. I., Bagbanly, I. L., Bagbanly, S. I., Guseinov, I. K., and Rustamov, N. Kh.**, Anion-exchange method for separating tellurium from heavy metals, *Zh. Anal. Khim.*, 25(7), 1374, 1970.
20. **Bagbanly, I. L., Bagbanly, S. I., and Guseinov, I. K.**, Reineckate method for the determination of tellurium in materials, *Azerb. Khim. Zh.*, No. 1, 126, 1971.
21. **Šušic, M. V.**, Separation of tellurium, ruthenium, caesium, and rare earths from one another, using anion-exchange resin Dowex-1, *Bull. Inst. Nucl. Sci. Belgrade*, 7, 39, 1957.
22. **Dreipa, E. F., Pakholkov, V. S., and Luk'yanov, S. A.**, Sorption of tellurium from aqueous solutions by anionites and amphoteric ion-exchange resins, *Zh. Prikl. Khim. (Leningrad)*, 54(5), 1040, 1981.
23. **Kimura, K., Ikeda, N., and Inarida, M.**, Separation of tellurite and tellurate by ion-exchange, *Jpn. Analyst*, 7(3), 174, 1958.
24. **Faris, J. P.**, Adsorption of the elements from hydrofluoric acid by anion-exchange, *Anal. Chem.*, 32, 520, 1960.
25. **Faix, W. G., Caletka, R., and Krivan, V.**, Element distribution coefficients for hydrofluoric acid/nitric acid solutions and the anion-exchange resin Dowex 1X8, *Anal. Chem.*, 53, 1719, 1981.
26. **Casella, V. R., Bishop, C. T., Glosby, A. A., Hiatt, M. H., Mathews, N. F., Bunce, L. A., and Hahn, P. B.**, Separation of polonium by ion-exchange chromatography, *Radiochem. Radioanal. Lett.*, 55(5-6), 279, 1983.
27. **Orvini, E. and Gallorini, M.**, Speciation problems solved with neutron-activation analysis. Some actual cases for Hg, V, Cr, As and Se, *J. Radioanal. Chem.*, 71(1-2), 75, 1982.

28. **Strel'nikova, N. P. and Lÿstsova, G. G.**, Separation of tellurium from platinum and non-ferrous metals by means of a cationite, *Zavod. Lab.*, 26(2), 142, 1960.
29. **Chen, S.**, Separation of selenium and tellurium by ion exchange and their determination by induced reaction, *Fenxi Huaxue*, 10(6), 342, 1982.
30. **Elson, C. M. and Macdonald, A. S.**, Determination of selenium in pyrite by an ion-exchange-electrothermal atomic absorption spectrometric method, *Anal. Chim. Acta*, 110, 153, 1979.
31. **Yamashige, T., Ohmoto, Y., and Shigetomi, Y.**, Ion-exchange separation and determination of selenium in ambient particulates by heated quartz cell-atomic absorption spectrophotometry, *Jpn. Analyst*, 27(10), 607, 1978.
32. **Chau, Y. K. and Riley, J. P.**, Determination of selenium in seawater, silicates and marine organisms, *Anal. Chim. Acta*, 33, 36, 1965.
33. **Tzeng, J. and Zeitlin, H.**, Separation of selenium from seawater by adsorption colloid flotation, *Anal. Chim. Acta*, 101, 71, 1978.
34. **Ishizaki, M.**, Determination of selenium in biological materials by flameless atomic-absorption spectrometry using a carbon-tube atomiser, *Jpn. Analyst*, 26(3), 206, 1977.
35. **Ishizaki, M.**, Simple method for determination of selenium in biological materials by flameless atomic-absorption spectrometry using a carbon-tube atomiser, *Talanta*, 25, 167, 1978.
36. **Lau, H. K. Y. and Lott, P. F.**, Indirect atomic absorption method for the determination of selenium, *Talanta*, 18, 303, 1971.
37. **Chiou, K. Y. and Manuel, O. K.**, Determination of tellurium and selenium in atmospheric aerosol samples by graphite furnace atomic absorption spectrometry, *Anal. Chem.*, 56, 2721, 1984.
38. **Hiraki, K., Yoshii, O., Hirayama, H., Nishikawa, Y., and Shigemastu, T.**, Fluorimetric determination of selenium in seawater, *Jpn. Analyst*, 22(6), 712, 1973.
39. **Kirchnawy, F., Kainz, G., and Sontag, G.**, Fluorimetric determination of selenium in Austrian spa and mineral waters, *Ernaehrung (Vienna)*, 6(6), 267, 1982.
40. **Nixon, G. S. and Myers, V. B.**, Estimation of selenium in human dental enamel by activation analysis, *Caries Res.*, 4(2), 179, 1970.
41. **Oyamada, N. and Ishizaki, M.**, Determination of trimethyl selenonium ions and total selenium in human urine by graphite-furnace atomic-absorption spectrometry, *Jpn. Analyst*, 31(1), 17, 1982.
42. **Nahapetian, A. T., Young, V. R., and Janghorbani, M.**, Measurement of trimethylselenonium ion in human urine, *Anal. Biochem.* 140(1), 56, 1984.
43. **Weinert, C. H. S. W., Strelow, F. W. E., and Böhmer, R. G.**, Cation exchange in thiourea-hydrochloric acid solutions, *Talanta*, 30, 413, 1983.
44. **Bÿkov, I. E. and Gorshkova, L. S.**, Separation of selenium and tellurium by the use of a cationite, *Tr. Inst. Metall. Ural. Fil. Akad. Nauk SSSR*, No. 1, 151, 1957.
45. **Babayan, G. G., Kapantsyan, E. E., and Organesyan, E. N.**, Ion-exchange chromatographic separation of selenium, tellurium and bismuth from nitric acid solutions, *Arm. Khim. Zh.*, 25(4), 291, 1972.
46. **Babayan, G. G., Kapantsyan, E. E., and Organesyan, E. N.**, Ion-exchange chromatographic separation of selenium, tellurium and bismuth in sulfuric acid medium, *Uch. Zap. Erevan. Gos. Univ. Estestv. Nauk*, 3(121), 46, 1972.
47. **Darbinyan, M. V. and Gaibakyan, D. S.**, Ion-exchange separation of selenium and tellurium. II. Separation of selenium from tellurium in solutions of some complexing acids, *Izv. Akad. Nauk Arm. SSR Khim. Nauk*, 16(5), 443, 1963.
48. **Strelow, F. W. E.**, Separation of tellurium from gold(III), indium, cadmium and other elements by cation exchange chromatography in hydrochloric acid-acetone, *Anal. Chem.*, 56, 2069, 1984.
49. **Abrão, A.**, Rapid radiochemical ion-exchange separation of iodine from tellurium: a novel radio-iodine-132 generator, Report IEA-371, Relat. Inst. Energ. At. (São Paulo) 1975.
50. **Varand, V. L.**, Use of cationites in the determination of selenium and tellurium, *Nauchn. Tr. Irkutsk. Nauchno Issled. Inst. Redk. Met.*, No. 10, 97, 1961.
51. **Okashita, H.**, Rapid separation of radionuclides by spontaneous electro-deposition on metallic mercury. II. Metallic mercury-metal ion systems, *Radiochim. Acta*, 7(2-3), 85, 1967.
52. **Uzumasa, Y., Hikime, S., Hayashi, K., and Yoshida, H.**, Separation of tellurium by ion exchange, *Jpn. Analyst*, 11(1), 78, 1962.
53. **Russell, B. G., Lubbe, W. V., Wilson, A., Jones, E., Taylor, J. D., and Steele, T. W.**, Determination of selenium and tellurium in silver chloride and uranium oxide, *Talanta*, 14, 957, 1967.
54. **Zelyanskaya, A. I. and Gorshkova, L. S.**, Polarographic determination of tellurium in copper and lead compounds, *Tr. Inst. Metall. Ural. Fil. Akad. Nauk SSSR*, No 5, 141, 1960.
55. **Zelyanskaya, A. I. and Gorshkova, L. S.**, Determination of small amounts of selenium in anodic copper, *Tr. Inst. Metall. Ural. Fil. Akad. Nauk SSSR*, No. 5, 137, 1960.
56. **Münze, R.**, Separation of tellurium and iodine in different oxidation states by oxidation-reduction resins, *J. Prakt. Chem.*, 7, 262, 1959.

57. **Radhakrishna, P.,** Ion exchange separations of radioelements, *J. Chim. Phys.,* 51(7-8), 354, 1954.
58. **Bowling, J. L. and Dean, J. A.,** Determination of selenium in natural waters using the centrifugal photometric analyzer, *Anal. Let.,* 7(3), 205, 1974.
59. **Magin, G. B., Thatcher, L. L., Rettig, S., and Levine, H.,** Suggested modified method for colorimetric determination of selenium in natural water, *J. Am. Water Works Assoc.,* 52(9), 1199, 1960.
60. **Henn, E. L.,** Determination of selenium in water and industrial effluents by flameless atomic absorption, *Anal. Chem.,* 47, 428, 1975.
61. **Andrews, R. W. and Johnson, D. C.,** Determination of selenium(IV) by anodic-stripping voltammetry in a flow system with ion exchange separation, *Anal. Chem.,* 48, 1056, 1976.
62. **Kudrin, A. N., Krasnyuk, I. I., and Efremenko, O. A.,** Kinetik determination of selenium in rat blood and organs in experimental myocardial infarction, *Farmatsiya (Moscow),* 34(1), 25, 1985.
63. **Inui, T., Terada, S., Tamura, H., and Ichinose, N.,** Determination of selenium by hydride generation with reducing tube, followed by graphite furnace atomic-absorption spectrometry, *Fresenius' Z. Anal. Chem.,* 311, 492, 1982.
64. **Lau, C. M., Ure, A. M., and West, T.,** Determination of selenium by atom-trapping atomic-absorption spectrometry, *Anal. Chim. Acta,* 141, 213, 1982.
65. **Cresser, M. S. and West, T. S.,** Studies in the analytical chemistry of selenium: absorptiometric determination with 2-mercaptobenzoic acid, *Analyst (London),* 93, 595, 1968.
66. **West, P. W. and Ramakrishna, T. V.,** Catalytic method for determining traces of selenium, *Anal. Chem.,* 40, 966, 1968.
67. **Lederer, M. and Kertes, S.,** Chromatography on paper impregnated with ion-exchange resins. II. Separation of selenite and tellurite, *Anal. Chim. Acta,* 15, 226, 1956.
68. **Hamilton, T. W., Ellis, J., and Florence, T. M.,** Determination of selenium and tellurium in electrolytic copper by anodic stripping voltammetry at a gold-film electrode, *Anal. Chim. Acta,* 110, 87, 1979.
69. **Nakahara, T., Wakisaka, T., and Musha, S.,** Determination of tellurium by non-dispersive atomic-fluorescence spectrometry using the hydride-generation technique, *Spectrochim. Acta Part B,* 36(7), 661, 1981.
70. **Howard, A. G., Gray, M. R., Waters, A. J., and Oromiehie, A. R.,** Determination of selenium(IV) by differential pulse polarography of 4-chloro-o-phenylenediamine piazselenol, *Anal. Chim. Acta,* 118, 87, 1980.
71. **Nakayama, M., Itoh, K., Chikuma, M., Sakurai, H., and Tanaka, H.,** Anion-exchange resin modified with bismuthiol II, as a new functional resin for selective collection of selenium(IV), *Talanta,* 31, 269, 1984.
72. **Fodor, P. and Barnes, R. M.,** Determination of some hydride-forming elements in urine by resin complexation and inductively coupled plasma atomic spectroscopy, *Spectrochim. Acta,* 38(1-2), 229, 1983.
73. **Barnes, R. M. and Genna, J. S.,** Concentration and spectrochemical determination of trace metals in urine with poly(dithiocarbamate) resin and inductively coupled plasma atomic emission spectrometry, *Anal. Chem.,* 51, 1065, 1979.
74. **Blotcky, A. J., Hansen, G. T., Opelanio-Buencamino, L. R., and Rack, E. P.,** Determination of trimethylselenonium ion in urine by ion exchange chromatography and molecular neutron activation analysis, *Anal. Chem.,* 57, 1937, 1985.
75. **Ikeda, M.,** Determination of selenium by atomic absorption spectrometry with miniaturized suction flow hydride generation and on-line removal of interferences, *Anal. Chim. Acta,* 170, 217, 1985.
76. **Itoh, K., Nakayama, M., Chikuma, M., and Tanaka, H.,** Separation and determination of selenium(IV) in environmental water samples by an anion exchange resin modified with bismuthiol-II and diaminonaphthalene fluorophotometry, *Fresenius' Z. Anal. Chem.,* 321, 56, 1985.

THE HALOGENS: FLUORINE, CHLORINE, BROMINE, AND IODINE

These elements are usually present as anions and, hence, are adsorbed on anion exchange resins, e.g., resins of the strongly basic type on which the adsorbability of the monovalent anionic species increases in the order: fluoride (F^-) < chloride (Cl^-) < bromide (Br^-) < iodide (I^-). For the oxyanions in which the halogen is present in the +5 oxidation state the affinity of the resins increases in the order IO_3^- < BrO_3^- < ClO_3^-. No adsorption is observed on cation exchange resins so that this type of ion exchanger is usually employed to separate the anionic halogen species from the cationic constituents of the sample solutions. Very little relevant information is available on astatine, so this element will not be treated here.

Procedures based on ion chromatographic separations of the halogens are presented in the chapter "Special Analytical Techniques Using Ion Exchange Resins", Volume I.

FLUORINE

ANION EXCHANGE RESINS

General

Fluoride is readily adsorbed on strongly basic anion exchange resins in the hydroxide,[1-12] chloride,[1,13-21] and acetate[22,23] forms from solutions adjusted to pH 5 to 9.[1,2,7,14,20,21,24-26]

Thus, fluoride can be adsorbed directly from samples of natural waters,[15-17,22] air,[18] and ammoniacal solutions.[1] Examples for this adsorption principle are presented in Tables 1 to 5*. Coadsorbed with fluoride are other anions as, for example, sulfate, phosphate, nitrate, chloride, bromide, and iodide, but not the cationic constituents of the sample solutions. Fluoride can also be adsorbed on anionic resins when present as an anionic fluoride complex, e.g., as SiF_6^{2-} (see Table 3)[5] or BF_4^- (see chapter on Boron). On the other hand, elution of fluoride as BeF_4^{2-} can be effected by using 0.1 M acetic acid - 0.005 M Be(II) solution as eluent (see Table 2).[22]

Separation from most coadsorbed anions, especially from sulfate, phosphate, and nitrate, can be achieved by elution of the adsorbed fluoride with 0.025 to 0.5 M ammonium chloride solutions adjusted with ammonia to pH ~ 9 (see Tables 1 to 5)[1,15,16,20,21] or with 0.1 to 0.5 M NaOH solutions.[7-9,12,26,27] With these alkaline eluents only fluoride is eluted, so that very selective separations of fluoride from essentially all elements can be obtained following removal of the cations during the sorption step from the media mentioned above. Subsequent to elution with 0.1 or 0.2 M NaOH the eluate may be passed through a cation exchange resin in the H^+ form and the free hydrofluoric acid in the effluent is determined enthalpimetrically with NaOH in a flow calorimeter.[12,27] Other eluents that have been used for the elution of adsorbed fluoride include 1[4,6] and 2 M NaOH,[3] 0.1[2,23] and 0.2 M Na acetate,[23,28,29] 0.05[30] and 0.1 M Na nitrate,[31] 4% KNO_3 solution,[10] 5% NaCl solution,[18] 0.1 M KCl,[17] 0.3 M Na_2SO_4,[14] and 1 M Na oxalate.[32] With the 0.1 M Na nitrate eluent sulfate is not coeluted (see Table 5). No elution of fluoride is required when this element can be determined directly on the resin using a radiometric procedure (see Table 3).

The fact that fluoride is not adsorbed on anionic resins in the presence of high concentrations of chloride, as, for example, from 10 to 12 M hydrochloric acid (HCl) (see Table 1)[13] or 0.5 M NaCl at pH 6.0 (see Table 4),[19] can be employed to separate it from elements forming anionic chloride complexes (e.g., Fe[III], Ga, Co, Cu, and Mn) or from oxyanions of metals (e.g., Mo, W, and Re), respectively.

---

* Tables for this chapter appear at the end of the text.

Applications

The separation principles discussed above have variously been used in connection with the determination of fluoride in geological materials (see Tables 1 and 2), biological matrices (see Table 3), and industrial products (see Table 4). To isolate fluoride from synthetic mixtures with anions of other elements the procedures illustrated in Table 5 or other methods[11,33] can be employed.

The anion exchange resin Dowex® 1-X1 impregnated with colored metal chelates, as, for example, Ce(III)-alizarin complexan, Be(II)-Chrome Azurol S, and Al(III)-hematoxylin, has been utilized for the microchemical detection of microgram amounts of fluoride using resin spot tests.[34]*

CATION EXCHANGE RESINS

General

On strongly acid cation exchange resins fluoride is not retained from acid, neutral, or alkaline media. Therefore, fluoride can readily be separated from cationic constituents of the sample solutions provided that the adsorption of the latter is performed under suitable experimental conditions, as, for example, from very dilute acid media, e.g., at pH 2[35] or from 0.3 M nitric acid,[36] essentially neutral systems (e.g., water samples),[37-43] and dilute alkali carbonate[44,45] or alkali hydroxide solutions.[46-52]

From all these systems Fe, Al, and many other elements are adsorbed on the cation exchangers mainly used in the H form, while fluoride together with other anions such as halogen anions, sulfate, phosphate, nitrate, etc. are not retained. Examples for separations of this type are presented in Tables 6 to 8. To prevent evolution of CO_2 in the resin bed (when carbonates are present) the retention of cations can be effected by application of a batch process which may be combined with the slurry column technique (see Table 7).[44,45] Following adsorption of the cations, residual fluoride remaining in the resin is usually eluted with water.[35,36,38,40,41,43-49,51-53] For the same purpose ethanol may also be employed.[42]

Batch adsorption of cationic constituents can also be effected by equilibrating the resin, e.g., Dowex® 50, with an aqueous suspension of the sample, e.g., MgOH, on which fluoride was preconcentrated by coprecipitation from a natural water (seawater).[54]

The separation of fluoride from accompanying elements on cation exchange resins may be incomplete in the presence of metals forming stable anionic fluoride complexes such as Al and Zr. On the other hand, this complex formation can be utilized to completely retain fluoride on a resin, e.g., Dowex® 50, loaded with Zr(IV) (see Table 6).[55] The adsorption separates fluoride from accompanying cations (which are less strongly adsorbed than Zr) and also from anions. This type of separation has been termed as "inorganic affinity chromatography".

Applications

From Tables 6 and 7 it is seen that cation exchange separations of fluoride from accompanying elements have variously been used in connection with the determination of this element in geological, biological, and industrial matrices which include natural waters,[37,38,54-60] effluents,[58] phosphate rocks,[36] potato tops,[40] cryolite and other products of the Al industry,[44,45,61] chrome plating baths,[41] fluorides of U[46-49,62] and Pu,[62] petroleum products,[42] synthetic organic compounds,[43,50-52,63-67] cordites,[68] and electroslags.[69]

To separate fluoride from synthetic mixtures with other elements the procedures outlined in Table 8[35,53,70] can be employed. Similar methods have been used for the separation of fluoride from interfering metals[29,71-79] which include Pb,[76] Ca and Mg,[77] Al and lanthanides,[78] and Li.[79]

---

* See also Table 80 in chapter on Copper, Volume III.

Table 1
DETERMINATION OF FLUORIDE IN GEOLOGICAL MATERIALS AFTER ANION EXCHANGE SEPARATION[a]

Material	Ion exchange resin, separation conditions, and remarks	Ref.
Rock phosphates	Deacidite® FF (−60 + 100 mesh BSS; hydroxide form) Column: 20 × 1 cm conditioned with eluent (c) (50 mℓ) a) Ammoniacal sample solution (10-mℓ aliquot) (adsorption of fluoride; accompanying elements pass into the effluent) b) Distilled water (10 mℓ) (as a rinse) c) 0.025 M NH$_4$Cl adjusted to pH 9.2 with NH$_3$ solution (115 mℓ) (elution of fluoride; the first 15 mℓ of eluate is discarded) Fluoride is determined titrimetrically Before the separation, the sample (≯0.2 g) is digested with 1 M HCl (6 mℓ) for 24 hr and then alkaline EDTA solution (3 mℓ) (∼1 M in EDTA and 4 M in NH$_3$) is added and the mixture is heated to 70°C for a few minutes; after filtration, the filtrate is diluted with water to 250 mℓ to prepare solution (a)	1
Rocks and minerals	Deacidite® FF (100-200 mesh; OH$^-$ form) Column of 1 cm ID containing 750 mg of the resin a) Distillate (300—400 mℓ containing HF) (adsorption of F$^-$) b) 0.1 M Na acetate (25 mℓ) (elution of F$^-$) Fluoride is determined spectrophotometrically Before the separation, the sample (200 mg) is fused with Na$_2$CO$_3$ (1 g) and F$^-$ is steam-distilled from H$_2$SO$_4$ medium using a form of Willard-Winter distillation	2
Fe ore and apatite	Dowex® 2-X10 (Cl$^-$ form) Column: 18.5 × 2.3 cm conditioned with 10 M HCl a) ≃10—12 M HCl sample solution (50 mℓ) (adsorption of Fe[III], Mn[II], and other elements; into the effluent pass F$^-$ and PO$_4^{3-}$) b) 10 M HCl (50 mℓ) (elution of remaining F$^-$) Fluoride is determined spectrophotometrically; before the separation the Fe ore (0.2—1 g) or the apatite concentrate (0.1—0.5 g) is dissolved in HCl	13
Atmospherically polluted soils	Amberlite® IRA-400 (20-50 mesh; Cl$^-$ form) The sample (1 g) is shaken in water (50 mℓ) with the resin (5 mℓ) for 16 hr to adsorb F$^-$; then the resin is separated from the soil by sieving (60 mesh) and F$^-$ is eluted with 0.3 M Na$_2$SO$_4$ (50 mℓ) at a rate of 2.5 mℓ/min Fluoride is determined with a specific ion electrode	14

[a] See also Table 6 in the chapter on Phosphorus.

Table 2
DETERMINATION OF FLUORIDE IN NATURAL WATERS AND AIR AFTER ANION EXCHANGE SEPARATION

Material	Ion exchange resin, separation conditions, and remarks	Ref.
Fluoridated drinking water	Dowex® 1-X10 (50-100 mesh; chloride form) Column: 15 × 1 cm operated at a flow rate of 2 mℓ/min a) Water sample (100—130 mℓ; containing ~1 mg fluoride per liter) (adsorption of fluoride and sulfate; matrix constituents pass into the effluent) b) Water (10 mℓ) (as a rinse) c) 0.1 M NH$_4$Cl (40 mℓ) adjusted to pH 9.2 with NH$_3$ followed by water (10 mℓ) (elution of fluoride) Fluoride is determined spectrophotometrically (Zr and Eriochrome cyanine R) With the ion exchange method interfering sulfate is separated from fluoride because sulfate is not eluted with eluent (c)	15,16
Potable waters	Dowex® 1-X4 (50-100 mesh; chloride form) Column: 10 × 1 cm operated at a flow rate of 1 mℓ/min a) Water sample (1 ℓ containing fluoride at the 1 mg level) (adsorption of fluoride; matrix components including interfering ions pass into the effluent) b) Distilled water (as a rinse) c) 0.1 M KCl (45 mℓ) (elution of fluoride discarding the first 5 mℓ of eluate) Fluoride is determined titrimetrically with Th-nitrate; down to 1 ppm of fluoride can be determined to within ±1%	17
	Dowex® 2-X8 (50-100 mesh; acetate form) Column of 1 cm ID containing 25 mℓ of a 1:1 slurry of the resin and distilled water and operated at a flow rate of ~3 mℓ/min a) Water sample (50 mℓ) (adsorption of fluoride; cationic constituents pass into the effluent) b) Distilled water (100 mℓ) (removal of residual nonadsorbed elements) c) 0.1 M acetic acid which is 0.005 M in Be(II) (elution of fluoride as BeF$_4^{2-}$) Fluoride is determined spectrophotometrically (Zr-SPADNS method)	22
Atmospheric fluorides	Duolite® A41 a) Aqueous sample solution (adsorption of F$^-$) b) 0.1 M NaOH followed by 0.01 M NaOH (elution of F−) Fluoride is determined titrimetrically Before the separation, the air sample (with F$^-$ contents of 1—10 parts per 10^9) is passed through a wash bottle containing water or dilute HCl to absorb F$^-$	26
Air	Amberlite® IRA 400 (Cl$^-$ form) The air sample is passed through a column of the resin at 28 ℓ/min for 30 min (for 10—100 parts per 10^9 of F$^-$) or for 2 hr (for 1—10 parts per 10^9) Subsequently, F$^-$ is eluted with 5% NaCl solution (10 mℓ) and determined spectrophotometrically No interference is caused by S^{2-}, Cl$^-$, or SO$_2$ at the 200-μg level	18

Table 3
DETERMINATION OF FLUORIDE IN BIOLOGICAL MATERIALS AFTER ANION EXCHANGE SEPARATION

Material	Ion exchange resin, separation conditions, and remarks	Ref.
Urine, plant material (including rice), and soil	Dowex® 1-X8 (50-100 mesh; washed with 1 M NH$_3$ solution and then with water) Column: 3 × 1 cm washed with 10% ascorbic acid solution (100 mℓ) and then with distilled water a) Sample solution (urine) adjusted to pH 6—7 (adsorption of fluoride using a flow rate of ≈20 mℓ/min) b) The resin is taken from the column, dried at 110°C, and then irradiated in a linear accelerator (photonuclear activation) c) HF is steam distilled from the irradiated resin and after precipitation of fluoride with MgCl$_2$ the activity due to ^{18}F is determined radiometrically At the level of 2 ppm of fluoride, the coefficient of variation is ≈8% (ten determinations); the limit of detection is 0.01 μg of F$^-$; before the ion exchange separation, the sample (100 mℓ) is treated with conc NH$_3$ (5 mℓ) and the supernate containing the fluoride is used to prepare solution (a) In the analysis of plants and soils the sample solution is passed through a column (5 × 1 cm) containing Dowex® 50-X8 (100-200 mesh; to adsorb cations) connected in series with a column (2 × 1 cm; lower column) filled with the anion exchange resin (flow rate: 0.5 mℓ/min); then the columns are disconnected and the anionic resin is washed with distilled water (50 mℓ); subsequently, the resin is dried at 40°C and subjected to activation analysis; before the separation the sample (0.5 g) is decomposed by fusion with Na$_2$O$_2$ + NaOH	24, 25
Urine, tea, fossilized bone, and organic fluoro compounds	Dowex® 1-X8 (50-100 mesh; OH$^-$ and Cl$^-$ forms) A) First column operation Column of 0.7 cm ID containing 3.8 mℓ of the resin in the OH$^-$ form and operated at a flow rate of ≯0.75—1.5 mℓ/min a) Neutral sample solution (containing 5—100 μg of fluoride and ≯100 mg of sulfate; a total of anions ≯3 meq) (adsorption of fluoride and sulfate) b) Water (until effluent is neutral) B) Second column operation Column of 0.7 cm ID containing 3 mℓ of the resin in the Cl$^-$ form is placed underneath the column used for separation (A) and through both columns 0.1 M NH$_4$Cl (adjusted to pH 9.2 with NH$_3$) is passed (elution of fluoride from both columns and adsorption of sulfate on the lower column; the first 13 mℓ of eluate is discarded in the next 15 mℓ the fluoride is determined colorimetrically); before the ion exchange separation the organic matter is oxidized by incineration of the sample, covered with WO$_3$, at up to 650°C in a stream of moist O$_2$; fluoride is collected in dilute NaOH	80
Urine	Merck® Ion Exchanger III (OH$^-$ form) Column: 16 × 1 cm operated at a flow rate of 5 mℓ/min a) Urine sample (10 mℓ) (adsorption of F$^-$; matrix constituents pass into the effluent) b) Water (20—30 mℓ) (as a rinse) c) 2 M NaOH (25 mℓ) (elution of F$^-$) Fluoride is determined spectrophotometrically; coefficients of variation are 4.1, 6.3, and 15.3% for amounts of F$^-$ of 1.59, 0.6, and 0.16 μg/mℓ (11 determinations)	3

Table 3 (continued)
DETERMINATION OF FLUORIDE IN BIOLOGICAL MATERIALS AFTER ANION EXCHANGE SEPARATION

Material	Ion exchange resin, separation conditions, and remarks	Ref.
Blood serum and plasma	Dowex® 1-X16 (100-200 mesh; OH⁻ form) Column: 4.6 × 0.6 cm operated at a flow rate of ≃0.7 mℓ/min a) Mixture (5 mℓ) consisting of serum (3 mℓ) and water (2 mℓ) (adsorption of fluoride, chloride, and other anions; Na and other cations pass into the effluent) b) Water (4 mℓ) (as a rinse) c) 1 M NaOH (elution of fluoride) Subsequently, a diffusion is performed and the neutralized diffusate is passed through a column (1.7 × 0.6 cm) of Dowex® 50-X8 (100-200 mesh; H⁺ form) to remove Na⁺; finally, fluoride is determined spectrophotometrically using a method which is based on aniline-isobutanol extraction of a ternary fluoro complex with La-alizarin complexan	4
Animal tissues	Dowex® 1-X8 (200-400 mesh; acetate form) Column: ≃2.5 × 0.9 cm operated at a flow rate of 6—12 drops/min a) Aqueous sample solution (≥250 mℓ; e.g., distillate) containing up to 10 μg of F⁻ (adsorption of F⁻; matrix components pass into the effluent) b) 0.1, 0.2, and 3 M Na-acetate solutions (10 mℓ each passed in this order) (elution of F⁻) F is determined spectrophotometrically; common ions in biological extracts do not interfere and a precision of ±5% on 1—10-μg amounts of F⁻ is obtained The method can also be used for determining F in atmospheric samples	23
Bone	Dowex® 1-X8 (200-400 mesh; OH⁻ form) Column of 1.1 cm ID containing ≃5 mℓ of the wet resin The sample (50 mg) is irradiated with neutrons and then ^{18}F is distilled as H_2SiF_6 at 145°C; the distillate (at pH ≃ 4) is passed through the resin bed to adsorb SiF_6^{2-} and ^{18}F is determined radiometrically directly on the resin The radiochemical yield is ≃98%; the lower limit of determination is ≃0.1 μg	5

Table 4
DETERMINATION OF FLUORIDE IN INDUSTRIAL PRODUCTS AFTER ANION EXCHANGE SEPARATION

Material	Ion exchange resin, separation conditions, and remarks	Ref.
Mo, W, and Re	Dowex® 1-X8 (20-50 mesh; chloride form). Column: 70 cm × 2 cm² conditioned with 0.5 M NaCl (200 mℓ) and operated at a flow rate of 1—2 mℓ/min a) Sample solution (50 mℓ) of pH 6.0 and which is 0.5 M in NaCl (adsorption of Mo, W, and Re; the effluent is discarded) b) 0.5 M NaCl (250 mℓ) (elution of fluoride) Fluoride is determined spectrophotometrically or by pF-electrode measurements Before the separation, the sample (0.12—1.5 g) is decomposed at 550°C by treatment with NaOH (1 g) and NaNO$_3$ (0.6 g)	19
Etching solutions	Dowex® 1-X8 (50-100 mesh; chloride form) Column: ~15 × ~0.9 cm operated at a flow rate of ≃2 mℓ/min a) Sample solution (10-mℓ aliquot) (adsorption of fluoride and sulfate) b) Water (20 mℓ) (as a rinse) c) 0.1 M NH$_4$Cl (25 mℓ) (elution of fluoride; sulfate is not coeluted) Fluoride is determined titrimetrically Before the separation, the sample (10 g) is diluted with water to 150 mℓ and filtered, and NaOH solution is added to adjust the pH to 5—6.5; then solution (a) is prepared by further dilution to 1 ℓ	20
Al salts	Amberlite® IRA-400 (OH⁻ form) Column: 54 cm × 1 cm² operated at a flow rate of 2—2.5 mℓ/min a) 0.2 M NaOH sample solution (~50 mℓ) containing ≮100 ppm of F⁻ and ≃100 mg of Al (adsorption of F⁻; Al passes into the effluent) b) 0.2 M NaOH (280—300 mℓ) (elution of remaining Al) c) 1 M NaOH (100—250 mℓ) (elution of F⁻) Fluoride is determined colorimetrically; the error is ±0.5—2% for 1—10 mg of F⁻	6
Organic compounds (4-fluorobenzoic acid in admixture with phenylphosphonic acid)	Dowex® 2-X10 (100-200 mesh; OH⁻ form) Column of 18 cm length operated at a flow rate of 1.5 mℓ/min a) Aqueous sample solution at pH 7 (adsorption of F⁻ and PO$_4^{3-}$) b) 0.1 M NaOH (elution of F⁻ in ≃90 min; PO$_4^{3-}$ is not coeluted) Fluoride is determined spectrophotometrically Before the separation, the sample is subjected to O-flask combustion	7

Table 5
ANION EXCHANGE SEPARATION OF FLUORIDE FROM SYNTHETIC MIXTURES WITH OTHER ELEMENTS

Ions separated	Ion exchange resin, separation conditions, and remarks	Ref.
F^-, oxalate, PO_4^{3-}, Cl^-, SO_4^{2-}, Br^-, $S_2O_3^{2-}$, $Fe(CN)_6^{4-}$, I^-, $Fe(CN)_6^{3-}$, and SCN^-	Anionite AV-17 (NO_3^- form) Column: 12 × 1 cm a) Aqueous sample solution (adsorption of the anions) b) Water (20 mℓ) (as a rinse) c) 0.05 M $NaNO_3$ (15 mℓ) (elution of F^-) d) 0.1 M $NaNO_3$ (20 mℓ) (elution of oxalate and PO_4^{3-}) e) 0.2 M $NaNO_3$ (20 mℓ) (elution of Cl^- and SO_4^{2-}) f) 0.5 M $NaNO_3$ (20 mℓ) (elution of Br^-) g) 1 M $NaNO_3$ (20 mℓ) (elution of $S_2O_3^{2-}$ and $Fe[CN]_6^{4-}$) h) 2 M $NaNO_3$ (20 mℓ) (elution of I^- and $Fe[CN]_6^{3-}$) i) 0.1 M $FeCl_3$ (20 mℓ) (elution of SCN^-)	30
F^-, Cl^-, Br^-, and I^- ($\simeq$3 meq each)	Anion exchange resin Wofatit® SBW Column containing the resin to a height of 150 cm a) 1 M Na oxalate (elution of F^- followed by Cl^-) b) 1 M $NaNO_3$ (elution of Br^- followed by I^-)	32
F^- from interfering ions	Amberlite® IRA-410 (100 mesh; OH^- form); when the amount of F^- is > 0.1 mg/mℓ, the Cl^- form can also be used and the column is eluted with NaCl solution Column: 15 × 0.7 cm operated at a flow rate of 1 mℓ/min a) Dilute acid sample solution (adsorption of F^-) b) 0.1 M NaOH (<100 mℓ) (elution of F^-; this elution can also be effected when F was adsorbed as SiF_6^{2-} which is decomposed with NaOH) Fluoride is determined colorimetrically; before the anion exchange separation, the sample solution is passed through a column of a cation exchange resin (H^+ form) to remove cations	8
Fluoride from nitrate	Amberlite® IRA-400 (20-50 mesh; chloride form). Column: 75 × 1 cm conditioned with 0.5 M NH_4Cl (100 mℓ) and operated at a flow rate of 7—10 mℓ/min a) Sample solution (50 mℓ) adjusted to pH 9 with HCl (adsorption of fluoride and nitrate) b) 0.5 M NH_4Cl adjusted to pH 9 (elution of fluoride; the first 20 mℓ of eluate is discarded and fluoride is collected in the next 100 mℓ; nitrate is not coeluted) Before the separation, the sample (containing from 100—500 μmol of fluoride) is pyrolyzed (at 900 °C) in the presence of WO_3 and the distillate is collected in NaOH solution; this solution which contains fluoride and nitrate is used to prepare solution (a)	21
Fluoride from sulfate	Amberlite® IRA-400 (30-50 mesh; nitrate form) Column: 12 × 1.5 cm operated at a flow rate of 1.5 mℓ/min a) Neutral sample solution (50—200 mℓ) (adsorption of fluoride and sulfate) b) 0.1 M $NaNO_3$ (35 mℓ) (elution of fluoride; sulfate is not coeluted)	31
Fluoride from phosphate	Dowex® 1-X10 (100-200 mesh; OH^- form) Column: 10.5 × 1 cm operated at a flow rate of 8—9 drops/min With 0.5 M NaOH as the eluent, fluoride (25 μg) is eluted before the phosphate (12.5 mg)	9

Table 5 (continued)
ANION EXCHANGE SEPARATION OF FLUORIDE FROM SYNTHETIC MIXTURES WITH OTHER ELEMENTS

Ions separated	Ion exchange resin, separation conditions, and remarks	Ref.
F^- from Br^-	Anionite AV-17 (OH^- form) Column: 9 × 1.6 cm operated at a flow rate of 5 mℓ/min a) Aqueous sample solution (20 mℓ containing ≃2 mg of each of the anions) (adsorption of F^- and Br^-) b) Water (15 mℓ) (as a rinse) c) 4% KNO_3 solution (80 mℓ) (elution of F^-) d) 1% KNO_3 solution (330 mℓ) (elution of Br^-)	10

Table 6
DETERMINATION OF FLUORIDE IN GEOLOGICAL AND BIOLOGICAL MATERIALS AFTER CATION EXCHANGE SEPARATION[a]

Material	Ion exchange resin, separation conditions, and remarks	Ref.
River-, rain-, and tapwater	Dowex® 50W-X8 (100-200 mesh; Zr[IV] form) Column of 1.2 cm ID containing ≃8—10 mℓ of the resin and operated at a flow rate of 5 mℓ/min., the Zr form is prepared by passing 1 M HNO$_3$ containing 1% Zr-nitrate until saturation; excess of this solution is removed by a water wash a) Sample (e.g. 1 ℓ) (adsorption of fluoride through complex formation with the Zr; accompanying elements pass into the effluent) b) Deionized water (few milliliters) (as a rinse) c) 2 M NaOH (~40 mℓ) (elution of fluoride; Zr is not coeluted) Fluoride is determined potentiometrically using an ion selective electrode; yields are 95 ± 5% for concentrations of fluoride <1 mM; interference by chloride and nitrate is negligible even at concentrations of ≃1 M, and millimolar concentrations of sulfate and phosphate cause only slight interference; only Al causes severe interference (due to anionic complex formation with fluoride in the sorption solution), but this can be masked (at concentrations of <1 mM) by the addition of 1,2-diaminocyclohexanetetraacetic acid or citrate; this method is also applicable to the determination of fluoride in reagent chemicals, cooking salts, and samples from lake salt-beds	55
Potable water	Dowex® 50W (H$^+$ form) Column: 40 × 3 cm The sample (100 mℓ) is passed through the resin bed to adsorb interfering cations such as Al and Fe(III) and in the effluent fluoride is determined spectrophotometrically (La-alizarin complexan) A similar procedure has been described[39] using a 20 × 1-cm column of the cation exchange resin Serdolit® (Serwa-Heidelberg Nr. 45540, Germany) (H$^+$ form); samples containing >1 mg/ℓ of Al have to be adjusted to pH 9 to achieve complete adsorption of this element	37
Mineral waters	Dowex® 50-X8 (50-100 mesh; Na$^+$ form) Column: 10 × 0.8 cm operated at a flow rate of 0.5 ± 0.1 mℓ/min a) Sample (~10 mℓ) (adsorption of Ca, Mg, and other multiple charge cations; NaF passes into the effluent) b) Distilled water (4 × 2 mℓ) (elution of residual NaF) Fluoride is determined potentiometrically (fluoride selective electrode) The recovery of fluoride is >97.9%	38
Phosphate rocks	Amberlite® IR-120 (H$^+$ form) Column: 100-mℓ buret filled with the resin to the 50-mℓ mark and operated at a flow rate of ~1 drop/sec a) ~0.3 M HNO$_3$ sample solution (10 mℓ, containing 5 mg of the rock sample) (adsorption of cations; fluoride passes into the effluent) b) Water (2 × 10 mℓ) (elution of residual fluoride) Fluoride is determined spectrophotometrically	36

Table 6 (continued)
DETERMINATION OF FLUORIDE IN GEOLOGICAL AND BIOLOGICAL MATERIALS AFTER CATION EXCHANGE SEPARATION[a]

Material	Ion exchange resin, separation conditions, and remarks	Ref.
Potato tops	Amberlite® CG-120 (H$^+$ form) a) Aqueous sample solution (adsorption of cationic constituents; F$^-$ passes into the effluents) b) Water (elution of remaining F$^-$) Fluoride is determined with an ion-specific electrode Before the separation, the dried sample (1 g) is extracted with water (25 mℓ) for 30 min (steam bath), the extract is filtered, and from the filtrate, solution (a) is prepared by dilution with water	40

[a] See also Table 3 in the chapter on Calcium, Volume V.

Table 7
DETERMINATION OF FLUORIDE IN INDUSTRIAL PRODUCTS AFTER CATION EXCHANGE SEPARATION[a]

Material	Ion exchange resin, separation conditions, and remarks	Ref.
Al fluoride, cryolite (Na$_3$AlF$_6$), and fluorspar (CaF$_2$)	Amberlite® IR-120 (H$^+$ form) Column: ~35 × 1.9—2.0 cm operated at a flow rate of 25 mℓ/min To an aliquot of the sample solution, a batch (10 g) of the resin is added and the mixture is allowed to stand 15 min swirling frequently (evolution of liberated CO$_2$); then the resin (containing the adsorbed cations) and the solution are transferred to the column using water as a rinse; in the effluent fluoride is determined titrimetrically Before the ion exchange separation, the fluoride in cryolite and fluorspar is solubilized by fusion with Na$_2$CO$_3$-K$_2$CO$_3$ mixture in the presence of SiO$_2$; the melt is extracted with water and after filtration the sample solution is obtained	44
Products of the Al industry (e.g., cryolite)	Cationite KU-2 (H$^+$ form) The sample (0.25 g) is fused with silica (0.25 g) and KNaCO$_3$ (6 g), the melt is leached with hot water, and the solution is diluted with water to 250 mℓ; an aliquot (25 mℓ, containing up to 15 mg of F) is stirred with 30 mℓ of the resin for 10 min, the resin is filtered off, and washed with water; in the filtrate and washings F$^-$ is determined titrimetrically	45
Chrome plating baths	Diaion® SK (H$^+$ form) a) Aqueous sample aliquot (5 mℓ) treated with 2 mℓ ethanol and 1 M HCl (3 mℓ) to reduce Cr$_2$O$_7^{2-}$ to Cr^{3+} (adsorption of Cr[III]; F$^-$ passes into the effluent) b) Water (20 mℓ) (elution of remaining F$^-$) Fluoride is determined colorimetrically using the Al-aluminon method Before the separation, the sample (10 mℓ), containing 0.2—5 mg of F$^-$ per liter is diluted with water to 750 mℓ	41
U tetrafluoride	Amberlite® IR-120 (16-50 mesh; H$^+$ form) Column: Jones reductor tube containing the resin to a height of 11—12 cm and operated at a flow rate of ~5 drops/sec a) Sample solution (~100 mℓ, containing 0.4—0.7 g of UF$_4$ dissolved in 25% NaOH (5 mℓ) and 30% H$_2$O$_2$ (10 mℓ, to oxidized U [IV] to U [VI]) (adsorption of U [VI] and Na; HF passes into the effluent) b) Water (150 mℓ) (elution of residual HF) HF is determined by titration with 0.2 M NaOH The method is also applicable to samples of UF$_6$ or UO$_2$F$_2$	46, 47
U tetrafluoride and NaUF$_5$	Amberlite® IR-120 (H$^+$ form) Column of 0.5 in. ID containing 100 mℓ of the resin and operated at a flow rate of 1 drop/sec a) Alkaline sample solution (80 mℓ) (adsorption of U[VI] and Na; HF passes into the effluent) b) Distilled water (400 mℓ) (elution of remaining HF) Fluoride is determined titrimetrically with Th-nitrate Before the separation, the sample (0.25 g) is decomposed with NaOH (0.5 g) and 30% H$_2$O$_2$ (1 mℓ) and then diluted with water to prepare solution (a)	48
U tetrafluoride	Katex® KU-2 (H$^+$ form) Column: 30 × 2 cm a) Alkaline sample solution (100 mℓ) (adsorption of U and Na; HF passes into the effluent) b) Water (250 mℓ) (elution of remaining HF) HF is determined titrimetrically with 0.2 M NaOH	49

Table 7 (continued)
DETERMINATION OF FLUORIDE IN INDUSTRIAL PRODUCTS AFTER CATION EXCHANGE SEPARATION[a]

Material	Ion exchange resin, separation conditions, and remarks	Ref.
Petroleum products (hydrocarbons)	Before the separation, the sample (0.3—0.5 g) is treated with 25% NaOH solution (5 mℓ) and 30% H_2O_2 solution (10 mℓ), and after removal of excess of the latter by boiling, the solution is diluted with water to prepare solution (a) Dowex® 50W-X8 (H^+ form) Column: 10 × 1.5 cm a) Aqueous extract (60 mℓ) (adsorption of Na^+; F^- passes into the effluent) b) 50% Ethanol (100 mℓ) (elution of remaining F^-) Fluoride is determined titrimetrically; the detection limit on samples of catalytically processed hydrocarbons in the C_{12} boiling range is 0.2 ± 0.1 ppm of F^- and the standard deviation at the 1-ppm level is ±0.087 ppm	42
Organic compounds	Before the ion exchange separation, the sample (100—200 mℓ) is mixed with Na biphenyl (45 mℓ) and toluene (100 mℓ) and after 10 min the mixture is extracted with water (4 × 15 mℓ) to prepare solution (a) Amberlite® IR-120 (H^+ form) Column of 18 cm height containing 20 g of the resin a) Aqueous sample solution (25 mℓ) (adsorption of Na^+ and other cations; F^- and Cl^- pass into the effluent) b) Water (elution of residual F^- and Cl^-) F^- and Cl^- are determined titrimetrically	43
Organic fluoro compounds	Before the separation, the sample (containing 10 mg each of F and Cl) is decomposed by fusion with Na Amberlite® IR-112 (H^+ form) Column: ≈50 × 3 cm operated at a flow rate of 35 drops/min The sorption solution (50 mℓ) is percolated through the resin bed to adsorb Na^+; the HF passing into the effluent is collected in 0.2 M NaOH and then determined titrimetrically	81
Organic polymers (fluoro and chloro-fluoro compounds)	Before the ion exchange separation, the sample (20—60 mg) is mixed with sucrose (200 mg) and decomposed by ignition in the presence of Na_2O_2 (4 g); the sorption solution is obtained by extracting the residue with water (50 mℓ) Cationite KU-1 (or KU-2 or SBS) (H^+ form) Column: 50-mℓ buret filled with the resin and operated at a flow rate of 8—10 mℓ/min a) Sample solution (25 mℓ) (adsorption of Na^+; HF passes into the effluent) b) Water (300 mℓ) (elution of residual HF) HF is determined titrimetrically with 0.05 M NaOH; before the ion exchange separation, the sample (0.2—0.3 g, containing 37—76% of F) is fused with Na (four- to fivefold amount)	51
Organofluorosilicon compounds	Cationite SDV-3 Column: 8 × 1.3 cm a) Sample solution (10 mℓ) (adsorption of Na^+; F^- passes into the effluent) b) Water (20 mℓ) (elution of remaining F^-) Fluoride is determined titrimetrically; before the separation the sample is decomposed by fusion with alkali	52

[a] See also Table 6 in the chapter on Boron and Table 10 in the chapter on Sulfur.

Table 8
CATION EXCHANGE SEPARATION OF FLUORIDE FROM SYNTHETIC MIXTURES WITH OTHER ELEMENTS

Elements separated	Ion exchange resin, separation conditions, and remarks	Ref.
Fluoride from Al, Fe(III), and other cations	Amberlite® IR-120 (H⁺ form) Column of ~3.1 cm ID containing the resin to a height of 6 in. and operated at a flow rate of 60 drops/min The sample solution (225 mℓ) is passed through the resin bed to adsorb the cations and in the effluent (the first 75 mℓ are discarded) fluoride is determined spectrophotometrically	70
F⁻ from Ca, Cd, Co, Cu, Ni, Pb, Zn, etc.	Amberlite® IR-120 (H⁺ form) Column: 25-mℓ buret containing 10 mℓ of the resin and operated at a flow rate of 1 mℓ/min a) HCl sample solution of pH ≃2 (adsorption of cations; F⁻ passes into the effluent) b) Deionized water (10 mℓ) (elution of remaining F⁻)	35
Fluoride from NaF	Zeo-Karb® 225 (40-80 mesh; H⁺ form) Column: 15 × 1.8 cm a) 0.1 M NaF (10 mℓ) (adsorption of Na⁺; HF passes into the effluent) b) Distilled water (until effluent is neutral) (elution of HF) HF is determined titrimetrically with NaOH The same procedure is applicable for the separation of chloride from NaCl	53

REFERENCES

1. **Newman, A. C. D.**, Separation of fluoride ions from interfering anions and cations by anion-exchange chromatography, *Anal. Chim. Acta*, 19, 471, 1958.
2. **Evans, W. H. and Sergeant, G. A.**, Determination of small amounts of fluorine in rocks and minerals, *Analyst (London)*, 92, 690, 1967.
3. **Rausa, G. and Trivello, R.**, Spectrophotometric method for the determination of fluorine in urine, *Ig. Mod.*, 63, 89, 1970.
4. **Cox, F. H. and Backer Dirks, O.**, The determination of fluoride in blood serum, *Caries Res.*, 2(1), 69, 1968.
5. **Van der Mark, W. and Das, H. A.**, Determination of fluorine in bone by fast neutron activation analysis, *J. Radioanal. Chem.*, 13(1), 107, 1973.
6. **Coursier, J. and Saulnier, J.**, Separation of trace amounts of fluoride from tervalent aluminum salts by ion exchange, *Anal. Chim. Acta*, 14, 62, 1956.
7. **Poirier, M.**, Ion-exchange separation of fluoride from phosphate for the micro determination of fluorine in the presence of phosphorus, *Talanta*, 22, 607, 1975.
8. **Funasaka, W., Kawase, M., Kojima, T., and Matsuda, Y.**, Separation of fluoride ions by the use of ion exchange resin, *Jpn. Analyst*, 4 (8), 514, 1955.
9. **Zipkin, I., Armstrong, W. D., and Singer, L.**, Chromatographic separation of fluoride and phosphate, *Anal. Chem.*, 29, 310, 1957.
10. **Eristavi, D. I., Shatirishvili, I. Sh., and Mekvabishvili, L. F.**, Anion-exchange chromatographic separation of fluorine from bromine, *Tr. Gruz. Politekh. Inst.* 4(132), 33, 1969.
11. **Starobinets, G. L. and Zakhartseva, E. P.**, Chromatographic separation of halide and thiocyanate ions, *Izv. Akad. Nauk B. SSR Ser. Khim. Nauk*, No. 6, 105, 1970.
12. **Johansson, C. E.**, Enthalpimetric determination of fluoride, *Talanta*, 17, 739, 1970.
13. **Glasö, Ö. S.**, Determination of fluorine in iron ore and apatite, *Anal. Chim. Acta*, 28, 543, 1963.
14. **Supharungsun, S. and Wainwright, M.**, Determination, distribution and adsorption of fluoride in atmospherically polluted soils, *Bull. Environ. Contam. Toxicol.*, 28(5), 632, 1982.
15. **Dirks, O. B. and Cox, F. H.**, Fluoride determination on fluoridated drinking water, *Caries Res.*, 1(4), 295, 1967.
16. **Fresen, J. A., van Gogh, H., and van Pinxteren, J. A. C.**, Determination of fluoride in drinking water with the aid of an ion-exchange resin, *Pharm. Weekbl.*, 95(2), 33, 1960.
17. **Light, T. S., Mannion, R. F., and Fletcher, K. S.**, Determination of fluoride in potable water by ion exchange and potentiometric titration, *Talanta*, 16, 1441, 1969.
18. **Paez, D. M. and Vasquez, J. A.**, Use of ion-exchange resins for the ultra-micro determination of fluoride in air, *Revta Asoc. Bioquim. Argent.*, 33, 152, 1968.
19. **Raby, B. A.**, The direct determination of fluoride in molybdenum, rhenium and tungsten, Report UCRL-50522, University of California, Lawrence Radiation Laboratory, October 9, 1968.
20. **Nickl, J. J. and von Braunmuehl, C.** Separation and determination of sulfate and fluoride ions in etching solutions, *Fresenius' Z. Anal. Chem.*, 246, 313, 1969.
21. **Maeck, W. J., Booman, G. L., Elliott, M. C., and Rein, J. E.**, Radiometric extraction method for fluoride. Pyrolysis-ion-exchange separation, *Anal. Chem.*, 32, 922, 1960.
22. **Kelso, F. S., Matthews, J. M., and Kramer, H. P.**, Ion-exchange method for determination of fluoride in potable water, *Anal. Chem.*, 36, 577, 1964.
23. **Nielsen, H. M.**, Determination of microgram quantities of fluoride, *Anal. Chem.*, 30, 1009, 1958.
24. **Ohno, S., Suzuki, M., Sasajima, K., and Iwata, S.**, Determination of fluorine in urine by photonuclear-activation analysis, *Analyst (London)*, 95, 260, 1970.
25. **Ohno, S., Suzuki, M., Kadota, M., and Yatazawa, M.**, Determination of trace fluorine in biological materials by photonuclear activation analysis, *Mikrochim. Acta*, No. 1, 61, 1973.
26. **Nielsen, J. P. and Dangerfield, A. D.**, Use of synthetic resin ion-exchange materials for the determination of atmospheric fluorides, *Arch. Ind. Health*, 11(1), 61, 1955.
27. **Johansson, C. E.**, Fluoride determination with a micro flow-calorimeter, *Talanta*, 19, 1349, 1972.
28. **Read, J. I. and Collins, R.**, Estimation of total fluoride in foodstuffs, *J. Assoc. Public Anal.*, 20(4), 109, 1982.
29. **Brunzie, G. F. and Pflaum, R. T.**, Direct spectrophotometric determination of fluoride ion, *Proc. Iowa Acad. Sci.*, 69, 186, 1963.
30. **Boiko, V. F., Zalevskaya, T. L., and Matusevich, L. A.**, Chromatographic separation of anions on the anion exchanger'AV-17, *Khim. Khim. Tekhnol. (Minsk)*, 10, 98, 1976.
31. **Mehra, M. C. and Lambert, J. L.**, Colorimetric determination of fluoride ions with a solid analytical reagent, *Microchem. J.*, 18(3), 226, 1973.
32. **Holzapfel, H. and Gürtler, O.**, Analytical investigations of the anion exchanger Wofatit SBW. II. Separation of halide ions, *J. Prakt. Chem.*, 35(3-4), 113, 1967.

33. **Zalevskaya, T. L. and Starobinets, G. L.**, Chromatographic separation of mixture of halide ions, *Zh. Anal. Khim.*, 24(5), 721, 1969.
34. **Fujimoto, M., Nakayama, H., Ito, M., Yanai, H., and Suga, T.**, Micro-analysis with use of ion-exchange resins. XXVII. Detection of traces of fluoride by decolorisation of enriched metal chelates and by color changes of lanthanoid(III)-alizarin complexans, *Mikrochim. Acta*, No. 2, 151, 1974.
35. **Deane, S. F., Leonard, M. A., McKee, V., and Svehla, G.**, Sulphonated alizarin fluorine blue (AFBS). IV. Critical comparison of the use of AFBS against alizarin fluorine blue (AFB) and the fluoride electrode for the determination of low fluoride concentrations; interferences with the AFBS method and their removal, *Analyst (London)*, 103, 1134, 1978.
36. **Shapiro, L.**, Rapid determination of fluorine in phosphate rocks, *Anal. Chem.*, 32, 569, 1960.
37. **Hidalgo de Cisneros, J. L., Bueno Garesse, E., and Perez-Bustamante, J. A.**, Determination of fluoride in potable water from Cadiz province, *An. Bromatol.*, 34(1), 13, 1983.
38. **Gorenc, B. and Babic, J.**, Potentiometric determination of fluoride in mineral waters by means of an ion-selective electrode, *Acta Pharm. Jugosl.*, 25(3), 171, 1975.
39. **Kempf, T.**, Photometric determination of fluoride in potable water with lanthanum-alizarin complexan after separation of interfering ions by ion-exchange, *Fresenius' Z. Anal. Chem.*, 244, 113, 1969.
40. **McElfresh, P. M.**, Ion-specific-electrode analysis of fluoride in potato tops using an ion-exchange pre-treatment, *J. Agric. Food Chem.*, 26(1), 276, 1978.
41. **Higashino, T. and Musha, S.**, Determination of fluorine in chrome-plating baths containing fluorides, *Jpn. Analyst*, 4(1), 3, 1955.
42. **Miller, M. and Keyworth, D. A.**, Determination of traces of organic fluorine in hydrocarbons, *Talanta*, 14, 1287, 1967.
43. **Banks, R. E., Cuthbertson, F., and Musgrave, W. K. R.**, Semimicro determination of fluorine, chlorine and nitrogen in organic compounds. II. The use of cation-exchange resins, *Anal. Chim. Acta*, 13, 442, 1955.
44. **Shehyn, H.**, Acidimetric determination of fluorine after ion exchange. Application to aluminum fluoride, cryolite and fluorspar, *Anal. Chem.*, 29, 1466, 1957.
45. **Zinov'eva, L. D. and Gladysheva, K. F.**, Rapid determination of fluorine in products of the aluminum industry, *Zavod. Lab.*, 39(1), 23, 1973.
46. **Sporek, K. F.**, Determination of total fluoride content in uranium tetrafluoride using ion-exchange columns, *Anal. Chem.*, 30, 1030, 1958.
47. **Sporek, K. F.**, New analytical methods in the technology of uranium, *Chem. Can.*, 11(4), 66, 1959.
48. **Vogliotti, F.**, Use of thorium nitrate in the titration of fluorine after its separation from uranium by ion exchange resins, *Energ. Nucl.*, 6, 661, 1959.
49. **Singer, E.**, Analysis of uranium tetrafluoride, *Chem. Listy*, 59(2), 225, 1965.
50. **Lundgren, D. P. and Loeb, N. P.**, Automation of ion-exchange chromatographic analysis of condensed phosphate mixtures, *Anal. Chem.*, 33, 366, 1961.
51. **Gurvich, D. B. and Balandina, V. A.**, Determination of fluoride ion in polymers by electrometric titration and ion-exchange chromatography, *Plast. Massy*, No. 6, 54, 1962.
52. **Klimova, V. A. and Vitalina, M. D.**, Application of a cationite in the determination of fluorine in organo-fluorosilicon compounds by thorimetric titration, *Izv. Akad. Nauk SSSR Otd. Khim. Nauk*, No. 12, 2245, 1962.
53. **Aynsley, E. E.**, The determination of fluoride and chloride by an ion-exchange method, *Sch. Sci. Rev.*, 135, 270, 1957.
54. **Fukushi, K. and Hiiro, K.**, Determination of fluoride ion in seawater by capillary-type isotachophoresis after coprecipitation enrichment, *Jpn. Analyst*, 34(4), 205, 1985.
55. **Kokubu, N., Kobayasi, T., and Yamasaki, A.**, Preconcentration of low-level fluoride ions in natural water samples with ziconium-loaded cation-exchange resin, *Jpn. Analyst*, 29(2), 106, 1980.
56. **Jeffery, P. G. and Williams, D.**, Determination of fluorine in deposite-gange samples, *Analyst (London)*, 86, 590, 1961.
57. **Fong, C. C. and Huber, C. O.**, Fluoride determination by magnesium atomic-absorption inhibition-release effects, *Spectrochim. Acta B*, 31(3), 113, 1976.
58. **Zolotavin, V. L. and Kazakova, V. M.**, Absorptiometric determination of fluorine in natural waters and effluents, *Zavod Lab.*, 31(3), 297, 1965.
59. **Iglesias, M. A., Garcia-Vargas, M., and Perez-Bustamente, J. A.**, Fluorimetric determination of fluoride traces using an ion-exchange method and the aluminum-Eriochrome red B complex, *Quim. Anal.*, 3(4), 280, 1984.
60. **Brownley, F. I. and Howle, C. W.**, Spectrophotometric determination of fluoride in water, *Anal. Chem.*, 32, 1330, 1960.
61. **Ashratova, Sh. K.**, Determination of fluoride ion in cryolite by means of ion-exchange chromatography, *Zavod. Lab.*, 23(9), 1064, 1957.

62. **Benz, R., Douglas, R. M., Kruse, F. H., and Penneman, R. A.**, Preparation and properties of several ammonium uranium(IV) and ammonium plutonium(IV) fluorides, *Inorg. Chem.*, 2(4), 799, 1963.
63. **Volodina, M. A., Gorshkova, T. A., and Terentèv, A. P.**, Ammonia in organic analysis. I. Ammonia method for determination of fluorine in organic compounds, *Zh. Anal. Khim.*, 24(7), 1121, 1969.
64. **Eger, C. and Yarden, A.**, Determination of fluorine in organic fluoro compounds, *Bull. Res. Counc. Isr.*, 4(3), 305, 1954.
65. **Johncock, P., Musgrave, W. K. R., and Wiper, A.**, Semi-micro determination of fluorine and chlorine in organic compounds. III. The decomposition of fluoro and chloro-fluoro compounds by the diphenyl-sodium-dimethoxy-ethane complex, *Analyst (London)*, 84, 245, 1959.
66. **Kijima, R. and Ueno, H.**, Determination of chlorine and fluorine in fluorocarbon plastics by fusion with potassium, *J. Chem. Soc. Jpn. Ind. Chem. Sect.*, 61(2), 270, 1958.
67. **Schröder, E. and Waurick, U.**, Determination of the fluorine content of high polymers, *Plaste Kautsch.*, 7(1), 9, 1960.
68. **Gallacher, E. J. and Wright, H.**, Micro-determination of cryolite in cordites by the oxygen-flask technique, *Mikrochim. Acta*, No. 3, 535, 1967.
69. **Mitchell, A.**, Simple method for determination of fluoride in electroslag melts, *J. Iron Steel Inst.*, 203(4), 378, 1965.
70. **Yasuda, S. K. and Lambert, J. L.**, Cellulose-supported thorium-alizarin red S reagent for fluoride-ion determination, *Anal. Chem.*, 30, 1485, 1958.
71. **Henrion, G., Marquardt, D., and Stoecker, B.**, Determination of halogens by flame emission of metal halides, *Z. Phys. Chem. (Leipzig)*, 260(6), 1053, 1979.
72. **Wayman, D. H.**, Determination of hydrofluoric acid in nitric-hydrofluoric acid mixtures. Development of a field test, *Anal. Chem.*, 28, 865, 1956.
73. **Nordmann, F. and Engelmann, C.**, Separation of fluorine-18 and fluorine-17 from sodium, *Radiochem. Radioanal. Lett.*, 16(1), 17, 1973.
74. **Hensley, A. L. and Barney, J. E., II**, Spectrophotometric determination of fluoride with thorium chloranilate, *Anal. Chem.*, 32, 828, 1960.
75. **Abe, S. and Kikuchi, H.**, Application of adsorption-barrier technique in the ring-oven method. II. Micro-determination of fluoride and oxalate, *Mikrochim. Acta*, 1(4), 379, 1975.
76. **Mader, C.**, Removal of lead interference in fluoride analysis by cation-exchange resin. *Chemist Analyst*, 44(4), 86, 1955.
77. **Ramasamy, J. and Lambert, J. L.**, Colorimetric determination of fluoride with an insoluble inverse basic beryllium-carboxylate dye complex, *Anal. Chem.*, 51, 2044, 1979.
78. **Bond, A. M. and O'Donnell, T. A.**, Determination of fluoride by atomic-absorption spectrophotometry, *Anal. Chem.*, 40, 560, 1968.
79. **de Kleijn, J. P. and van Zanten, B.**, Paper-chromatographic behavior of reactor-produced fluorine-18, *J. Radioanal. Chem.*, 45(1), 195, 1978.
80. **van Gogh, H.**, Determination of microgram amounts of fluorine in biological material, *Pharm. Weekbl. Ned.*, 101(40), 881, 1966.
81. **Eger, C. and Yarden, A.**, Semi-micro determination of fluorine in organic fluoro compounds, *Anal. Chem.*, 28, 512, 1956.

CHLORINE

ANION EXCHANGE RESINS

General

Adsorption of chloride on strongly basic anion exchange resins from very dilute acid and neutral media[1-11] as well as from alkali hydroxide solutions is a very effective means to separate Cl from virtually all accompanying cationic elements. For this purpose chloride is usually adsorbed on resins in the nitrate form[1,2,8,9,11-14] or hydroxide form.[10,15,16] Retention of chloride by a sulfate-form resin has also been utilized.[17] Coadsorbed with chloride from the above systems are other anions such as perchlorate,[18,19] chlorate,[20-22] chlorite,[22-24] bromide, iodide, phosphate, and sulfate.

Elution of the adsorbed chloride is, in most cases, performed with 0.5 M solutions of Na nitrate[1-3,9,25] or ammonium nitrate[9,16,26] or with 0.5, 1, and 1.5 M solutions of the former eluent in 10,[27] or 20,[28] >20,[29] and 50%[8] acetone, respectively. Other eluents that have been used for this elution include 0.75 M NH$_4$NO$_3$,[7] 1.5 M NaNO$_3$,[12] 1.8[30] and 2 M KNO$_3$,[11] 5 M NH$_4$NO$_3$,[5,6] 0.01 M Na acetate (pH ≃ 5.4),[10] and 1 M NaOH.[4] After activation with neutrons adsorbed chloride can be determined radiometrically directly on the resin.[15]

Adsorbed perchlorate can be eluted with 1 M trichloroacetate at pH 7.3[18] or 1.5 M Na iodide.[19] For the elution of chlorate 2 M Na nitrate,[20] 1 M NaCl,[22] and 0.1 M K sulfate[21] have been used. Coadsorbed chlorite is eluted before the chlorate using 0.1 M NaCl as eluent.[22-24]

Applications

Adsorption of chloride on anion exchange resins has been employed in connection with the determination of this ion in rocks (see Table 1*),[1,15] waters,[14,27,28] and in a variety of industrial products (see Table 3).[3-7,14,16,26] From Table 2 it is seen that anion exchange on basic resins has also been applied for the assay of perchlorate in biological materials and seawater.[18,19]

For the mutual separation of chloride, bromide, and iodide the procedures outlined in Table 4[8-11] and also other methods based on anion exchange chromatography can be used.[12,25,29,30]

Anion exchange of oxyanions of halogens has not only been used for their mutual separation (see Table 5),[20-22] but also in connection with the determination of chlorate and chlorite in spent bleaching liquors.[24]

CATION EXCHANGE RESINS

General

Chloride and other anionic Cl species are not retained by strongly acidic cation exchange resins from weakly acid to neutral solutions[16,31-37] and ammoniacal systems (see Tables 6 and 7).[2,17,38] This nonadsorbability from acid and neutral solutions makes it possible to separate Cl from accompanying cations which are retained by the H$^+$-form resins under these conditions. Adsorption of the cationic amine complex of Ag, i.e., Ag(NH$_3$)$_2^+$ from ammoniacal media allows to isolate chloride from AgCl after precipitation of this compound and its subsequent dissolution in ammonia solution (see Table 7).[2,17,38] Freshly precipitated AgCl also reacts quantitatively with a strong acid ion exchange resin (H$^+$ form) liberating hydrochloric acid (HCl) which can be determined titrimetrically.[39]

---

* Tables for this chapter appear at the end of the text.

After adsorption of the cationic constituents of the sample solution the resin is usually washed with water to elute residual chloride remaining in the ion exchange column.[2,16,32-34,36,38,40,41]

Coeluted with chloride are other halogen species as well as all anionic constituents.

Applications

Cation exchange separations of Cl species, mainly chloride, have been used in connection with the determination of Cl in waters (see Table 6),[37,42-45] beer (see Table 6),[32] plants (see Table 6),[31] and in industrial materials of the type listed in Table 7,[2,33-36,38,46] as well as in organic compounds,[39,47] pharmaceutical preparations,[40] flue dusts (or bonded deposits),[41] paper extracts,[48] and pulping liquors.[49]

Cation exchange methods have also been described for the isolation of chloride from synthetic mixtures with other elements,[50-53] and for its microchemical detection (see Table 17 in the chapter on Sulfur).

Table 1
DETERMINATION OF CHLORIDE IN GEOLOGICAL MATERIALS AFTER ANION EXCHANGE SEPARATION

Material	Ion exchange resin, separation conditions, and remarks	Ref.
Silicate rocks	Dowex® 1-X10 (100-200 mesh; nitrate form) Column: 8 × 0.7 cm operated at a flow rate of 0.85 mℓ/min a) Neutral sample solution (10-mℓ aliquot) (adsorption of chloride; matrix constituents pass into the effluent) b) 0.5 M NaNO$_3$ (elution of chloride) Chloride is determined potentiometrically using a Cl$^-$-sensitive electrode Before the separation, the sample (0.5 g) is fused with Na$_2$CO$_3$ (4 g) and the melt is dissolved in water; the solution is neutralized with conc HNO$_3$ and digested for 2—3 hr, then the digest is filtered and the filtrate is diluted to 50 mℓ (= solution [a])	1
Micas and amphiboles (in granitic rocks)	Amberlite® IRA-400 (OH$^-$ form) After pyrohydrolysis of the sample, the condensate is stirred for 5 min with 500 mg of the resin to adsorb chloride; then the resin is irradiated with neutrons and ^{38}Cl is determined by γ spectrometry	15

Table 2
DETERMINATION OF PERCHLORATE IN BIOLOGICAL MATERIALS AND SEAWATER AFTER ANION EXCHANGE SEPARATION

Material	Ion exchange resin, separation conditions, and remarks	Ref.
Urine, silver beet, cabbage, and seawater	Dowex® 2-X8 (200-400 mesh; trichloroacetate form) Column: 12.5 × 1.2 cm a) Aqueous sample solution (adsorption of ClO$_4^-$) b) Water (as a rinse) c) 1 M ammonium trichloroacetate of pH 7.3 (200 mℓ) (elution of perchlorate which is contained in the second 100-mℓ fraction of eluate) Then the perchlorate is coprecipitated with tetraphenylarsonium perrhenate and determined by IR spectrophotometry; the coefficient of variation is 14% (11 samples), and the limit of detection is ≃5 μg of KClO$_4$	18
Urine and serum	Amberlite® IR-45 (weakly basic resin) Column filled with the resin to a height of 2.5 cm a) Aqueous sample solution (adsorption of ClO$_4^-$) b) Distilled water (small amount) (as a rinse) c) 1.5% NaI solution (18 mℓ) (elution of ClO$_4^-$) Perchlorate is determined spectrophotometrically; the mean recovery of ClO$_4^-$ (80 μg) is 97 ± 1% and the limit of detection is 5 μg ClO$_4^-$ per milliliter	19

Table 3
DETERMINATION OF CHLORIDE IN INDUSTRIAL PRODUCTS AFTER ANION EXCHANGE SEPARATION

Material	Ion exchange resin, separation conditions, and remarks	Ref.

NaOH, Na metal, and S	Amberlite® CG-400 (100-200 mesh; OH$^-$ form) and Amberlite® CG-120 (100-200 mesh; NH$_4^+$ form)
	I) Analysis of NaOH and Na metal
	A) First column operation (anion exchange resin)
	Column: 5.5 × 1.15 cm operated at a flow rate of 1 mℓ/min
	  a) ≃5 M NaOH sample solution (containing 10 g of the NaOH or 5 g of the metal sample) (adsorption of Cl$^-$; the matrix passes into the effluent)
	  b) Water (50 mℓ) (removal of remaining NaOH)
	  c) 0.5 M NH$_4$NO$_3$ (20 mℓ) (elution of Cl$^-$ together with coadsorbed species of Cu[II])
	Al and other elements, which after neutralization of the eluate with 1 M HNO$_3$, are removed by the following cation exchange separation procedure
	B) Second column operation (cation exchange resin)
	Column: 5 × 0.75 cm operated at a flow rate of 0.5 mℓ/min
	  a) Neutralized eluate obtained from eluate (c) (adsorption of Cu and Al; Cl$^-$ passes into the effluent)
	  b) Water (5 mℓ) (elution of remaining Cl$^-$)
	Chloride is determined spectrophotometrically
	II) Analysis of S
	A) First column operation (same resin as above under [I.A])
	Column: 2.8 × 1.4 cm operated at a flow rate of 1 mℓ/min
	  a) <3 M NaOH sample solution (adsorption of Cl$^-$)
	  b) 3 M NaOH followed by water (50 mℓ) (as rinses)
	  c) 0.5 M NH$_4$NO$_3$ (30 mℓ) (elution of Cl$^-$ and coadsorbed elements)
	B) Second column operation (same as described above under [I.B.])
	Before the first column operation the sample (1 g) is dissolved in 3 M NaOH + H$_2$O$_2$
		16

Al	Amberlite® CG-400 (100-200 mesh; OH$^-$ form)
	Column: 5 × 1.2 cm
	  a) Alkaline sample solution (50 mℓ) (adsorption of Cl$^-$; into the effluent passes Al as aluminate)
	  b) 3 M NaOH (20 mℓ) (elution of remaining Al)
	  c) Water (20 mℓ) (as a rinse)
	  d) 0.5 M NH$_4$NO$_3$ (30 mℓ) (elution of Cl$^-$)
	Cl is determined spectrophotometrically
	Before the separation, the sample (1 g, containing ≪1 ppm of Al) is dissolved in 3 M NaOH (35 mℓ) and 3% H$_2$O$_2$ (1 mℓ) and the solution is diluted to 50 mℓ to prepare solution (a)
		26

Organic compounds	Dowex® 1-X10 (100-200 mesh; NO$_3^-$ form)
	Column: 10 cm × 3 cm^2
	  a) Neutral sample solution (≃8 mℓ, which is < 3 M in NaNO$_3$) (adsorption of Cl$^-$ and Br$^-$)
	  b) 0.5 M NaNO$_3$ (70 mℓ) (elution of Cl$^-$)
	  c) 2 M NaNO$_3$ (70 mℓ) (elution of Br$^-$)
	Before the separation, the sample (10 mg) is decomposed with Na$_2$O$_2$ in a microbomb and then treated with HNO$_3$ and water
		3

Sulfite pulp waste liquors	Strongly basic anion exchange resin Dowex® SAR (20-50 mesh; nitrate form) Column: 17 × 3 cm operated at a flow rate of 2—5 mℓ/min	5

Table 3 (continued)
DETERMINATION OF CHLORIDE IN INDUSTRIAL PRODUCTS AFTER ANION EXCHANGE SEPARATION

Material	Ion exchange resin, separation conditions, and remarks	Ref.
	a) Sample solution of pH > 4 (adsorption of chloride; organic material and other constituents of the liquor pass into the effluent) b) Methanol (100 mℓ) (as a rinse) c) 5 M NH$_4$NO$_3$ (elution of chloride) Chloride is determined potentiometrically	
Sulfite waste liquor	Dowex® 2 (grain size: 0.12—0.3 mm; NO$_3^-$ form) Column: 15 × 1 cm operated at a flow rate of 2.5 mℓ/min a) Sample solution (50 mℓ) (adsorption of Cl$^-$, thiosulfate, and polythionate) b) Water (50 mℓ) (as a rinse) c) 5 M NH$_4$NO$_3$ (100 mℓ) (elution of Cl$^-$; coeluted are thiosulfate and polythionate) After destruction of the thio- and polythionates by oxidation with H$_2$O$_2$ in alkaline medium, Cl$^-$ is determined by potentiometric titration with AgNO$_3$	6
Alcohol-soluble matter in detergents	Strongly basic anion exchange resin (OH$^-$ form) Column: 25 × 1.4 cm operated at a flow rate of 2 mℓ/min a) Sample solution (diluted to contain ≃0.01 M Cl$^-$) (adsorption of Cl$^-$) b) Water (15 mℓ) (as a rinse) c) 1 M NaOH (500 mℓ) (elution of Cl$^-$) Chloride is determined titrimetrically	4
Purified water	Anionite AV-17 (NO$_3^-$ form) (2.8 g) in a column operated at a flow rate of 24 ℓ/hr a) Water sample (10 ℓ) (adsorption of Cl$^-$; cationic impurities such as Na$^+$ pass into the effluent) b) 0.75 M NH$_4$NO$_3$ (20 mℓ) (elution of Cl$^-$) Chloride is determined potentiometrically or nephelometrically; when the sample contained 100 μg of Cl$^-$, the separation is complete and the error ±10%	7

Table 4
MUTUAL SEPARATION OF CHLORIDE, BROMIDE, AND IODIDE, BY ANION EXCHANGE

Elements separated	Ion exchange resin, separation conditions, and remarks	Ref.
Chloride, bromide, and iodide	High performance TSK LS-222 anion exchange resin (grain size: 6 μm) PTFE column: 5.9 × 0.05 cm a) Sample solution (injection volume of 0.2—0.9 μℓ containing 6.9 μg of Cl$^-$, 0.057 μg of Br$^-$, and 0.1 μg of I$^-$) b) 1:1 Mixture (152 μℓ) of acetone and 1.5 M NaNO$_3$ (sequential elution of Cl$^-$, Br$^-$, and I$^-$ within 19 min) Iodide (0.08 μg) could be separated in 4.5 min from Br$^-$ (34 μg) and Cl$^-$ (41 μg) on a 1.2-cm-long column of the same resin, with use of a 1:3 mixture of acetone and 1.5 M NaNO$_3$	8
Cl$^-$, Br$^-$, and I$^-$	Dowex® 1-X10 (100-200 mesh; NO$_3^-$ form) Column: 6.7 cm × 3.4 cm^2 conditioned with 0.5 M NaNO$_3$ and operated at a flow rate of 1 mℓ/min a) 0.5 M NaNO$_3$ sample solution (2 mℓ containing ≯2.6 meq of any one halide) (adsorption of Cl$^-$, Br$^-$, and I$^-$) b) 0.5 M NaNO$_3$ (55 mℓ) (elution of Cl$^-$) c) 2 M NaNO$_3$ (elution of Br$^-$ which is collected in the first 55 mℓ of eluate followed by I$^-$ contained in the third fraction of 260 mℓ)	9
	Anionite MG-36 (OH$^-$ form) Column operated at a flow rate of 25 mℓ/min After adsorption of the anions (≃25 meq) their fractionation is achieved by using the following eluents a) 0.01 M Na acetate at pH 5.45 (200—300 mℓ) (elution of Cl$^-$) b) 0.025 M Na acetate (200—300 mℓ) (elution of Br$^-$) c) 0.1 M NaOH (150—200 mℓ) (elution of I$^-$) The error caused by the separation is within the limits ≃1.16 to +2.01%	10
Cl$^-$ from I$^-$	Amberlite® IRA-401 (NO$_3^-$ form) Column: 26 × 0.6 cm Following adsorption of both ions, Cl$^-$ is separated from up to a 100-fold excess of coadsorbed I$^-$ by elution with 2 M KNO$_3$ (30 mℓ); in the eluate chloride is determined by mercurimetric, potentiometric titration	11

Table 5
MUTUAL SEPARATION OF HALOGEN OXYANIONS BY ANION EXCHANGE

Elements separated	Ion exchange resin, separation conditions, and remarks	Ref.
ClO_3^-, BrO_3^-, and IO_3^-	Dowex® 2-X8 (100-200 mesh; nitrate form) Column: 28 cm × 6.15 cm² operated at a flow rate of 2 mℓ/min After adsorption of the anionic species the following eluents are used for their fractionation a) 0.5 M NaNO₃ (270 mℓ) (elution of iodate and bromate; the former is contained in the eluate fraction from 50—200 mℓ, while bromate is partly contained in the next 70 mℓ) b) 2 M NaNO₃ (elution of residual bromate and of chlorate which is present in the 410—620-mℓ fraction) Similar separations can be achieved on the resin Dowex® 21K	20
	Dowex® 2 Column: 50 × 1 cm a) Sample solution (≃10 mℓ) (adsorption of the halogenates at a flow rate of ≃0.3 mℓ/min) b) 0.2 M KOH (elution of IO_3^- and BrO_3^- in this order at a rate of ≃0.2 mℓ/sec) c) 0.1 M K₂SO₄ (elution of ClO_3^- at a flow rate of ≃0.2 mℓ/sec)	54
Chlorite, chlorate, and bromate	Dowex® 1-X8 (200-400 mesh; chloride form) Column: 10 × 1 cm a) Aqueous solution of the anionic species (adsorption of ClO_2^-, ClO_3^-, and BrO_3^-) b) 0.1 M NaCl (desorption of ClO_2^- and BrO_3^-, which are eluted with 65 mℓ each; ClO_2^- is eluted first) (flow rate: 2 mℓ/min) c) 1 M NaCl (100 mℓ) (elution of ClO_3^- at a rate of 1 mℓ/min)	22

Table 6
DETERMINATION OF Cl SPECIES IN PLANTS, BEER, AND SEAWATER AFTER CATION EXCHANGE SEPARATION[a]

Material	Ion exchange resin, separation conditions, and remarks	Ref.
Plant materials	Dowex® 50W-X8 (20-50 mesh; Na$^+$ form)	31
	The dried sample (20—100 mg) is burnt in a Schöniger flask containing 20 mℓ of water; after 30 min, 0.2 g of the resin is added to remove interfering cations	
	In a portion (10 mℓ) of the supernate, chloride is determined spectrophotometrically	
	The method is suitable for chloride concentrations in the range of 0.05—5%	
Beer	Amberlite® IR-120 (H$^+$ form)	32
	Column containing the resin to a height of 12 in. and operated at a flow rate of 1 drop/sec	
	a) Degassed sample (20 mℓ) (adsorption of cationic constituents; Cl$^-$ passes into the effluent)	
	b) Water (65 mℓ) (elution of remaining Cl$^-$)	
	Chloride is determined titrimetrically; the recovery of Cl$^-$ is 97.3—102.2%	
Seawater	Amberlite® IR-120 (H$^+$ form)	37
	Column: 20 × 1.5 cm	
	The sample (250 mℓ containing 50 mg of standard labeled KClO$_4$) is passed through the resin bed to adsorb cationic constituents, and in the effluent ^{36}Cl is determined radiometrically	
	This isotope dilution method can be used for the determination of ClO$_4^-$ in seawater samples	

[a] See also Tables 1 and 4 in the chapter on Sulfur.

Table 7
DETERMINATION OF HALOGENS IN INDUSTRIAL MATERIALS AFTER CATION EXCHANGE SEPARATION[a]

Material	Ion exchange resin, separation conditions, and remarks	Ref.
Irradiated Fe, Ti, and Ca targets	A) First column operation (Bio-Rad® AG50W-X8, 200-400 mesh; NH_4^+ form) a) 6 M NH_3 sample solution (adsorption of $Ag[NH_3]_2^+$; Cl^- passes into the effluent) b) 6 M NH_3 solution (20 mℓ) (elution of residual chloride) Following removal of most of the NH_3 by evaporation, the eluates (a) and (b) are subjected to the following separation (B) B) Second column operation (same resin as in [A] but in the H^+ form and of 400 mesh) a) Evaporated eluates (a) and (b) obtained by (A) (adsorption of NH_3 and any residual $Ag(NH_3)_2^+$; chloride passes into the effluent) b) Water (30—40 mℓ) (elution of remaining chloride) C) Third column operation (Bio-Rad® AG1-X10, 200-400 mesh; nitrate form) a) Combined eluates (a) and (b) obtained by (B) (adsorption of chloride) b) Water (4—5 column volumes) (as a rinse) c) 0.5 M $NaNO_3$ (80 mℓ) (elution of chloride) Then chloride is precipitated as Hg_2Cl_2 and ^{36}Cl is determined radiometrically Before the first ion exchange separation step, the irradiated sample (+1 mℓ of 1 M NaCl as carrier) is dissolved by heating with HNO_3 (for Fe), H_3PO_4 (for Ti), or H_2SO_4 (for CaF_2), and the resulting HCl is swept by N_2 into $AgNO_3$ solution (in HNO_3); the AgCl produced is dissolved in conc NH_3 solution (5 mℓ) and diluted with water to prepare solution (a) for (A)	2
Zinc sulfate electrolyte	Amberlite® IR-120 (180 mesh; NH_4^+ form) Column: 10 × 0.9 cm a) Ammoniacal sample solution (adsorption of Ag and Zn as the cationic complexes $Ag[NH_3]_2^+$ and $Zn[NH_3]_4^{2+}$; Cl^- passes into the effluent) b) Water (30 mℓ) (elution of remaining Cl^-) Chloride is determined titrimetrically Before the ion exchange separation step the Cl^- is preconcentrated from 40 mℓ of the electrolyte by coprecipitation of AgCl with Ag_2SO_4; the precipitate is dissolved in NH_3 solution to prepare solution(a)	38
Ni electrolyte	Cationite SBS (H^+ form) Column operated at a flow rate of 0.5—1.5 mℓ/min a) Sample solution (5 mℓ of electrolyte diluted to 50 mℓ) (adsorption of Ni, Cu, Co, Na, and other cations; Cl^- passes into the effluent) b) Water (150—200 mℓ) (elution of remaining Cl^-) Chloride is determined titrimetrically	33
Na dithionite	Catex® FN (H^+ form) Column operated at a flow rate of 1.5 mℓ/min a) Aqueous sample solution (~60 mℓ) (adsorption of Cu[II]; Cl^- passes into the effluent) b) Water (200 mℓ) (elution of remaining Cl^-) Chloride is determined titrimetrically; before the separation, the sample (10 g) is mixed with conc NH_3 (90 mℓ), water (50 mℓ), 10% Cu sulfate solution (25 mℓ) and dithionite is de-	34

Table 7 (continued)
DETERMINATION OF HALOGENS IN INDUSTRIAL MATERIALS AFTER CATION EXCHANGE SEPARATION[a]

Material	Ion exchange resin, separation conditions, and remarks	Ref.
	composed by boiling for 30 min; after dilution with water to 100 mℓ the solution is filtered, and a 50-mℓ aliquot is oxidized with 10% H_2O_2 (10 mℓ) to prepare solution (a)	
Corrosion products	Zerolit® 225 (H^+ form) Column: 60 × 2.5 cm conditioned with HNO_3 From the sample solution of pH 4.7 (containing 1—50 µg of Cl^-) Cu and Fe are adsorbed on the resin, while Cl^- passes into the effluent where it is determined spectrophotometrically	35
Binary mixtures containing $HgCl_2$, $HgBr_2$, and HgI_2	Cationite KU2-X8 (H^+ form) Column: 30 × 0.3 cm operated at a flow rate of 0.25 mℓ/min a) 1 M H_2SO_4 solution containing > 20 ng Cl^- and ≈6—20 µg Hg per milliliter (adsorption of uncomplexed Hg[II]; into the effluent passes $HgCl_2$) b) 0.5 M H_2SO_4 solution containing > 50 ng Br^- per milliliter and the same amount of Hg as in (a) (adsorption of uncomplexed Hg[II]; into the effluent passes $HgBr_2$) c) 0.3 M H_2SO_4 solution containing > 70 ng I^- per milliliter and the same amount of Hg as in (a) (adsorption of uncomplexed Hg[II]; into the effluent passes HgI_2) These separation procedures can be applied to the indirect atomic absorption determination of Cl^-, Br^-, and I^- via measurement of the Hg contained in the effluents (a), (b), and (c), respectively, which corresponds to the amount of halide in the sample The limits of detection were 0.2 ng of Cl^-, 0.5 ng of Br^-, and 0.8 ng of I^-	46
$KClO_4$ in the presence of $AlCl_3$ and HCl	Dowex® 50-X8 (50-100 mesh; H^+ form) Columns: 7 × 2 cm (for samples < 20% of $KClO_4$) and 34 × 2 cm (for samples with higher contents of $KClO_4$) and operated at a flow rate of 4 mℓ/min a) Sample solution (adsorption of K and Al; $HClO_4$ + HCl pass into the effluent) b) Water (~170 mℓ) (elution of remaining $HClO_4$ and HCl) After evaporation of the eluate (to 6—7 mℓ) $HClO_4$ is determined titrimetrically in the presence of residual HCl using nonaqueous Na acetate solution	36

[a] See also Table 8 in the chapter on Nitrogen; Table II in the chapter on Sodium, Volume V; and Table 9 in the chapter on Sulfur.

REFERENCES

1. **Akaiwa, H., Kawamoto, H., and Hasegawa, K.**, Determination of chlorine in silicate rocks by ion-exchange chromatography and direct potentiometry with an ion-selective electrode, *Talanta*, 26, 1027, 1979.
2. **Lagarde-Simonoff, M., Baklouti, M., Regnier, S., and Simonoff, G. N.**, Chemical separation of chlorine-36 formed by nuclear reactions in iron, titanium and calcium targets, *J. Inorg. Nucl. Chem.*, 39(9), 1710, 1977.
3. **Kondo, A.**, Organic elementary analysis with a micro-bomb. IV. Separation and micro-determination of chlorine and bromine in an organic compound by ion-exchange chromatography, *Jpn. Analyst*, 7(4), 232, 1958.
4. **Pomeranz, J.**, Determination of chloride in alcohol-soluble matter in detergents, *Chemist Analyst*, 43(4), 89, 1954.
5. **Kuenstner, W., Huepfl, J., and Pfeiffer, S.**, Chloride determination in sulphite pulp waste liquors, *Papier (Darmstadt)*, 32(10), 441, 1978.
6. **Lagerström, O. and Samuelson, O.**, Determination of chloride in sulphite waste liquor, *Sven. Papperstidn.*, 62(19), 679, 1959.
7. **Nikitina, N. G., Galkina, N. K., and Senyavin, M. M.**, Choice of conditions for the ion-exchange concentration and determination of trace impurities in high-purity substances, *Zh. Anal. Khim.*, 21(10), 1165, 1966.
8. **Ishii, D., Hirose, A., and Horiuchi, I.**, Studies on micro high-performance liquid chromatography. VI. Application of micro scale liquid-chromatographic technique to anion-exchange separation of halide ions, *J. Radioanal. Chem.*, 45(1), 7, 1978.
9. **DeGeiso, R. C., Rieman, W., III, and Lindenbaum, S.**, Analysis of halide mixtures by ion-exchange chromatography, *Anal. Chem.*, 26, 1840, 1954.
10. **Nabivanets, B. I.**, Chromatographic separation of chlorides, bromides and iodides, *Ukr. Khim. Zh.*, 22(6), 816, 1956.
11. **Brajter, K., Janowski, A., and Jachimowicz, E.**, Application of anion-exchange in the mercurimetric potentiometric titration of chloride in the presence of excess of iodide, *Chem. Anal. (Warsaw)*, 15(3), 657, 1970.
12. **Liu, M., Fan, B., and Hu, Z.**, High performance ion-exchange chromatographic separation and determination of chloride, bromide and iodide, *Gaodeng Xuexiao Huaxue Xuebao*, 3(1), 48, 1982.
13. **Bhatnager, R. P. and Mishra, D. D.**, Anion-exchange selectivity studies of halate ions on Amberlite IRA-400 (nitrate form) in aqueous acetone media, *J. Indian Chem. Soc.*, 58(8), 804, 1981.
14. **Howard, G. A. and Gjertsen, P.**, The determination of chloride in beer, wort and water, *J. Inst. Brew.*, 83, 161, 1977.
15. **Gillberg, M.**, Halogens and hydroxyl contents of micas and amphiboles, *Geochim. Cosmochim. Acta*, 28(4), 495, 1964.
16. **Fukasawa, T., Kano, S., Ono, H., and Mizuike, A.**, Ion-exchange in basic media and its analytical applications. I. Anion-exchange in sodium hydroxide medium and its application to the spectrophotometric determination of chloride in sodium hydroxide, sodium metal and sulphur, *Jpn. Analyst*, 19(10), 1417, 1970.
17. **Ubaldini, I. and Capizzi Maitan, F.**, Analysis of mixtures of chlorides and bromides, *Ann. Chim. (Roma)*, 48(3), 209, 1958.
18. **Loach, K. W.**, Determination of low-concentrations of perchlorate in natural materials, *Nature*, 196, 754, 1962.
19. **Weiss, J. A. and Stanbury, J. B.**, Spectrophotometric determination of micro amounts of perchlorate in biological fluids, *Anal. Chem.*, 44, 619, 1972.
20. **Skloss, J. L., Hudson, J. A., and Cummiskey, C. J.**, Quantitative separation of halate mixtures by ion-exchange chromatography, *Anal. Chem.*, 37, 1240, 1965.
21. **Fratz, D. D.**, Automated determination of salts in water-soluble certifiable color additives by ion chromatography, *J. Assoc. Off. Anal. Chem.*, 63(4), 882, 1980.
22. **Denis, M. and Masschelein, W. J.**, Determination of mixtures of chlorite, chlorate and bromate ions in aqueous solution, *Analusis*, 11(2), 79, 1983.
23. **Cauquis, G. and Limosin, D.**, Determination of chloride ions and mixed species containing oxygen and chlorine in aqueous media, *Analusis*, 5(2), 70, 1977.
24. **Eriksson, B. and Sjostrom, L.**, Determination of inorganic chlorine compounds and total chlorine in spent bleaching liquors. I. Volumetric method, *Sven. Papperstidn.*, 79(17), 570, 1976.
25. **Burriel-Marti, F. and Alvarez Herrero, C.**, Quantitative separation of halides in a mixture by ion-exchange chromatography, *Inf. Quim. Anal.*, 17(3), 77, 1963.
26. **Fukasawa, T. and Katagiri, K.**, Ion-exchange in basic media and its analytical applications. III. Determination of traces of chlorine in aluminum, *Jpn. Analyst*, 21(4), 480, 1972.

27. **Takata, Y.**, Determination of traces of chloride ions in water by ion-exchange chromatography, *Jpn. Analyst*, 24(8), 531, 1975.
28. **Takata, Y., Miyagi, H., Hirota, K., and Arikawa, Y.**, Trace analysis of water with mini-column preconcentration liquid chromatography, *Jpn. Analyst*, 26, 752, 1977.
29. **Muto, G., Takata, Y., and Tsuda, H.**, Ion-exchange separation of halide ions with mixed eluents, *J. Chem. Soc. Jpn. Pure Chem. Sect.*, 88(4), 432, 1967.
30. **Roehse, W., Roewer, G., Boran, R., and Hellmig, R.**, Determination of small concentrations of chloride, bromide and iodide ions by anion-exchange chromatography, *Z. Chem.*, 22(6), 226, 1982.
31. **Likussar, W., Huber, H., Raber, H., and Grill, D.**, Rapid method for the spectrophotometric determination of chloride in plant material, *Mikrochim. Acta*, 2(5-6), 467, 1976.
32. **West, D. B. and Lautenbach, A. F.**, Chlorides in beer by ion exchange, *Proc. Am. Soc. Brew. Chem.*, 87, 1959.
33. **Chernukha, G. N. and Churkina, K. M.**, The mercurimetric determination of chlorides in nitrate baths, electroplating solutions and other materials, in *Sovrem Metody Anal. Metall.*, M., *Metallurgizdat*, report of symposium, 1955, 220.
34. **Kuśák, O.**, Determination of sodium chloride in sodium dithionite, *Chem. Prum.*, 11(1), 26, 1961.
35. **Mor, E. D., Beccaria, A. M., and Poggi, G.**, Spectrophotometric determination of traces of chloride in corrosion products, *Anal. Chim. Acta*, 99, 361, 1978.
36. **Albaugh, E. W., Buhlert, J. E., and Pearson, R. M.**, Determination of potassium perchlorate in the presence of chloride ion by ion exchange and non-aqueous titration, *Anal. Chem.*, 35, 153, 1963.
37. **Johannesson, J. K.**, Determination of perchlorate by isotopic dilution with potassium perchlorate-chlorine-36, *Anal. Chem.*, 34, 1111, 1962.
38. **Kojima, M.**, Rapid determination of chlorine in zinc sulfate electrolyte, *Jpn. Analyst*, 6(5), 309, 1957.
39. **Makineni, S., McCorkindale, W., and Syme, A. C.**, Direct titration method for determining chlorine in organic compounds after Carius combustion, *J. Appl. Chem.*, 8(5), 310, 1958.
40. **Jarzebinski, J.**, Application of synthetic ion-exchange resins in quantitative analysis of pharmaceutical preparations. I. Analysis of inorganic compounds, *Farm., Pol.*, 26(3), 191, 1970.
41. **Grant, J.**, Improved methods of deposit analysis. I. Silicon, aluminum, calcium, magnesium, and chloride, *J. Appl. Chem. (London)*, 14(12), 525, 1964.
42. **Johannesson, J. K.**, The radiochemical determination of trace amounts of chloride, *J. Radioanal. Chem.*, 6(1), 27, 1970.
43. **Franks, M. C. and Pullen, D. L.**, Technique for determination of trace anions by the combination of a potentiometric sensor and liquid chromatography, with particular reference to the determination of halides, *Analyst (London)*, 99, 503, 1974.
44. **Moskvin, L. N., Krasnoperov, V. M., Fokina, R. G., and Vilkov, N. Ya.**, Continuous control of pH and chloride concentration in the aqueous coolant of nuclear reactors, *At. Energ.*, 38(3), 143, 1975.
45. **Konstantinov, B. P., Oshurkova, O. V., and Starobina, I. Z.**, Analysis of natural water and brine for chloride, nitrate and sulfate, *Zh. Prikl. Khim. (Leningrad)*, 40(6), 1260, 1967.
46. **Chuchalina, L. S., Yudelevich, I. G., and Chinenkova, A. A.**, Indirect atomic absorption determination of microgram amounts of halides. Determination of chloride, bromide, or iodide in pure solutions, *Zh. Anal. Khim.*, 36(5), 920, 1981.
47. **Shah, R. A. and Jabbar, S. A.**, Micro-determination of chlorine in chloro-organic compounds, *Pak. J. Sci. Ind. Res.*, 5(3), 162, 1962.
48. **Heathcote, C., Dunk, R., and Mostyn, R. A.**, Determination of chloride, bromide and iodide by cool-flame emission spectroscopy, Report of Ministry of Defense (London), QAD (MATS), No. 196, 1972.
49. **Papp, J.**, Potentiometric determination of chloride in pulping liquors with a chloride ion-selective electrode, *Sven. Papperstidn.*, 75(16), 677, 1972.
50. **Bardin, V. V., Ivanov, Yu. M., and Shartukov, O. F.**, Use of flowthrough concentration cell with presaturation for potentiometric detection in liquid chromatography of anions, *Zh. Anal. Khim.*, 33(9), 1732, 1978.
51. **Starobinets, G. L. and Mechkovskii, S. A.**, Partition chromatography on ion-exchange resins. II. Separation of halogenate and halogenide, *Zh. Anal. Khim.*, 18(2), 255, 1963.
52. **Starobinets, G. L. and Mechkovskii, S. A.**, Chromatographic separation of halide ions, *Zh. Anal. Khim.*, 16(3), 319, 1961.
53. **Barney, J. E. and Bertolacini, R. J.**, Colorimetric determination of chloride with mercuric chloranilate, *Anal. Chem.*, 29, 1187, 1957.
54. **Kikindai-Cassel, M.**, Separation of halogenates by ion exchangers, *Ann. Chim. (Paris)*, 3(1-2), 5, 1958.

BROMINE

ANION EXCHANGE RESINS

General

Bromide ion is strongly retained by anion exchange resins, e.g., of the strongly basic type, from weakly acid media, e.g., of pH 4[1] to neutral solutions.[2-6] This adsorption takes place on resins in the nitrate,[2,3,6] chloride,[4] hydroxide,[1] and acetate[7] forms and allows Br to be separated from essentially all cationic constituents of the solutions. Coadsorbed with the bromide are, among other anions, bromate, chloride, iodide, and other anionic species of the halogens as well as phosphate and sulfate. This adsorption principle has been utilized for most separations outlined in Tables 1 to 3*. Removal of coadsorbed chloride can be achieved by use of 0.06,[8] 0.1,[2,6] and 0.16 M Na nitrate as eluents which do not elute the bromide. For the subsequent elution of the adsorbed bromide the following eluents have been employed: 0.4,[2,8] 0.5,[3,5,6] and 2 M[9] Na nitrate and 2 M Na perchlorate.[4] No elution of the adsorbed bromide is required whenever this ion can be determined directly on the resin, e.g., by use of X-ray fluorescence spectroscopy (see Table 2)[1] or radiometric assay (see Table 3).[10-12]

Also adsorbed on basic resins are other Br species such as bromate (see Table 3[7] and in Table 5 in the chapter on Chlorine) and elemental Br. The latter is retained by a mechanism of polybromide formation which may involve direct addition of Br molecules onto adsorbed bromide ions according to the reaction: $R_s^+ Br^- + 3Br_2 \rightleftarrows R_s^+ Br_7^-$ (R_s = resin matrix).[13] On a chloride-form resin the Br is adsorbed as polychlorobromide ion. This adsorption of Br is of analytical interest inasfar as it occurs from systems containing this element as a holding oxidant (see in the chapter on Thallium).

Applications

Adsorption of bromide on basic resins is the basis of separations that have been used in conjunction with the determination of Br in rocks (see Table 1),[2,3] waters (see Table 2),[1,4,6,14] and in biological matrices (see Table 3) such as foodstuffs and cereals,[9] plants,[5] and body fluids.[8,10-12] Included in Table 3 is a separation method which can be used for the determination of bromate in food.[7]

CATION EXCHANGE RESINS

Like the other halogen anions, bromide is not retained by strongly acid cation exchange resins, a fact which has been utilized for its separation from cationic constituents contained in natural waters (see Table 4)[15,16] and urine (see Table 4).[17] Included in Table 4 is a method which is based on reactive ion exchange of bromide with the Ag form of a strongly acid resin.[18]

In Table 5 a procedure is presented which can be employed for the microchemical detection of Br on Dowex® 50 using a resin spot test.[19]

---

* Tables for this chapter appear at the end of the text.

Table 1
DETERMINATION OF BROMIDE IN ROCKS AFTER ANION EXCHANGE SEPARATION

Material	Ion exchange resin, separation conditions, and remarks	Ref.
Standard rocks	Bio-Rad® AG1-X10 (200-400 mesh; nitrate form) Column: 17 × 0.5 cm a) Aqueous sample solution (~18 mℓ) (adsorption of bromide and chloride) b) 0.1 M NaNO$_3$ (45 mℓ) (elution of chloride) c) 0.4 M NaNO$_3$ (elution of bromide which is contained in the eluate fraction from 50—60 mℓ) Bromide is determined electrochemically by means of an ion-selective electrode Before the separation, the sample (~0.5 g) is decomposed by fusion at 960° C with a 2:1 mixture (2 g) of Na$_2$CO$_3$ and ZnO and the melt is shaken with water (10 mℓ) for several hours; a few drops of ethanol are added to the hot suspension (to reduce Mn to MnO$_2$) and the residue is filtered off and washed with ~8 mℓ of water (filtrate = solution [a])	2
Rocks (e.g., siltstone, shale, and pond sediment)	Dowex® 1-X10 (100-200 mesh; nitrate form) Column: 8 × 0.7 cm a) Sample solution (10-mℓ aliquot) (adsorption of bromide) b) 0.5 M NaNO$_3$ (elution of bromide at a flow rate of 1.5 mℓ/min) Bromide is determined potentiometrically using an ion-selective electrode; at the 4.3-ppm level, the coefficient of variation is 4% (six determinations) Before the separation, the sample (0.5 g, e.g., containing 3—15 ppm of Br) is fused with Na$_2$CO$_3$ (4 g) at 1000° C for 30 min, the melt is dissolved in water, and the solution is neutralized with conc HNO$_3$ and then heated at 60—70° C for 2—3 hr The digest is filtered and the filtrate is diluted to 50 mℓ to prepare solution (a)	3

Table 2
DETERMINATION OF BROMIDE IN WATERS AFTER ANION EXCHANGE SEPARATION

Material	Ion exchange resin, separation conditions, and remarks	Ref.
Lake and waste waters	Dowex® 1-X8 (100-200 mesh; Cl⁻ form) Column: 4 × 0.8 cm containing a 2-mℓ bed of the resin and operated at a flow rate of ~100 mℓ/hr a) Water sample (≯1 ℓ containing up to 10 μg of Br) made just acid to methyl orange (adsorption of Br⁻; matrix components pass into the effluent) b) 2 M NaClO₄ (3 × 3 mℓ) (elution of Br) After oxidation to BrO_3^- the Br is determined spectrophotometrically The Br recovery is 90—100%; the limit of determination is 0.4 μg of Br per liter Bicarbonate interferes with the adsorption of Br⁻, but this interference is eliminated by acidifying the sample with HCl	4
River-, supply, and waste water	Amberlite® CG-400 (nitrate form) Column: 7 × 1 cm operated at a flow rate of 2 mℓ/min a) Sample (100 mℓ) (adsorption of bromide and chloride; into the effluent pass matrix constituents and interfering organic materials such as humic acid and phenols) b) Distilled water (few milliliters) (as a rinse) c) 0.1 M NaNO₃ (70 mℓ) (elution of chloride) d) 0.5 M NaNO₃ (25 mℓ) (elution of bromide) Afterwards bromide is treated with citric acid and KMnO₄ to form pentabromoacetone, which is converted into CHBr₃ for gas chromatographic determination of Br	6
Natural waters	Filter paper disk (3.5 cm radius) loaded with Amberlite® SB-2 anion exchange resin To prepare the OH⁻ form of the resin the disk is soaked in 5% NaOH solution for 30 min and then mounted in a filtration column and washed by passage of 30 mℓ of 5% NaOH solution and two 50-mℓ portions of water; subsequently, the sample (50 mℓ), adjusted to pH 4 with acetic acid, is applied to the filter during 130 sec to adsorb Br; after three successive filtrations the disk is dried at 35° C and Br is determined by X-ray fluorescence spectroscopy; the error and detection limit are 5% and 0.05 ppm of Br, respectively	1

Table 3
DETERMINATION OF Br SPECIES IN BIOLOGICAL MATERIALS AFTER ANION EXCHANGE SEPARATION

Material	Ion exchange resin, separation conditions, and remarks	Ref.
Food (bread and fish paste products)	Amberlite® IRA-47 (acetate form) Column: 5 × 1 cm operated at a flow rate of 3 mℓ/min a) Sample solution (300 mℓ) which is 50% in acetic acid (adsorption of bromate) b) 50% Acetic acid (200 mℓ) (as a rinse) c) Water (200 mℓ) (as a rinse) d) 0.5 M K-acetate - 0.05 M KOH (45 mℓ) (elution of bromate) Br is determined spectrophotometrically using the o-toluidine method Before the ion exchange separation, the sample (e.g., 5 g of bread) is extracted for 30 min with water (60 mℓ), and after filtration Celite® 545 (3 g) is added to the filtrate; the adsorbent is filtered off and to the filtrate (150 mℓ) acetic acid (150 mℓ) is added to prepare solution (a) Recoveries were 74—83% from bread and 79—88% from fish paste products; the detection limit was 1 ppm	7
Plant material (lettuce)	Dowex® 1-X10 (100-200 mesh) Column: 6.7 × 1 cm a) Sample solution which is 0.5 M in NaNO$_3$ (adsorption of Br$^-$ and other constituents) b) 0.5 M NaNO$_3$ (elution in this order of soluble vegetable substances [e.g., chlorophylls], Cl$^-$ and Br$^-$) c) 2 M NaNO$_3$ (elution of I$^-$) Bromide is determined by use of an ion-selective electrode Before the separation, the dried and ground sample (containing Br$^-$ as a pesticide residue) is extracted with 0.5 M NaNO$_3$	5
Urine	Dowex® 2-X8 (200-400 mesh) Column containing 150 mg of the resin to a height of ~1 cm and washed until saturation with 1 M NH$_4$Br solution containing 2% of Br$_2$ a) Sample (1 mℓ) (adsorption of ^{80}Br; into the effluent pass ^{24}Na and ^{38}Cl) b) 5% NaCl solution (10 mℓ) (as a rinse) c) ^{80}Br is determined directly on the resin by radiation measurements Before the separation, the sample (0.1 mℓ) is diluted with water (0.9 mℓ) and then irradiated with neutrons	10
Serum and urine	Amberlite® IRA-400 (20-50 mesh; chloride form) Column: 8 × 1 cm containing ~ 2 g of the resin Serum or urine (2 mℓ each) is passed through the resin column to adsorb ^{82}Br Into the effluent pass ^{22}Na and ^{42}K; after washing with water (1.5 mℓ), the resin is removed from the column and ^{82}Br is determined radiometrically	11

Table 4
DETERMINATION OF BROMIDE IN WATERS, URINE, AND ZINC SULFATE AFTER CATION EXCHANGE SEPARATION

Material	Ion exchange resin, separation conditions, and remarks	Ref.
River- and tapwater	Strongly acidic cation exchange resin of the sulfonic acid type (50 mesh; H$^+$ form) Column: 6 × 0.5 in. operated at a flow rate of ~10 mℓ/min The sample (100 mℓ) is passed through the resin bed to adsorb NH$_4^+$, Fe, and other cations, and in the effluent (the first 40 mℓ are discarded) Br is determined spectrophotometrically	15
Urine	Dowex® 50W-X8 (100-200 mesh; NH$_4^+$ form) Column of 4 cm ID containing the resin (200 g) to a height of 16 cm a) Urine sample (10 mℓ) (adsorption of cations and conversion of anions into NH$_4^+$ salts) b) Water (500 mℓ) (elution of bromide as NH$_4$Br) Bromide is determined by flame photometry	17
Zn sulfate	Sulfonic acid cation exchange resin (Ag$^+$ form) a) Sorption solution adjusted to pH 11 with 1 M NaOH (formation of Ag^{82}Br, Ag$_3^{32}$PO$_4$, and Ag$_2^{35}$SO$_4$ in the resin bed) b) Water (elution of Ag sulfate) c) 5% HNO$_3$ (elution of Ag phosphate) d) 1:1 NH$_3$ solution (elution of Ag bromide) Bromide is determined radiometrically; as little as 1.71 μg of Br$^-$ per gram ZnSO$_4$ was detected Before the separation, the sample is irradiated with neutrons, and after dissolution in water (in the presence of carriers) Zn and other cationic constituents are removed by passage through a column of the resin in the H$^+$ form; from the effluent, the sorption solution (a) is prepared	18

Table 5
MICROCHEMICAL DETECTION OF BROMIDE BY A RESIN SPOT TEST

Ion exchange resin	Experimental conditions and remarks	Ref.
Dowex® 50W-X8 (50-100 mesh; Na$^+$ form)	A drop of the reagent is placed on a spot plate with a bead of the resin and mixed; after several minutes the resin is washed two or three times with water and then treated with the Br-containing test solution, when, according to the amount of Br present, the resin assumes a violet color, readily observed through a lens The method has a detection limit of 1.2 μg with a limiting concentration of 1 in 1 × 10^5 for bromides The reagent is 2% fuchsine solution (1 mℓ) treated with 6 M HCl (0.3 mℓ) and diluted to ≃ 5 mℓ with water then treated dropwise with 30% NaHSO$_3$ solution until the color is discharged; after the addition of 6 M HCl (0.1 mℓ), the mixture is diluted to 10 mℓ	19

REFERENCES

1. **Radcliffe, D.,** Rapid analysis for traces (1 p.p.m.) of bromide in natural water, *Anal. Lett.,* 3(11), 573, 1970.
2. **Heumann, K. G., Schrödl, W., and Weiß, H.,** Mass-spectrometric and electrochemical trace determination of bromide in geochemical standard reference materials, *Fresenius' Z. Anal. Chem.,* 315, 213, 1983.
3. **Akaiwa, H., Kawamoto, H., and Hasegawa, K.,** Determination of a trace amount of bromine in rocks by ion-exchange chromatography and direct potentiometry with an ion-selective electrode, *Talanta,* 27, 909, 1980.
4. **Lundstrom, U., Olin, A., and Nydahl, F.,** Determination of low levels of bromide in fresh water after chromatographic enrichment, *Talanta,* 31(1), 45, 1984.
5. **Nangniot, P., Agneessens, R., Zenon-Roland, L., and Berlemont-Frennet, M.,** Comparison between gas-liquid chromatography and selective ion exchange for determination of bromide in plant materials, *Analusis,* 12(4), 197, 1984.
6. **Ando, M. and Sayato, Y.,** Micro-determination of bromide in water by gas-chromatography, *Water Res.,* 17(12), 1823, 1983.
7. **Hidaka, T., Kirigaya, T., Kamijo, M., Suzuki, Y., and Kawamura, T.,** Studies on potassium bromate in foods. I. Determination of potassium bromate by anion-exchange resin, column separation followed by colorimetry, *Shokuhin Eiseigaku Zasshi,* 24(4), 376, 1983.
8. **Heurtebise, M. and Ross, W. J.,** Application of bromine-80m to semi-automated analysis for bromine in biological fluids, *J. Radioanal. Chem.,* 8(1), 5, 1971.
9. **Banks, H. J., Desmarchelier, J. M., and Elek, J. A.,** Determination of bromide-ion content of cereals and other foodstuffs by specific ion electrode, *Pestic. Sci.,* 7(6), 595, 1976.
10. **Heurtebise, M. and Ross, W. J.,** Radiochemical separation with halogenated resins, *Anal. Chem.,* 44, 596, 1972.
11. **Rovner, D. R. and Conn, J. W.,** A simple and precise method for the simultaneous measurement in man of plasma volume, radio-bromide space, exchangeable potassium and exchangeable sodium, *J. Lab. Clin. Med.,* 62(3), 492, 1963.
12. **Rovner, D. R. and Conn, J. W.,** Simple and precise method for the simultaneous measurement in man of plasma volume, radiobromide space, exchangeable potassium and exchangeable sodium, *J. Lab. Clin. Med.,* 62(3), 497, 1963.
13. **Aveston, J. and Everest, D. A.,** The adsorption of bromine and iodine by anion-exchange resins, *Chem. Ind.,* p. 1238, September 14, 1957.
14. **Leddicotte, G. W. and Navarrete Tejero, M.,** Separation and determination of traces of bromine in potable water by activation analysis, *Revta Soc. Quim. Mex.,* 12(5), 223A, 1968.
15. **Zitomer, F. and Lambert, J. L.,** Spectrophotometric determination of bromide ion in water, *Anal. Chem.,* 35, 1731, 1963.
16. **Basel, C. L., Defreese, J. D., and Whittemore, D. O.,** Interferences in automated phenol red method for determination of bromide in water, *Anal. Chem.,* 54, 2090, 1982.
17. **Gutsche, B. and Herrmann, R.,** Flame-photometric determination of bromine in urine, *Analyst (London),* 95, 805, 1970.
18. **Koch, H. and Grossmann, K. D.,** Bromide determination in zinc sulfate by activation analysis, *Kernenergie,* 6(11), 651, 1963.
19. **Fujimoto, M.,** Micro analyses with ion-exchange resins. IV. The detection of small quantities of bromides with the Schiff reagent, *Bull. Chem. Soc. Jpn.* 29(5), 571, 1956.

IODINE

ANION EXCHANGE RESINS

General

Iodide is very strongly retained by anion exchange resins from dilute acid, neutral, and alkaline solutions in which this I species shows distribution coefficients in the order of 10^3. For example, iodide is completely adsorbable, e.g., at pH 6.5,[1] and from neutral media[2,3] and from alkaline systems,[4-6] e.g., at pH 8.4[4] and 10.5[6] (see Table 1*), as well as from samples of natural waters (see Table 2),[7-10] milk (see Table 3),[1,11-22] blood serum (see Table 3),[23-29] and urine (see Table 3).[26,27,29-31] In water samples I can exist in several different valency states (e.g., as iodide, elemental I, and iodate). A preliminary oxidation-reduction step is therefore required to convert all the I to the strongly adsorbed iodide form. For this purpose all forms of I are first oxidized to iodate (IO_3^-) by using 5% Na hypochlorite (NaOCl) solution and slowly acidifying with nitric acid. Subsequently, iodate is reduced to iodide with hydroxylamine hydrochloride or Na bisulfite.[1,4] When using the column technique to adsorb iodide from milk samples (see Table 3), both fresh milk and a fast flow rate are essential, because milk will clog the resin bed if it is allowed to turn sour. Preservatives such as formaldehyde should be avoided since they complex the I and inhibit its adsorption on the resin. Na metabisulfite is a much more suitable preservative. In milk approximately 95% of the radionuclides ^{128}I and ^{131}I are present as iodide so that adsorption of the latter directly from the milk samples is the preferred method for the isolation of radioactive I. With this technique, which is also applicable to serum and urine, iodide is not only separated from all principal radionuclides, but also from all other cationic constituents of the samples including alkali metals and alkaline earth elements. Coadsorbed with iodide are elemental I, iodate, periodate (IO_4^-), and some organic iodo compounds as well as anions of other elements, e.g., phosphate, sulfate, nitrate, chloride, bromide, and fluoride. Elemental I is strongly retained by weakly and strongly basic resins (e.g., in the iodide form; see example in Table 4).[32-34] A polyiodide I_7^- is probably the principal ionic species present in the resin (compare with analogous behavior shown by elemental Br; see in the chapter on Bromine). This adsorption of I can be utilized for its isolation from solutions (see Table 4) and from gases. The elution of the elemental I can be achieved with 1 M solutions of reducing agents, e.g., Na sulfite or Na thiosulfate.[34] Elution of iodate (and also periodate) can be effected with 0.1 M ammonia solution (see Table 2),[8] 0.1 M ammonium chloride,[35] and 1 M K nitrate of pH 10.5.[6] These eluents do not elute iodide, which is also not desorbed when using 0.2 M hydrochloric acid (HCl) as eluent for iodopeptides, iodotyrosine, and di-iodotyrosine[5] or on application of 2 M[1] or 5%[2] NaCl solutions as column rinses (see Table 1). Coadsorbed chloride and bromide can be removed by elution with 0.1 M Na nitrate,[3] 0.2 M ammonium nitrate (see Table 3),[30] 0.4 M Na nitrate,[28,36] and 2 M K nitrate.[37] The latter eluent first elutes bromide and then iodide.

The most effective eluent for adsorbed iodide is sodium hypochlorite (NaOCl).[1,4,17,19,38] It is usually applied as 5 or 2% aqueous solution which, due to hydrolysis of this compound, shows basic and strong oxidizing properties. Thus, hypochlorite oxydizes the adsorbed iodide and also any coadsorbed elemental I to iodate which is only negligibly retained by strongly basic resins, e.g., Dowex® 1, under these conditions, so that rapid and complete elutions can be achieved. Any periodate that might be formed during this oxidative elution is coeluted with the iodate. Other eluents that have been used for the elution of iodide include: 2[7] and 4 M[16] Na nitrate, 0.5 M ammonium nitrate,[22] 1[23] and 2 M[39] Na perchlorate, 2 M K nitrate,[10,40]

---

* Tables for this chapter appear at the end of the text.

62% solution of Mg nitrate,[28,36] 4 M ammonium chloride,[9,41] and 0.5 M Na salicylate.[5] Additional eluents for iodide are listed in Table 4 in the chapter on Chlorine.

Frequently the adsorbed radioisotopes of I, i.e., ^{128}I and ^{131}I, are not eluted, but determined directly on the resin using radiometric procedures (see Tables 1, 3, and 4).[2,3,11-15,18,26,27,30,42,43]

Adsorbed iodide can also be determined on the resin by means of X-ray fluorescence spectroscopy (see Table 3).[20]

Applications

Adsorption of iodide and other I species on anion exchange resins is the basis of numerous separation methods which have been used in connection with the determination of I in natural materials, i.e., soils and rocks (see Table 1),[2,4] waters (see Table 2),[7-10,43-46] biological fluids (milk, serum, urine, and saliva) (see Table 3),[1,11-20,23-28,30,31,36,39,47-51] cheese,[52] and rat thyroid,[5] as well as in industrial products (see Table 4).[3,32,53]

For the separation of I species the procedures outlined in Table 5 can be employed. A procedure for the microchemical detection of I by means of a resin spot test is described in Table 17 (see in the chapter on Sulfur).

CATION EXCHANGE RESINS

This type of ion exchangers has mainly been used to separate iodide and other I species from cationic constituents of samples such as rainwater,[54] serum,[55] animal tissues,[56] mixtures of fission products,[57] and Se.[58] These methods are based on the nonadsorbability of iodide and other anions on the resins, e.g., of the sulfonated polystyrene type, which, however, retain the accompanying cationic elements.

On Amberlite® IR-120 in the Ag form, iodide is adsorbed by reactive ion exchange forming insoluble Ag iodide. This principle has been used for the removal of radioactive I.[40,44,45] The resin adsorbs only ionic I.

* Tables for this chapter appear at the end of the text.

Table 1
DETERMINATION OF IODIDE IN SOILS AND ROCKS AFTER ANION EXCHANGE SEPARATION

Material	Ion exchange resin, separation conditions, and remarks	Ref.
Soils	Dowex® 1-X8 (chloride form) Column of 2 cm ID containing 25 mℓ of the resin a) Soil extract (~500 mℓ) (adsorption of ^{129}I + carrier; into the effluent pass coextracted matrix constituents) b) Hot water (until effluent is clear) c) 2 M NaCl (until effluent again becomes clear) d) 5% NaOCl solution (50 mℓ) (elution of I as IO_3^-) (flow rate: 2 mℓ/min) From the eluate I is extracted into CCl$_4$ and back into a bisulfite solution and is then precipitated as PdI$_2$; the dried precipitate is thermally decomposed, the liberated I is adsorbed on charcoal, and the charcoal is irradiated with neutrons Finally I is determined radiometrically Before the ion exchange separation, the sample (100 g + 30 mg of NaI carrier) is extracted by boiling for 1 hr with 250 mℓ of 10% Na$_2$CO$_3$ solution and 20 mℓ of 1 M NH$_2$OH·HCl; after decantation of the supernate the digestion process is repeated; then the two supernates are combined and filtered, thus obtaining the final soil extract (a) The method can also be applied to the determination of ^{129}I in natural waters and milk; to the water sample (4 ℓ + 30 mg of NaI-carrier), while stirring, 5% NaOCl solution (5 mℓ), 1 M NH$_2$OH·HCl (25 mℓ), and 1 M NaHSO$_3$ (10 mℓ) are added at intervals of a few minutes; then the pH is adjusted to 6.5 and after addition of a batch (25 mℓ) of the resin the mixture is stirred for 1 hr to adsorb I; afterwards the resin is allowed to settle for 20 min and the supernate is equilibrated with a further 25-mℓ batch of the resin; subsequently, the resin batches are combined and transferred to the empty column; through the resulting resin bed 250 mℓ of hot water followed by 100 mℓ of 2 M NaCl is passed at a rate of 10—20 mℓ/min; finally the adsorbed I is eluted with eluent (d); the milk sample (4 ℓ + 30 mg NaI-carrier) is not subjected to the oxidation-reduction reactions performed with the water sample; analysis is started with the batch-extraction process and then continued as outlined for waters and soils	4
Sedimentary rocks	Dowex® 2-X8 (200-400 mesh; chloride form) Column operated at a flow rate of 1—4 mℓ/min a) Sample solution (~10 mℓ) neutralized with 0.5 M HNO$_3$ (adsorption of I$^-$) b) 5% NaCl solution (5 mℓ) (elution of interfering activities such as ^{24}Na and ^{38}Cl) c) Radiometric determination of ^{128}I directly on the resin Before the separation, the sample is irradiated with neutrons and I is separated by volatilization in the presence of ^{131}I tracer; the I is swept from the furnace by air and collected in 0.1 M NaOH (~10 mℓ); the method is suitable for samples containing >0.01 ppm of I	2

Table 2
DETERMINATION OF I SPECIES IN NATURAL WATERS AFTER ANION EXCHANGE SEPARATION

Material	Ion exchange resin, separation conditions, and remarks	Ref.
Seawater [a]	Bio-Rad® AG1-X8 (100-200 mesh; nitrate form) Column: 11—12 × 1 cm operated at a flow rate of 2 mℓ/min a) Water sample (250 mℓ) (adsorption of I$^-$; matrix constituents pass into the effluent) b) Distilled water (~5 mℓ) (as a rinse) c) 2 M NaNO$_3$ (110 mℓ) (elution of I which is contained in the last 80 mℓ of the eluate) After precipitation as PdI$_2$ (with elementary Pd acting as a carrier) and neutron irradiation of the precipitate, the ^{128}I formed is determined radiometrically	7
Rainwater	Dowex® 2-X8 (OH$^-$ form) Column containing 40 ± 2 mg of the resin (dry weight) a) Sample (20—25 mℓ, filtered through a 0.45-μm membrane filter) (adsorption of iodate and iodide at a flow rate of 0.22 mℓ/min) b) Bidistilled water (3 mℓ) (as a rinse) c) 0.1 M NH$_3$ solution (≃10 mℓ) (elution of IO$_3^-$; iodide is not coeluted) (flow rate: 0.3 mℓ/min) Eluate (c) is irradiated with neutrons and ^{128}I formed is determined radiometrically The limit of determination is 0.02 μg of iodate per liter	8

[a] See also Reference 61.

Table 3
DETERMINATION OF IODIDE IN BIOLOGICAL FLUIDS AFTER ANION EXCHANGE SEPARATION[a]

Material	Ion exchange resin, separation conditions, and remarks	Ref.
Milk	Dowex® 1-X8 (50-100 mesh; chloride form) Column of 2 cm ID containing 20 mℓ of the resin a) Milk sample (4 ℓ, containing 30 mg of iodide-carrier) (adsorption of I at a flow rate of 20 mℓ/min) b) Hot distilled water (500 mℓ) (as a rinse) c) 2 M NaCl (100 mℓ) (as a rinse; flow rate: 4 mℓ/min) d) 2% NaOCl solution (50 mℓ) (oxidative elution of I) (flow rate: 2 mℓ/min) After a cleanup of the eluate by alternate extractions with CCl_4, water, toluene, and water, the purified I is finally extracted into toluene for measurement of the ^{129}I activity; the chemical recovery is 58 ± 3% for raw milk, and 80 ± 4% for pasteurized milk; the sensitivity is 0.3 pCi of ^{129}I per liter for a 4-ℓ sample The same method can be used for determining ^{129}I in natural waters; before the ion exchange separation step, the sample (4 ℓ) is treated for 2—3 min with 5% NaOCl (5 mℓ) (in the presence of 30 mg I^--carrier) to oxidize the entire I to higher valency states; subsequently, all positive oxidation states of I are reduced to iodide by the addition of 1 M $NH_2OH \cdot HCl$ (25 mℓ) and 1 M $NaHSO_3$ (10 mℓ) and adjustment of the pH to 6.5 using 50% NaOH or 10% HNO_3; the mixture is stirred for a few minutes (up to 45 min) and then filtered if necessary before passing it through the column using the procedure outlined above for milk; in place of eluent (b), 200 mℓ of distilled water is used for washing the resin bed; the average recovery of I is 74 ± 2%	1
	Dowex® 1-X8 (20-50 mesh; chloride form) Column: 3 × 1.5 in. operated at a flow rate of 30—40 mℓ/min a) Milk sample (1 gal, i.e., 3785 mℓ) (adsorption of I) b) Distilled water (30—40 mℓ) (as a rinse) c) The resin is removed from the column and the adsorbed ^{131}I is determined radiometrically directly on the resin; the average recovery of I is 98%	11
	Strongly basic anion exchange resin (20-50 mesh; chloride form) Column (ion exchange cartridge): polyethylene phial containing ≃40 mℓ of the resin and operated at a flow rate of 1 ℓ/20—30 min a) Milk sample (1 ℓ) (adsorption of ^{131}I) b) The resin is removed from the cartridge and the adsorbed ^{131}I is determined radiometrically directly on the resin	12
	Dowex® 2-X8 (20-50 mesh; chloride form) Column: 4.8 × 2 cm containing 15 mℓ of the resin and operated at a flow rate of 12—14 mℓ/min; before use the resin bed is washed with a solution of pH 6.6 containing NaCl, NaH_2PO_4, and trisodium citrate (the concentrations of the anions in this solution expressed in mM/ℓ are 4.03, 4.64, and 20.4, respectively) a) Milk sample (adsorption of ^{131}I) b) Distilled water (as a rinse)	13

Table 3 (continued)
DETERMINATION OF IODIDE IN BIOLOGICAL FLUIDS AFTER ANION EXCHANGE SEPARATION[a]

Material	Ion exchange resin, separation conditions, and remarks	Ref.
	^{131}I is determined radiometrically; the recovery is 98 ± 2% from milk samples ≥ 2 ℓ	
	Deacidite® FF (chloride form) Column of 3.7 cm ID containing 50 mℓ of the resin and operated at a flow rate of 20 mℓ/min a) Milk sample (1 ℓ, containing 20 mg I$^-$ [as KI] as a carrier) (adsorption of ^{131}I) b) Water (as a rinse) c) The resin is removed from the column and the adsorbed ^{131}I is determined radiometrically directly on the resin	14
	Deacidite® FF-SRA63-X3-5 (14-52 mesh; chloride form) Column of 1.5 cm ID containing 50 mℓ of the resin and operated at a flow rate of 5 mℓ/min a) Sample of fresh milk (1ℓ) (adsorption of ionic and protein-bound ^{131}I) b) The resin is removed from the column, washed with water (2 × 100 mℓ), and the adsorbed ^{131}I is determined radiometrically directly on the resin	15
	Dowex® 1 (Cl$^-$ form) Column: 8 × 1 cm a) Milk sample (adsorption of I$^-$; the milk matrix passes into the effluent) b) NaOCl solution (elution of I$^-$) ^{129}I is determined by X-ray spectrometry	17
	Dowex® 1-X8 (100-200 mesh; I$_3^-$ [I$^-$ + I$_2$] form which is prepared by equilibration of the chloride form with 1 M KI containing 2.5% of I$_2$) The milk sample (100 mℓ) is irradiated with neutrons and then stirred for 5 min with the resin (200 mg) to adsorb ^{128}I; immediately afterwards the ^{128}I is determined radiometrically directly on the resin The precision is ±5% for I concentrations >40 µg/ℓ	18
	Dowex® 1-X8 (200-400 mesh; chloride form) To the sample (1—10 ℓ), I carrier (1—2 mℓ) (10 mg I$^-$ per milliliter) and 50 mℓ of the resin are added and the mixture is stirred magnetically for ≤3 hr to adsorb I; then the resin is separated from the milk by decantation, washed with water (until the supernate is clear), and I is eluted by heating on the water bath for 10 min each time with three portions of 50 mℓ each of 5% NaOCl solution; subsequently, I is extracted with CCl$_4$, precipitated as PdI$_2$, and determined radiometrically (^{131}I) and gravimetrically (yield determination)	19
	Anion exchange resin paper Amberlite® SB-2; disks of 5 cm diameter Milk (20 mℓ) diluted with water (25 mℓ) is warmed at 60° C for 5 min, then 5% trichloroacetic acid (10 mℓ) is added; the mixture is filtered and then passed through a disk of the resin paper; the disk is washed with 5 M NaOH (3 mℓ) and with water (15 mℓ) and the adsorbed iodide is determined by X-ray fluorescence spectrometry; the limit of detection is 0.05 ppm of I$^-$	20

Table 3 (continued)
DETERMINATION OF IODIDE IN BIOLOGICAL FLUIDS AFTER ANION EXCHANGE SEPARATION[a]

Material	Ion exchange resin, separation conditions, and remarks	Ref.
Urine	Amberlite® IRA-400 (14-52 mesh; chloride form) Column of 1.5 cm ID containing 2 g of the resin conditioned with 1% NH$_4$Cl solution (100 mℓ) and operated at a flow rate of 1—2 mℓ/min a) Urine sample (100 mℓ) (adsorption of iodide, bromide, and chloride) b) Distilled water (50 mℓ) (as a rinse) c) 0.2 M NH$_4$NO$_3$ (40 mℓ) (elution of bromide and chloride) d) Distilled water (400 mℓ) (as a rinse) e) The resin is removed from the column, irradiated with neutrons, and then I is determined by measurement of the γ activity of the radionuclide ^{128}I formed	30
	Bio-Rad® AG1-X10 (100-200 mesh; nitrate form) Column: 2 × 0.8 cm a) Sample solution (1 mℓ) (adsorption of I$^-$) b) 0.1 M KNO$_3$ (20 mℓ) (elution of residual contrast agent) c) 2 M KNO$_3$ (25 mℓ) (elution of I) After extraction into CCl$_4$ and back extraction with arsenite solution, I is determined colorimetrically; before the ion exchange separation, the sample (24 hr urine) is diluted to a concentration of $\simeq$100—500 µg I per liter and 2 mℓ of diluted urine is hydrolyzed with β-glucuronidase, after which most of the contrast agent (e.g., iodamide and metrizoate; in the free-acid form) is extracted into CCl$_4$; the remaining aqueous solution (containing the urinary I as iodide) is the sample solution (a) This method can be used to study the excretion of I$^-$ following administration of a I-containing contrast agent	31
Serum	Dowex® AG1-X2 (200-400 mesh; ClO$_4^-$ form) Column: 1 × 0.3 cm a) Serum (5 mℓ) (adsorption of iodide; organic I compounds pass into the effluent) b) Deionized water (10 mℓ) (as a rinse) c) 1 M NaClO$_4$ (5 mℓ) (elution of iodide) Recovery of I$^-$ from the column is 97% L-Thyroxine is not adsorbed on the resin, but monoiodo- and diiodotyrosines and triiodothyronine are retained to the extent of 3, 8, and 5%, respectively; adsorption of iodide on a column (15 × 1.5 cm) of Dowex® 1 or 2 (OH$^-$ form) has also been used for its separation from organically bound I^{59}	23
	Dowex® 1-X8 or Amberlite® IRA-401 (20-50 mesh; chloride form) To separate inorganic I from protein-bound I the sample (1.5—2.0 mℓ) is shaken for ~5 min with the resin (300 mg) which retains iodide; in the supernate the protein-bound I is determined colorimetrically after wet or dry ashing of an aliquot (0.1 mℓ); the dry ashing is carried out in the presence of K$_2$CO$_3$	24, 25
Saliva, urine, and serum	Column of Dowex® 2-X8 (150 mg) conditioned with a nearly saturated solution of I in 1 M KI (15 mℓ) and then with water (1 mℓ) a) Irradiated sample solution (adsorption of ^{128}I; all other radioactive components pass into the effluent)	26, 27

Table 3 (continued)
DETERMINATION OF IODIDE IN BIOLOGICAL FLUIDS AFTER ANION EXCHANGE SEPARATION[a]

Material	Ion exchange resin, separation conditions, and remarks	Ref.
	b) Water (4 mℓ) (as a rinse) c) 2% NaCl solution (10 mℓ) (as a rinse) ^{128}I is determined radiometrically directly on the resin Before the anion exchange separation, the sample (saliva or urine [2 mℓ each] or serum [1 mℓ] is diluted with water (2 mℓ) and 0.05% $K_2S_2O_5$ (0.1 mℓ) and the fraction containing protein-bound I is separated by anion exchange; then the samples are irradiated with neutrons	

[a] See also Table 13 in the chapter on Zinc, Volume IV and Reference 62.

Table 4
DETERMINATION OF I SPECIES IN INDUSTRIAL PRODUCTS AFTER ANION EXCHANGE SEPARATION[a]

Material	Ion exchange resin, separation conditions, and remarks	Ref.
Lugol I solution and tincture of I	Anion exchange resin IMAC-A.17 (I⁻ form) A volume of the sample solution containing 100—300 mg of I (I_2) is diluted to 50 mℓ, the resin (2 g) is added, and the solution is allowed to remain in contact with the resin until the color of I disappears (20—30 min) (adsorption of free I_2); then the resin is filtered off and iodide (which is not retained by the I⁻-form resin) is determined titrimetrically; the method has an efficiency of ≃99% for the recovery of I⁻ in the presence of I_2	32
Fission product mixtures	Bio-Rad® AG1-X4 (100-200 mesh; nitrate form) Column of ≃0.3 g of the resin packed in a special polyethylene insert that fits into a polyethylene rabbit (flow rate: 1 drop/sec) a) Aqueous sample solution (5 mℓ) (adsorption of ^{129}I) b) Water (5 mℓ) (removal of cations) c) 0.1 M $NaNO_3$ (10 mℓ) (elution of Br) d) Radiometric determination of ^{129}I directly on the resin Before the ion exchange separation, I is preconcentrated by extraction into CCl_4 The back extraction of I is effected with water (5 mℓ) containing 1 M $K_2S_2O_5$ (few drops)	3
Irradiated Te target	Amberlite® IRA-400 (80-120 mesh; Cl⁻ form) Column: 5 × 1 cm operated at a flow rate of 1.5 mℓ/min a) 1 M HCl sample solution containing 1 mg Te per milliliter (adsorption of carrier-free ^{131}I; Te passes into the effluent) b) 6 M HCl (50 mℓ) (elution of ^{131}I with a radiochemical yield of ≃50%) Before the separation, the irradiated Te is fused with KOH and the melt is dissolved in HCl	53

[a] See also Reference 63.

Table 5
ANION EXCHANGE SEPARATION OF I SPECIES

Elements separated	Ion exchange resin, separation conditions, and remarks	Ref.
I^-, IO_3^-, and IO_4^-	Amberlite® IRA-400 (20-50 mesh; OH^- form) Column: 50 × 1.2 cm a) NaOH solution (200 mℓ, pH 10.5) (adsorption of the anions) b) 1 M KNO_3 of pH 10.5 (1 ℓ) (iodate and periodate are recovered in the 25th—70th-mℓ fraction of the eluate and the iodide in the 300th—1000th-mℓ portion)	6
IO_4^- from highly colored solutions	Amberlite® IRA-400 (acetate form) Column: 20 × 0.8—1.2 cm operated at a flow rate of 1—2 mℓ/min a) Colored sample solution (adsorption of periodate; the colored constituents, e.g., flavonol and quercetin, pass into the effluent) b) 5% NaOH solution (elution of periodate) The periodate recovery is ≃99.3%	60
^{125}I	Dowex® 2-X8 (20-50 mesh; chloride form) The sample solution (containing a known amount of I^--carrier) is stirred with ~50 mℓ of the resin for 30 min to adsorb ^{125}I and then rinsed a few times with hot deionized water; afterwards ^{125}I is eluted by stirring with two 50-mℓ portions of 5% NaOCl solution; the eluate containing iodate is first treated with conc HNO_3 and then iodate is reduced to I_2 which is extracted into CCl_4; following reduction to iodide with $NaHSO_3$, ^{125}I is determined radiometrically; the chemical yield is ≃75% The separation can also be effected on a column of 2.5 cm ID containing the same quantity of resin	38

REFERENCES

1. **Gabay, J. J., Paperiello, C. J., Goodyear, S., Daly, J. C., and Matsuszek, J. M.**, Method for determining iodine-129 in milk and water, *Health Phys.*, 26(1), 89, 1974.
2. **Hasanen, E. and Salmela, S.**, Determination of iodine in geological samples using neutron-activation analysis and an induction furnace, *Radiochem. Radioanal. Lett.*, 37(4-5), 207, 1979.
3. **Bate, L. C. and Stokely, J. R.**, Iodine-129 separation and determination by neutron-activation analysis, *J. Radioanal. Chem.*, 72, 557, 1982.
4. **Wilkins, B. T. and Stewart, S. P.**, Sensitive method for determination of iodine-129 in environmental materials, *Int. J. Appl. Radiat. Isot.*, 33(12), 1385, 1982.
5. **Meyniel, G., Blanquet, P., Berger, J. A., Croizet, M., and Gaillard, G.**, Separation of inorganic and organic iodo-compounds from the thyroid gland on Dowex 1-X8 anion-exchange resin, *Bull. Soc. Chim. Biol.*, 47(1), 99, 1965.
6. **Good, M. L., Purdy, M. B., and Hoering, T.**, The anion-exchange separation of iodine anions, *J. Inorg. Nucl. Chem.*, 6(1), 73, 1958.
7. **Wong, G. T. F. and Brewer, P. G.**, Determination of iodide in seawater by neutron-activation analysis, *Anal. Chim. Acta*, 81, 81, 1976.
8. **Luten, J. B., Woittiez, J. R. W., Das, H. A., and deLigny, C. L.**, Determination of iodate in rainwater, *J. Radioanal. Chem.*, 43(1), 175, 1978.
9. **Lesigang, M. and Hecht, F.**, Laboratory experiments concerning the wash-out effect of radio-iodine, *Mikrochim. Acta*, No. 1, 32, 1963.
10. **Brutovský, M. and Zaduban, M.**, Use of the anion exchange resin Wofatit SBW for the concentration of iodine-131, *Collect. Czech. Chem. Commun.*, 32, 505, 1967.
11. **Boni, A. L.**, Rapid determination of iodine-131 in milk, *Analyst (London)*, 88, 64, 1963.
12. **Johnson, R. H. and Reavey, T. C.**, Evaluation of ion-exchange cartridges for field sampling of iodine-131 in milk, *Nature*, 208, 750, 1965.
13. **Murthy, G. K., Gilchrist, J. E., and Campbell, J. E.**, Method for removing iodine-131 from milk, *J. Dairy Sci.*, 45(9), 1066, 1962.
14. **Bonnyman, J. and Duggleby, J. C.**, Iodine-131 concentrations in Australian milk resulting from 1968 French nuclear weapon tests in Polynesia, *Aust. J. Sci.*, 31(11), 389, 1969.
15. **Smith, H. and Whitehead, E. L.**, Rapid method for estimation of iodine-131 in milk, *Nature*, 199, 503, 1963.
16. **Braun, T., Ruiz de Pardo, C., and Salazar, E. C.**, Iodine-131 monitoring in milk by using an iodine membrane-electrode for chemical yield determination, *Radiochem. Radioanal. Lett.*, 3(5), 397, 1970.
17. **Giacomelli, R. and Spezzano, P.**, X-ray spectrometric determination of iodine-129 in milk samples using planar intrinsic germanium detector, *Com. Naz. Ric. Sviluppo Energ. Nucl. Energ. Altern. (Rapp. Tec.)*, (NEA-RT/PROT (Italy), ENEA-RT (PROT[82])14, 1982.
18. **Ohno, S.**, Simple and rapid determination of iodine in milk by radioactivation analysis, *Analyst (London)*, 105, 246, 1980.
19. **Latimer, J. N., Bush, W.E., Higgins, L. J., and Shay, R. S.**, Radiochemical determination of iodine-131 in Handbook of Analytical Procedures, Report RMO-3008, U.S. Atomic Energy Commission, February 16, 1970.
20. **Lawrence, J. F., Chadha, R. K., and Conacher, H. B. S.**, The use of ion-exchange filters for determination of iodide in milk by X-ray-fluorescence spectrometry, *Int. J. Environ. Anal. Chem.*, 15(4), 303, 1983.
21. **Murthy, G. K. and Campbell, J. E.**, Removal of radionuclides from milk, *J. Dairy Sci.*, 47(11), 1188, 1964.
22. **Jankowska, S. and Zajac, W.**, Rapid radiochemical determination of iodine-131 in milk, *Rocz. Inst. Przem. Mlecz.*, 19(1), 37, 1977.
23. **Sacks, B. I.**, Recovery of iodide from anion-exchange resins, *Nature*, 202, 899, 1964.
24. **Yee, H. Y., Katz, E. S., and Jenest, E. S.**, Rapid semi-micro method for determining protein-bound iodine, *Clin. Chem.* 13(3), 220, 1967.
25. **Juengst, D. and Strauch, L.**, Rapid method for the determination of protein-bound iodine, *Z. Klin. Chem. Klin. Biochem.*, 7(6), 636, 1969.
26. **Heurtebise, M. and Ross, W. J.**, Application of an iodide specific resin to the determination of iodine in biological fluids by activation analysis, *Anal. Chem.*, 43, 1438, 1971.
27. **Heurtebise, M. and Ross, W. J.**, Radiochemical separations with halogenated resins, *Anal. Chem.*, 44, 596, 1972.
28. **Heurtebise, M.**, Semi-automated determination of iodine in biological fluids by activation analysis, *J. Radioanal. Chem.*, 7(2), 227, 1971.
29. **Comoy, E.**, Automatic determination of iodine in serum and urine, *Rev. Fr. Etud. Clin. Biol.*, 12(2), 189, 1967.

30. **Morgan, D. J., Black, A., and Mitchell, G. R.**, Determination of inorganic iodide in urine by neutron-activation analysis, *Analyst (London)*, 94, 740, 1969.
31. **Backer, E. T.**, Chemical method for determination of urinary iodide in presence of iodine-containing contrast agents, *J. Mol. Med.*, 4(1-2), 235, 1980.
32. **Franchi, G. and Gorgeri, L.**, Quantitative determination of iodide in solutions containing iodine and iodide by using an ion-exchange resin, *Boll. Chim. Farm.*, 102(6), 393, 1963.
33. **Saber, T. M. H.**, Sorption of molecular iodine and triiodide ion by weak-base anion resin, *U.A.R. J. Chem.*, 13(1), 67, 1970.
34. **Bhat, T. R. and Rao, T. V.**, Use of ion-exchange resins in the removal of iodine from liquids and gases, *Indian J. Chem.*, 1(11), 497, 1963.
35. **Khym, J. X.**, Direct spectrophotometric determination of iodate following periodate oxidation of alpha-glycol groups: quantitative removal of iodate and periodate by ion-exchange materials or by solvent extraction, *Methods Carbohydr. Chem.*, 6, 87, 1972.
36. **Commissariat á l'Énergie Atomique**, Activation Analysis, British Patent, 1,103,995; date appl. April 15, 1966.
37. **Ozaki, T. and Nakayama, T.**, Studies of analysis with convection electrodes. II. Scanning of the effluent from a column of ion-exchange resin. Separation of iodide and bromide, *J. Chem. Soc. Jpn. Pure Chem. Sect.*, 81(10), 1567, 1960.
38. **Chandrasekaran, E. S.**, Measurement of iodine-125 by liquid-scintillation-counting method, *Health Phys.*, 40(6), 896, 1981.
39. **Kahn, B.**, Determination of pico-curie concentrations of iodine-131 in milk, *J. Agric. Food Chem.*, 13(1), 21, 1965.
40. **Hingorani, S. B. and Venkateswarlu, K. S.**, Removal of radioactive iodine and iodomethane by use of silver-impregnated resin, *Chem. Eng. World*, 12(5), 59, 1977.
41. **Lesigang, M. and Hecht, F.**, Micro-determination of iodine-131. Sorption on Amberlite IRA-400, *Mikrochim. Acta*, No. 1-2, 327, 1962.
42. **Mulvey, P. F., Cardarelli, J. A., Meyer, R. A., Cooper, R., and Burrows, B. A.**, Sensitivity of bremsstrahlung activation analysis for iodine determination, in Radioisotope Sample Meas. Tech. Med. Biol. Proc. Symp., Vienna, 1965, 249.
43. **Dobosz, E.**, Simple method for determination of radio-iodine, *Radiochem. Radioanal. Lett.*, 13(5-6), 381, 1973.
44. **Hingorani, S. B. and Venkateswarlu, K. S.**, Adsorption of iodine-131 by silver-impregnated Amberlite IR-120 resin, *Radiochem. Radioanal. Lett.*, 31(6), 383, 1977.
45. **Hingorani, S. B. and Venkateswarlu, K. S.**, Use of silver-impregnated resin for trapping of radioactive iodine, *Kerntechnik*, 18(5), 207, 1976.
46. **Jansta, V.**, Comparison of variously pretreated ion-exchange resin for iodine-131 separation from aqueous solutions, *J. Radioanal. Chem.*, 80(1-2), 81, 1983.
47. **Dahl, J. B., Johansen, O., and Steinnes, E.**, Activation analysis of iodide in biological fluids, *Inst. Atomenergi Rep.*, KR-80, 1964.
48. **Knapp, G. and Spitzy, H.**, New technique for automated protein-bound iodine determination, *Clin. Chim. Acta*, 30(1), 119, 1970.
49. **Zieve, L., Vogel, W. C., and Schultz, A. L.**, Determination of protein-bound radioiodine with an anion-exchange resin, *J. Lab. Clin. Med.*, 47(4), 663, 1956.
50. **Strauss, H. D. and Richards, H. K.**, Rapid method for determination of liquid-bound radio-iodine in blood, *Proc. Soc. Exp. Biol. Med.*, 100, 461, 1959.
51. **Morgan, A.**, Measurement of non-ionic iodine in the milk of dairy cows following oral administration of labelled sodium iodide, *J. Dairy Res.*, 27(3), 399, 1960.
52. **Nordbye, P. I. and Auguston, J. H.**, Determination of iodine in cheese by neutron activation combined with separation of iodine by selective retention on Dowex 2-X8, Report FF1-80/3003, Forsvarets Forskningsinst., 1980.
53. **Inarida, M.**, Carrier-free separation of iodine-131 from irradiated unit I-131, *J. Chem. Soc. Jpn. Pure Chem. Sect.*, 80(4), 400, 1959.
54. **Funahashi, S., Tabata, M., and Tanaka, M.**, Determination of sub-microgram amounts of iodide by its catalytic effect on the substitution reaction of mercury (II)-4-(2-pyridylazo) resorcinol complex with DCTA, *Anal. Chim. Acta*, 57, 311, 1971.
55. **Mantzos, J. D. and Malamos, B.**, Direct method for the chemical determination of serum inorganic iodine, *Clin. Chim. Acta*, 21(3), 501, 1968.
56. **Brutovský, M., Zaduban, M., Baňas, J., and Liptáková, G.**, Determination of iodine-131 by an extraction method, *Chem. Zvesti*, 19(6), 470, 1965.
57. **Maeck, W. J. and Rein, J. E.**, Determination of fission-product iodine. Cation-exchange purification and heterogeneous isotopic exchange, *Anal. Chem.*, 32, 1079, 1960.

58. **Ernst, O. and Szolnoki, G.,** Indirect photometric determination of small amounts of iodine in selenium, *Fresenius' Z. Anal. Chem.*, 258, 263, 1972.
59. **Blanquet, P., Meyniel, G., Mounier, J., and Tobias, C. A.,** The anion-exchange resins Dowex 1 and 2. Immediate separation of organic and inorganic iodine. Rapid estimation of iodinated amino acids labelled with ^{131}I, *Bull. Soc. Chim. Biol.*, 39(4), 419, 1957.
60. **Smith, M. A. and Willeford, B. R.,** Determination of periodic acid in highly coloured solutions, *Anal. Chem.*, 26, 751, 1954.
61. **Zhang, S. C. and Lieser, K. H.,** Ion exchange of chloride, bromide and iodide on the anion exchanger Amberlite IRA 458, *Fresenius' Z. Anal. Chem.*, 320, 265, 1985.
62. **Byrne, A. R., Dermelj, M., and Tusek-Znidaric, M.,** Study of the iodinated resin column for the determination of iodine in biological fluids by radiochemical neutron-activation analysis, *J. Radioanal. Nucl. Chem.*, 91(2), 315, 1985.
63. **Lawrence, J. F., Chadha, R. K., O'Brien, R., and Conacher, H. B. S.,** Novel method for determination of iodide in table salt by X-ray fluorescence, *Microchem. J.*, 31(2), 237, 1985.

APPENDIX

ION EXCHANGE RESINS

Tradename and manufacturer	Functional group(s)	Matrix
Strongly Acidic Cation Exchangers		
Bio-Rad AG 50 or AG50W (1)	$-SO_3H$	cop-1
Allasion CS; CS/AD(2)	$-SO_3H$	cop-1
Allasion CP(2)		phen
Allasion CX(2)	$-SO_3H$	cop-1
Amberlite IR-1 (3)	$-CH_2SO_3H$; $-OH$	phen
Amberlite IR-100 (3)	$-CH_2SO_3H$; $-OH$	phen
Amberlite IR-105 (3)	$-CH_2SO_3H$; $-OH$	phen
Amberlite IR-105-G (3)	$-CH_2SO_3H$; $-OH$	phen
Amberlite IR-112 (3)	$-SO_3H$	cop-1
Amberlite IR-112-H (3)	$-SO_3H$	cop-1
Amberlite IR-120 (3)	$-SO_3H$	cop-1
Amberlite IR-122 (3)	$-SO_3H$	cop-1
Amberlite IR-124 (3)	$-SO_3H$	cop-1
Amberlite XE-100 (3)	$-SO_3H$	cop-1
Amberlite 200 (3)	$-SO_3H$	cop-1
Amberlite 200-C (3)	$-SO_3H$	cop-1
Amberlite 252 (3)	$-SO_3H$	cop-1
Amberlyst 15 (3)	$-SO_3H$	cop-1
Bio-Rex 40 (1)	$-OH$; $-CH_2SO_3H$	phen
Diaion PK 204 (4)	$-SO_3H$	cop-1
Diaion PK 208 (4)	$-SO_3H$	cop-1
Diaion PK 212 (4)	$-SO_3H$	cop-1
Diaion PK 216 (4)	$-SO_3H$	cop-1
Diaion PK 220 (4)	$-SO_3H$	cop-1
Diaion PK 224 (4)	$-SO_3H$	cop-1
Diaion PK 228 (4)	$-SO_3H$	cop-1
Diaion SK-1A (4)	$-SO_3H$	cop-1
Diaion SK-1B (4)	$-SO_3H$	cop-1
Diaion SL-102 (4)	$-SO_3H$	cop-1
Diaion SK-103 (4)	$-SO_3H$	cop-1
Diaion SK-104 (4)	$-SO_3H$	cop-1
Diaion SK-106 (4)	$-SO_3H$	cop-1
Diaion SK-110 (4)	$-SO_3H$	cop-1
Diaion SK-112 (4)	$-SO_3H$	cop-1
Diaion SK-116 (4)	$-SO_3H$	cop-1
Diaion SK-1 AG (4)	$-SO_3H$	cop-1
Dowex 30 (5)	$-CH_2SO_3H$; $-OH$	phen
Dowex 50 (5)	$-SO_3H$	cop-1
Dowex 50W (5)	$-SO_3H$	cop-1
Duolite C-1 (7)	$-CH_2SO_3H$; $-OH$	phen
Duolite C-2 (7)	$-CH_2SO_3H$	phen
Duolite C-3 (7)	$-CH_2SO_3H$; $-OH$	phen
Duolite C-10 (7)	$-CH_2SO_3H$	phen
Duolite C-20 (7)	$-SO_3H$	cop-1
Duolite C-21 (7)	$-SO_3H$	cop-1
Duolite C-25 (2)	$-SO_3H$	cop-1
Duolite C-26 (2)	$-SO_3H$	cop-1
Duolite C-27 (2)	$-SO_3H$	cop-1
Duolite C-261 (2)	$-SO_3H$	cop-1
Duolite-Micro Kationex (2)	$-SO_3H$	cop-1
FN-Katex (8)	$-SO_3H$; $-OH$	phen

APPENDIX (continued)

ION EXCHANGE RESINS

Tradename and manufacturer	Functional group(s)	Matrix
Gamranityt FPC (47)	$-SO_3H$; $-OH$; $-COOH$	phen
Chempro C-12 (9)	$-SO_3H$	cop-1
Chempro C-20 (9)	$-SO_3H$	cop-1
Chempro C-26 (9)	$-SO_3H$	cop-1
Chempro C-200 (9)	$-SO_3H$	cop-1
Imac C-2 (10)	$-SO_3H$	cop-1
Imac D-8-P (10)	$-SO_3H$	cop-1
Imac C-10 (10)	$-SO_3H$	cop-1
Imac C-10-P (10)	$-SO_3H$	cop-1
Imac C-11 (10)	—	phen
Imac C-12 (10)	$-SO_3H$	cop-1
Imac C-14 (10)	$-SO_3H$	cop-1
Imac C-16 (10)	$-SO_3H$	cop-1
Imac C-16-P (10)	$-SO_3H$	cop-1
Imac C-19 (10)	$-SO_3H$; $-COOH$	—
Imac C-22 (10)	$-SO_3H$	cop-1
Imac C-26 (10)	$-SO_3H$	cop-1
Ionac C-200 (11)	$-SO_3H$; $-OH$	phen
Ionac C-240 (11)	$-SO_3H$	cop-1
Ionac C-242 (11)	$-SO_3H$	cop-1
Ionac C-244 (11)	$-SO_3H$	cop-1
Ionac C-249 (11)	$-SO_3H$	cop-1
Ionac C-250 (11)	$-SO_3H$	cop-1
Ionac C-251 (11)	$-SO_3H$	cop-1
Ionac C-252 (11)	$-SO_3H$	cop-1
Ionac C-253 (11)	$-SO_3H$	cop-1
Ionac C-255 (11)	$-SO_3H$	cop-1
Ionac C-257 (11)	$-SO_3H$	cop-1
Ionac C-258 (11)	$-SO_3H$	cop-1
Ionac C-259 (11)	$-SO_3H$	cop-1
Ionac C-280 (11)	$-SO_3H$	cop-1
Ionac C-281 (11)	$-SO_3H$	cop-1
Ionac CI-294 (11)	$-SO_3H$	cop-1
Ionac CI-295 (11)	$-SO_3H$	cop-1
Ionenaustauscher (12)	$-SO_3H$	cop-1
Kastel C-300 (13)	$-SO_3H$	cop-1
Kastel C-300P (13)	$-SO_3H$	cop-1
Kastel C-300AGR (13)	$-SO_3H$	cop-1
Kastel C-300 AGR-P (13)	$-SO_3H$	cop-1
KU-1 (cationite KU-1; Espatit-1) (14)	$-SO_3H$; $-OH$	phen
KU-2 (cationite KU-2) (14)	$-SO_3H$	cop-1
KU-3 (cationite KU-3) (14)	$-SO_3H$	cop
KU-4 (cationite KU-4) (14)	$-SO_3H$	cop
KU-5 (cationite KU-5) (14)	$-SO_3H$; $-OH$	phen
KU-6 (cationite KU-6) (14)	$-SO_3H$; $-COOH$	phen
KU-6F (cationite KU-6F) (14)	$-SO_3H$; $-COOH$	phen
KU-7 (cationite KU-7) (14)	$-SO_3H$; $-OH$	phen
KU-8 (cationite KU-8) (14)	$-SO_3H$; $-OH$	phen
KU-9 (cationite KU-9) (14)	$-SO_3H$; $-OH$	phen
KU-21 (cationite KU-21) (14)	$-SO_3H$	phen
KU-22 (cationite KU-22) (14)	$-SO_3H$	cop-1
Lewasorb A-10 (16)	$-SO_3H$	cop-1
Lewasorb A-11 (16)	$-SO_3H$	cop-1

APPENDIX (continued)

ION EXCHANGE RESINS

Tradename and manufacturer	Functional group(s)	Matrix
Lewasorb CA-9252 (16)	$-SO_3H$	cop-1
Lewasorb CNS (16)	$-SO_3H$; $-COOH$	phen
Lewatit KSN (16)	$-SO_3H$	phen
Lewatit PN (16)	$-SO_3H$	phen
Lewatit S-100 (16)	$-SO_3H$	cop-1
Lewatit S-112 (16)	$-SO_3H$	cop-1
Lewatit S-115 (16)	$-SO_3H$	cop-1
Lewatit S-120 (16)	$-SO_3H$	cop-1
Lewatit SP-100 (16)	$-SO_3H$	cop-1
Lewatit SP-112 (16)	$-SO_3H$	cop-1
Lewatit SP-120 (16)	$-SO_3H$	cop-1
Lewatit TSW-40 (16)	$-SO_3H$	cop-1
Liquonex CRM (15)	$-SO_3H$; $-OH$	phen
Liquonex CRP (15)	$-SO_3H$; $-OH$	phen
Liquonex CRQ (15)	$-SO_3H$; $-OH$	phen
Merck-Lewatit S-1020 (12)	$-SO_3H$	cop-1
Merck-Lewatit S-1080 (12)	$-SO_3H$	cop-1
Merck-Lewatit SP-1080 (12)	$-SO_3H$	cop-1
MK-2 (49)	$-SO_3H$	phen
MK-3 (49)	$-SO_3H$	phen
MSF (14)	$-SO_3H$; $-OH$	phen
Mükion FG (17)	$-CH_2SO_3H$; $-OH$	phen
Mükion P (17)	$-CH_2SO_3H$; $-OH$	phen
Mükion PS (17)	$-SO_3H$	cop-1
Mükion PSM (17)	$-SO_3H$	cop-1
Ostion KS (8)	$-SO_3H$	cop-1
Ostion KS-2 (8)	$-SO_3H$	cop-1
Ostion KSP (8)	$-SO_3H$	cop-1
Nalcite HCR (18)	$-SO_3H$	cop-1
Nalcite HCRW (18)	$-SO_3H$	cop-1
NSF (cationite NSF) (14)	$-SO_3H$	phen
Permutit C 50D (19)	$-SO_3H$	cop-1
Permutit Q (19)	$-SO_3H$	cop-1
Permutit RS (19)	$-SO_3H$	cop-1
Permutit RS-20 (19)	$-SO_3H$	cop-1
Permutit RS-40 (19)	$-SO_3H$	cop-1
Permutit RS-60 (19)	$-SO_3H$	cop-1
Permutit RS-90 (19)	$-SO_3H$	cop-1
Permutit RS-120 (19)	$-SO_3H$	cop-1
Permutit RSP-100 (19)	$-SO_3H$	cop-1
Permutit RSP-100I (19)	$-SO_3H$	cop-1
Permutit RSP-120 (19)	$-SO_3H$	cop-1
Relite CF (20)	$-SO_3H$	cop-1
Relite CFS (20)	$-SO_3H$	cop-1
Relite CFZ (20)	$-SO_3H$	cop-1
Relite CM (20)	$-SO_3H$; $-COOH$	cop-1
Resex (21)	$-SO_3H$	phen
Resex P (21)	$-SO_3H$	cop-1
Rexyn 101 (22)	$-SO_3H$	cop-1
SBS (cationite SBS) (14)	$-SO_3H$	cop
SDV (cationite SDV) (14)	$-SO_3H$	cop-1
Serdolit CS-1 (23)	$-SO_3H$	cop-1
Serdolit CS-2 (23)	$-SO_3H$	cop-1

APPENDIX (continued)

ION EXCHANGE RESINS

Tradename and manufacturer	Functional group(s)	Matrix
Serdolit CS-11 (23)	$-SO_3H$	cop-1
Serdolit Rot (23)	$-SO_3H$	cop-1
Sine-1 (40)	$-SO_3H$	cop-1
SM-12 (cationite SM-12) (14)	$-SO_3H$	cop
SN (cationite SN) (14)	$-SO_3H$; $-OH$	phen
SNF (cationite SNF) (14)	$-SO_3H$; $-OH$	phen
Staionit F extra (8)	$-CH_2SO_3H$; $-OH$	phen
Staionit FK (8)		phen
Varion KS (29)	$-SO_3H$	cop-1
Varion KSM (29)	$-SO_3H$	cop-1
Varion KS-P (29)	$-SO_3H$	cop-1
Varion KO (29)	$-SO_3H$	cop-1
Vionit CS-2 (24)	$-SO_3H$	cop-1
Vionit KS-21 (24)	$-SO_3H$	cop-1
Wofatit D (25)	$-SO_3H$; $-OH$	phen
Wofatit F (25)	$-SO_3H$; $-OH$	phen
Wofatit K, KS (25)	$-SO_3H$; $-OH$	phen
Wofatit KS-10 (25)	$-SO_3H$	cop-1
Wofatit KPS (Wofatit KPS-200) (25)	$-SO_3H$	cop-1
Wofatit P (25)	$-SO_3H$; $-OH$	phen
Wofatit S-1T (25)	$-SO_3H$	phen
Xenonit SD (48)	$-SO_3H$	cop-1
Zeo-Rex (26)	$-CH_2SO_3H$; $-OH$	phen
Zerolit 215 (Zeo-Karb 215) (27)	$-CH_2SO_3H$; $-OH$	phen
Zerolit 225 (Zeo-Karb 225) (27)	$-SO_3H$	cop-1
Zerolit 227 (27)	$-SO_3H$; $-COOH$	acrylate
Zerolit 325 (27)	$-SO_3H$	cop-1
Zerolit 425 (27)	$-SO_3H$	cop-1
Zerolit 525 (27)	$-SO_3H$	cop-1
Zerolit 625 (925) (27)	$-SO_3H$	cop-1

Weakly Acidic Cation Exchangers

Allasion CC (2)	$-COOH$	cop-2
Amberlite IRA-50 (3)	$-COOH$	cop-2
Amberlite IRC-84 (3)	$-COOH$	cop-2
Bio-Rex 70 (1)	$-COOH$	cop-2
Diaion WK-10 (4)	$-COOH$	cop-2
Diaion WK-11 (4)	$-COOH$	cop-2
Duolite Cation S Selector (7)	$-COOH$	phen
Duolite CC-2 (7)	$-COOH$	cop-2
Duolite CC-3 (7)	$-COOH$	cop-2
Duolite C-10H (7)	$-COOH$	cop-2
Duolite C-464 (7)	$-COOH$	cop-2
Duolite CS-101 (7)	$-COOH$	cop-2
Duolite CS-464 (7)	$-COOH$	cop-2
Gamranityt FHF (47)	$-COOH$; $-OH$	phen
Imac Z-5 (10)	$-COOH$	cop-2
Ionac C-270 (11)	$-COOH$	cop-2
CG-1 (cationite CG-1) (14)	$-COOH$	cop-2
Ionenaustauscher IV (12)	$-COOH$	cop-2
Kastel C-100 (13)	$-COOH$	cop-2
Kastel C-101 (13)	$-COOH$	cop-2

APPENDIX (continued)

ION EXCHANGE RESINS

Tradename and manufacturer	Functional group(s)	Matrix
KB-1 (cationite KB-1) (14)	–COOH	cop-2
KB-2 (cationite KB-2) (14)	–COOH	cop
KB-3 (cationite KB-3) (14)	–COOH	cop
KB-4 (cationite KB-4) (14)	–COOH	cop-2
KB-5 (cationite KB-5) (14)	–CH$_2$COOH; –OH	phen
KFU (cationite KFU) (14)	–COOH; –OH	phen
KM (cationite KM) (14)	–COOH	pol
KMD (cationite KMD) (14)	–COOH	pol
KMG (cationite KMG) (14)	–COOH	pol
KMT (cationite KMT) (14)	–COOH	pol
KMTA (cationite KMTA) (14)	–COOH	pol
KMTB (cationite KMTB) (14)	–COOH	pol
KN (cationite KN) (14)	–COOH	cop
KR (cationite KR) (14)	–COOH	pol
KFFU (cationite KFFU) (14)	–OCH$_2$COOH; –OH	phen
KRFU (cationite KRFU) (14)	–OCH$_2$COOH; –OH	phen
KRFFU (cationite KRFFU) (14)	–OCH$_2$COOH; –OH	phen
KS-1 (cationite KS-1) (14)	–COOH	cop
Lewatit C (16)	–COOH	—
Lewatit CA-9267 (16)	–COOH	cop-2
Lewatit CNO (16)	–COOH; –OH	phen
Lewatit CNP (16)	–COOH	cop-2
Lewatit CNP-80 (16)	–COOH	cop-2
Lewatit CA-9269HL (16)	–COOH	cop-2
Lewatit OC-1001HL (16)	–COOH	cop-2
Merck-Lewatit CP-3050 (12)	–COOH	cop-2
Mükion CP (17)	–COOH	cop-2
Mükion KMK (17)	–COOH	cop-2
Ostion KM (8)	–COOH	cop-2
Permutit C (19)	–COOH	cop-2
Permutit C-65 (19)	–COOH	cop-2
Permutit C-67 (19)	–COOH	cop-2
Permutit Q-210 (19)	–COOH	—
Permutit 216 (19)	–COOH	—
Rexyn 102 (22)	–COOH	cop-2
Relite CC (20)	–COOH	cop-2
Relite CCN (20)	–COOH	cop-2
ROA Katex (8)	–OCH$_2$COOH; –OH	phen
SG-1 (cationite SG-1) (14)	–COOH	phen
Serdolit CW-1 (23)	–COOH	cop-2
Varion KC (29)	–COOH	cop-2
Varion KCM (29)	–COOH	cop-2
Wofatit C (25)	–COOH; –OH	phen
Wofatit CA (25)	–COOH	cop-2
Wofatit CN (25)	–COOH; –OH	phen
Wofatit CP-300 (25)	–COOH	cop-2
Wofatit CV (25)	–COOH; –OH	phen
Zerolit 216 (27)	–COOH; –OH	phen
Zerolit 226 (Zeo-Karb 226) (27)	–COOH	cop-2
Zerolit 236 (27)	–COOH	cop-2

Strongly Basic Anion Exchangers

AG-1 (Bio-Rad AG-1) (1)	Type 1	cop-3

APPENDIX (continued)

ION EXCHANGE RESINS

Tradename and manufacturer	Functional group(s)	Matrix
AG-2 (Bio-Rad AG-2) (1)	Type 2	cop-3
AG-21K (1)	—	cop-3
Allasion AR 10 (2)	Type 1	cop-3
Allasion AR 17 (2)	Type 1	cop-3
Allasion AQ 20 (2)	Type 2	cop-3
Allasion AQ 27 (2)	Type 2	cop-3
Allasion DC-22 (2)	Type 2	cop-3
Amberlite IRA-400 (3)	Type 1	cop-3
Amberlite IRA-401 (3)	Type 1	cop-3
Amberlite IRA-401-S (3)	Type 1	cop-3
Amberlite IRA-402 (3)	Type 1	cop-3
Amberlite IRA-405 (3)	Type 1	cop-3
Amberlite IRA-410 (3)	Type 2	cop-3
Amberlite IRA-411 (3)	Type 2	cop-3
Amberlite IRA-425 (3)	Type 1	cop-3
Amberlite IRA-900 (3)	Type 1	cop-3
Amberlite IRA-904 (3)	Type 1	cop-3
Amberlite IRA-910 (3)	Type 2	cop-3
Amberlite IRA-911 (3)	Type 2	cop-3
Amberlite IRA-938 (3)	Type 1	cop-3
Amberlyst A-26 (3)	Type 1	cop-3
Amberlyst A-27 (3)	Type 1	cop-3
Amberlyst A-29 (3)	Type 2	cop-3
AV-15 (anionite AV-15) (14)	Type 1	cop-3
AV-17 (anionite AV-17) (14)	Type 1	cop-3
AV-19 (anionite AV-19) (14)	Type 1	cop
AV-21 (anionite AV-21) (14)	Type 1	cop
AV-27 (anionite AV-27) (14)	Type 2	cop
Bio-Rex 9 (1)	$-N^+$ pyridine	cop-3
Diaion PA 304 (4)	Type 1	cop-3
Diaion PA 306 (4)	Type 1	cop-3
Diaion PA 308 (4)	Type 1	cop-3
Diaion PA 310 (4)	Type 1	cop-3
Diaion PA 312 (4)	Type 1	cop-3
Diaion PA 314 (4)	Type 1	cop-3
Diaion PA 316 (4)	Type 1	cop-3
Diaion PA 318 (4)	Type 1	cop-3
Diaion PA 320 (4)	Type 1	cop-3
Diaion PA 404 (4)	Type 2	cop-3
Diaion PA 406 (4)	Type 2	cop-3
Diaion PA 408 (4)	Type 2	cop-3
Diaion PA 410 (4)	Type 2	cop-3
Diaion PA 412 (4)	Type 2	cop-3
Diaion PA 414 (4)	Type 2	cop-3
Diaion PA 416 (4)	Type 2	cop-3
Diaion PA 418 (4)	Type 2	cop-3
Diaion PA 420 (4)	Type 2	cop-3
Diaion SA 10A (4)	Type 1	cop-3
Diaion SA 10B (4)	Type 1	cop-3
Diaion SA 11A (4)	Type 1	cop-3
Diaion SA 11B (4)	Type 1	cop-3
Diaion SA 20A (4)	Type 2	cop-3
Diaion SA 20B (4)	Type 2	cop-3

APPENDIX (continued)

ION EXCHANGE RESINS

Tradename and manufacturer	Functional group(s)	Matrix
Diaion SA 21A (4)	Type 2	cop-3
Diaion SA 21B (4)	Type 2	cop-3
Diaion SA 100-AG (4)	Type 1	cop-3
Dowex 1 (5)	Type 1	cop-3
Dowex 2 (5)	Type 2	cop-3
Dowex 11 (5)	Type 1	cop-3
Dowex 21K (5)	Type 1	cop-3
Duolite A-40, 42 (7)	Type 2	cop-3
Duolite A-44 (7)	Type 2	cop-3
Duolite A-101 (7)	Type 1	cop-3
Duolite A-101 D (7)	Type 1	cop-3
Duolite A-102 D (7)	Type 2	cop-3
Duolite A-121 (7)	Type 1	cop-3
Duolite A-162 (7)	Type 2	cop-3
Duolite A-161 (7)	Type 1	cop-3
Duolite ESF-12 (7)	Type 1	cop-3
Imac S-5-40 (10)	Type 1	cop-3
Imac S-42 (10)	Type 2	cop-3
Imac S-5-50 (10)	Type 1	cop-3
Imac S-5-52 (10)	Type 2	cop-3
Ionac A-540 (11)	Type 1	cop-3
Ionac A-550 (11)	Type 2	cop-3
Ionac 553 (11)	Type 2	cop-3
Ionac 580 (11)	$-N^+$pyridine	cop-3
Ionac 590 (11)	$-N^+$pyridine	cop-3
Ionac A-935 (11)	Type 1	cop-3
Ionenaustauscher III (12)	Type 1	cop-3
Kastel A-300 (13)	Type 2	cop-3
Kastel A-300 P (13)	Type 2	cop-3
Kastel A-500 (13)	Type 1	cop-3
Kastel A-500 P (13)	Type 1	cop-3
L-anex (OAL) (8)	$-N^+$pyridine	phen
Lewatit CA 9263 (16)	Type 1	cop-3
Lewatit CA 9268 (16)	Type 2	cop-3
Lewatit M-500 (16)	Type 1	cop-3
Lewatit MP-500 (16)	Type 1	cop-3
Lewatit M-504 (16)	Type 1	cop-3
Lewatit M-600 (16)	Type 2	cop-3
Lewatit MP-600 (16)	Type 2	cop-3
Lewasorb A 50 (16)	Type 1	cop-3
Merck-Lewatit M-5020 (12)	Type 1	cop-3
Merck-Lewatit M-5080 (12)	Type 1	cop-3
Merck-Lewatit M-5080 G3 (12)	Type 1	cop-3
Merck-Lewatit MP-5080 (12)	Type 1	cop-3
Ostion AT (S-8-TM) (8)	Type 1	cop-3
Ostion ATP (8)	Type 1	cop-3
Ostion AD (S-8-D) (8)	Type 2	cop-3
Ostion ADP (8)	Type 2	cop-3
Permutit A-300 D (19)	Type 2	cop-3
Permutit E-3 (19)	Type 2	cop-3
Permutit EHP (19)	Type 1	cop-3
Permutit EHP-274 (19)	Type 1	cop-3
Permutit ES (19)	Type 2	cop-3

APPENDIX (continued)

ION EXCHANGE RESINS

Tradename and manufacturer	Functional group(s)	Matrix
Permutit ES-26 (19)	Type 2	cop-3
Permutit ES-32 (19)	Type 2	cop-3
Permutit ES-274 (19)	Type 2	cop-3
Permutit ES-274-K (19)	Type 2	cop-3
Permutit ESB-26 (19)	Type 1	cop-3
Permutit ESB-32 (19)	Type 1	cop-3
Permutit ESB-274 (19)	Type 1	cop-3
Permutit ESB-274-I (19)	Type 1	cop-3
Permutit S-1 (19)	Type 1	cop-3
Permutit S-2 (19)	Type 1	cop-3
Permutit SK (19)	$-N^+$pyridine	cop-3
Relite 2A (20)	Type 2	cop-3
Relite 3A (20)	Type 1	cop-3
Relite 2AS (20)	Type 2	cop-3
Relite 3AS (20)	Type 1	cop-3
Resenax HB (28)	Type 1	cop-3
Rexyn 201 (22)	Type 1	cop-3
Rexyn 202 (22)	Type 2	cop-3
Rexyn 204 (22)	$-N^+$pyridine	cop-3
Serdolit AS-1 (23)	Type 1	cop-3
Serdolit AS-2 (23)	Type 2	cop-3
Serdolit AS-3 (23)	Type 1	cop-3
Serdolit AS-4 (23)	Type 1	cop-3
Varion AD (29)	Type 2	cop-3
Varion AP (29)	$-N^+$pyridine	cop-3
Varion AT-400 (29)	Type 1	cop-3
Varion AT-660 (29)	Type 1	cop-3
Varion AD-P (29)	Type 2	cop-3
Varion AT-P (29)	Type 1	cop-3
Varion ATM (29)	Type 1	cop-3
Varion ADM (29)	Type 2	cop-3
Vionit AT1 (24)	Type 1	cop-3
Wofatit SBW (25)	Type 1	cop-3
Wofatit SBT (25)	Type 1	cop-3
Wofatit SBK (25)	Type 2	cop-3
Wofatit SBU (25)	$-N^+$pyridine	cop-3
Zerolit FF (Zeo-Karb FF; Deacidite FF) (27)	Type 1	cop-3
Zerolit FF-IP (27)	Type 1	cop-3
Zerolit FS (27)	Type 1	cop-3
Zerolit MPF (27)	Type 1	cop-3
Zerolit MPN (27)	Type 2	cop-3
Zerolit P (IP) (27)	Type 2	cop-3
Zerolit N (IP) (27)	Type 2	cop-3
Zerolit N (27)	Type 2	cop-3

Medium and Weakly Basic Anion Exchangers

AG-3 (Bio-Rad AG 3) (1)	$-NH_2$; $=NH$; $\equiv N$	cop
Allasion AW-2 (2)	Polyamine	—
Allasion AWB-3 (2)	Polyamine	—
Amberlite IR-4B (3)	$=NH$; $\equiv N$	phen
Amberlite IR-45 (3)	$-NH_2$; $=NH$; $\equiv N$	cop
Amberlite IRA-68 (3)	$\equiv N$	cop

APPENDIX (continued)

ION EXCHANGE RESINS

Tradename and manufacturer	Functional group(s)	Matrix
Amberlite IRA-93 (3)	≡N	cop
Amberlyst A-21 (3)	−NR$_3$	cop
AN-1 (anionite AN-1) (14)	−NH$_2$; =NH	phen
AN-2F (anionite AN-2F) (14)	=NH, ≡N	phen
AN-4K (anionite AN-4K) (14)	=NH; ≡N	pol
AN-7K (anionite AN-7K) (14)	=NH; ≡N	pol
AN-9 (anionite AN-9) (14)	=NH; ≡N	phen
AN-10 (anionite AN-10) (14)	−NH$_2$; ≡N	phen
AN-15 (anionite AN-15) (14)	−NH$_2$	cop
AN-17 (anionite AN-17) (14)	=NH	cop
AN-18 (anionite AN-18) (14)	≡N	cop
AN-19 (anionite AN-19) (14)	−NH$_2$; =NH; ≡N	cop
AN-20 (anionite AN-20) (14)	−NH$_2$; =NH	cop
AN-21 (anionite AN-21) (14)	−NH$_2$; =NH	cop
AN-22 (anionite AN-22) (14)	−NH$_2$; =NH	cop
AN-23 (anionite AN-23) (14)	≡N	cop
AN-25 (anionite AN-25) (14)	−NH$_2$; =NH	cop
AV-16 (anionite AV-16) (14)	=NH; ≡N; −$\overset{+}{N}$R$_3$	—
AV-18 (anionite AV-18) (14)	≡N; −$\overset{+}{N}$R$_3$	—
AV-20 (anionite AV-20) (14)	≡N; −$\overset{+}{N}$R$_3$	—
Bio-Rex 5 (1)	≡N; −$\overset{+}{N}$R$_3$	—
Dowex 3 (5)	−NH$_2$; =NH; ≡N	cop
Dowex 44 (5)	−NR$_2$	—
Duolite A-2 (7)	−NHR	phen
Duolite A-4 (7)	−NR$_2$	phen
Duolite A-6 (7)	≡N; −NR$_2$	phen
Duolite A-7 (7)	−NHR	phen
Duolite A-30 (7)	−NR$_2$; −$\overset{+}{N}$R$_3$	—
Duolite A-30B (7)	−NR$_2$; −$\overset{+}{N}$R$_3$	—
Duolite A-30G (7)	≡N; −$\overset{+}{N}$R$_3$	—
Duolite A-30T (7)	−NR$_2$; −$\overset{+}{N}$R$_3$	phen
Duolite A-33 (7)	≡N; −$\overset{+}{N}$R$_3$	—
Duolite A-41 (7)	≡N; −$\overset{+}{N}$R$_3$	—
Duolite A-43 (7)	≡N; −$\overset{+}{N}$R$_3$	—
Duolite A-47 (7)	≡N; −$\overset{+}{N}$R$_3$	—
Duolite A-303 (7)	−NR$_2$; −$\overset{+}{N}$≡	cop-1
Duolite A-368 (7)	−NR$_2$	cop-1
Duolite AS-31 (7)	−NR$_2$	phen
Diaion WA-10 (4)	=NH; −NR$_2$	cop
Diaion WA-11 (4)	=NH; −NR$_2$	cop
Diaion WA-20 (4)	=NH; −NH$_2$	cop
Diaion WA-21 (4)	=NH; −NH$_2$	cop
Diaion WA-30 (4)	−NR$_2$	cop
EDE-10P (anionite EDE-10P) (14)	=NH; ≡N; −$\overset{+}{N}$R$_3$	phen
Gamranityt FD (47)	−NH$_2$	phen

APPENDIX (continued)

ION EXCHANGE RESINS

Tradename and manufacturer	Functional group(s)	Matrix
Gamranityt LF (47)	=NH; −NHR	phen
Imac A-13T (10)	≡N; −$\overset{+}{N}R_3$	—
Imac A-17G, P (10)	Polyamino	phen
Imac A-20 (10)	=NH; ≡N	cop
Imac A-21 (10)	≡N	cop
Imac A-27 (10)	−NH$_2$; =NH; −$\overset{+}{N}$≡; ≡N	—
Imac A-293 (10)	=NH; ≡N	—
Ionac A-260 (11)	=NH; ≡N	—
Ionac A-300 (11)	≡N; −$\overset{+}{N}R_3$	—
Ionac A-310 (XL) (11)	≡N; −$\overset{+}{N}R_3$	—
Ionac A-330 (11)	≡N	phen
Ionenaustauscher II (12)	Aliphatic amino group	cop
Kastel A-100 (13)	Epoxyamino ≡N; =NH	phen
Kastel A-101 (13)	=NH	cop
Lewatit MIH-59 (16)	=NH	phen
Lewatit MP-62 (16)	−NH$_2$	cop
Lewatit MP-64 (16)	=NH	cop
Merck-Lewatit MP-7080 (18)	−NR$_2$	cop
MFD (8)	−NH$_2$	phen
MMG-1 (anionite MMG-1) (14)	=NH; ≡N	phen
MN (anionite MN) (14)	=NH; ≡N	phen
Mükion G (17)	≡N guanidine	phen
Mükion PP (17)	−NH$_2$; =NH; ≡N	—
N (anionite N) (14)	=NH; ≡N	phen
NO (anionite NO) (14)	=NH; ≡N	phen
Ostion AW (PA anex) (8)	≡N; −NR$_2$	cop
Ostion AWP (8)	≡N; −NR$_2$	cop
PEK (anionite PEK) (14)	−NH$_2$; =NH; ≡N; −$\overset{+}{N}R_3$	—
Permutit E (19)	=NH; ≡N	—
Permutit EM-13 (19)	≡N	cop
Permutit EM-13-I (19)	≡N	cop
Relite 4 MS (20)	−NR$_2$	cop
Relite MS-170 (20)	−NR$_2$; =NR	cop
Rexyn 203 (22)	−NH$_2$; =NH	cop
Rexyn 205 (22)	≡N; −$\overset{+}{N}R_3$	—
Rexyn 206 (22)	=NH; ≡N	—
Rexyn 208 (22)	=NH	—
Serdolit AW-1 (23)	≡N	cop
Serdolit AW-2 (23)	=NH; ≡N	cop
Varion ADA (29)	−$\overset{+}{N}R_3$	cop
Varion ADAM (29)	−$\overset{+}{N}R_3$	cop
Wofatit AD-41 (25)	−$\overset{+}{N}R_3$	cop
Wofatit L-150 (25)	=NH; ≡N; −$\overset{+}{N}R_3$	—
Wofatit L-160 (25)	=NH; ≡N; −$\overset{+}{N}R_3$	—
Wofatit L-165 (25)	≡N; −NR$_3$	—
Wofatit M (25)	=NH; ≡N	phen
Wofatit Y-13 (25)	−NH$_2$; =NH	cop

APPENDIX (continued)

ION EXCHANGE RESINS

Tradename and manufacturer	Functional group(s)	Matrix
Zerolit E (27)	Aliphatic polyamino	phen
Zerolit G (27)	$-N(C_2H_5)_2$	cop
Zerolit H (27)	$-N(CH_3)_2; -\overset{+}{N}R_3$	cop
Zerolit H-IP (27)	$-NR_2; -\overset{+}{N}R_3$	cop
Zerolit J (27)	$-NH_2; =NH; \equiv N$	cop
Zerolit M (27)	$-NH_2; =NH; \equiv N$	cop
Zerolit M-IP (27)	$-NH_2; =NH$	cop

Chelating Resins

Diaion CR-10 (4)	$-CH_2N(CH_2COOH)_2$	cop-1
Dowex A-1 (5)	$-CH_2N(CH_2COOH)_2$	cop-1
Chelex-100 (1)	$-CH_2N(CH_2COOH)_2$	cop-1
KT-1 (14)	$-N(CH_2COOH)_2$	cop-1
KT-2(ANKB-50) (14)	$-CH_2N(CH_2COOH)_2$	cop-1
Lewatit TP-207 (16)	$-CH_2N(CH_2COOH)_2$	cop-1
Varion CH (29)	$-CH_2N(CH_2COOH)_2$	cop-1
Wofatit CM-50 (25)	$-CH_2N(CH_2COOH)_2$	cop-1

Resins Containing Phosphorus and Arsenic

Bio-Rex 63 (1)	$-PO(OH)_2$	cop-1
Duolite C-62 (C-60) (7)	$-P(OH)_2$	cop
Duolite C-63 (C-61) (7)	$-PO_3H_2$	cop-1
Duolite C-65 (C-62) (7)	$-OPO_3H_2; OH$	phen
Duolite ES-463 (7)	$-PO_3H_2$	cop-1
FV (cationite FV) (14)	$-PO_3H_2$	phen-cop
KF-1 (cationite KF-1) (14)	$-PO_3H_2$	cop
KF-2 (cationite KF-2) (14)	$-CH_2PO_3H_2$	cop
KF-3 (cationite KF-3) (14)	$-PO_3H_2$	cop
KF-4 (cationite KF-4) (14)	$-CH_2PO_3H_2$	cop
Nalcite X-219 (38)	$-PO_3H_2$	—
Permutit XP (19)	$-P(OH)_2$	—
PM-52 (38)	$-P(OH)_2$	—
RF (cationite RF) (14)	$-OPO_3H_2; -OH$	phen
AR (cationite AR) (14)	$-AsO_3H_2; -OH$	phen
SF (cationite SF) (14)	$-PO_3H_2$	cop-1

ABBREVIATIONS USED IN THE APPENDIX

phen:	polycondensate
cop:	copolymer
cop-1:	copolymer of styrene with divinylbenzene (DVB)
cop-2:	copolymer of acrylic (methacrylic) acid with DVB
cop-3:	chloromethylated copolymer of polystyrene with DVB
pol:	polymer
type 1:	see Table 1 in the first chapter of Volume I
type 2:	see Table 1 in the first chapter of Volume I

MANUFACTURERS AND DISTRIBUTORS OF ION EXCHANGERS

1. Bio-Rad Laboratories, 32nd and Griffin Ave., Richmond Calif.
2. Dia-Prosim, 107, Rue Edith Cavell, Vitry-sur-Seine, Seine, France
3. Rohm and Haas Co., Philadelphia, Pa., European Division: Rohm and Haas, France S. A. La Tour de Lyon, 185 Rue de Berey, Paris, France
4. Mitsubishi Chemical Industries Ltd., Tokyo, Japan
5. The Dow Chemical Co., Midland, Mich.
6. Reanal, Budapest 70, Hungary
7. Diamond Alkali Co., Western Division, 1901 Spring St., Redwood City, Calif.
8. Spolek pro chemickou a hutni výrobu, Ústi n. Labem, Czechoslovakia
9. Chemical Process Co., Redwood City, Calif.
10. AKZO Chemie Verkoopkantoor bV., James Wattstraat 100, Amsterdam, The Netherlands
11. Ionac Chemical Co., Birmingham, N.J.
12. E. Merck, Darmstadt, West Germany
13. Montecatini-Società Generale per l'Industria Mineraria a Chimica, Via F. Tutari 18 Milano, Italy
14. U.S.S.R. product
15. Liquid Cond. Corp., Linden, N.Y.
16. Farbenfabrik Bayer, Leverkusen, West Germany
17. MUKY, Hungaria Korüt 114, Budapest, Hungary
18. National Aluminate Corp., 6216 West 66 Place, Chicago, Ill.
19. Permutit Co., 330 W. 42nd St., New York, N. Y., Permutit AG, Auguste-Victoria Strasse 62, Berlin, Schmangendorf, West Germany
20. Resindion S.p.A., Division of Sybron Corp., Via Roma, Binasco, Italy
21. J. Crosfield and Sons Ltd., Warrington, Lancs., England
22. Fisher Scientific Co., 1458 N. Lamon Ave., Chicago, Ill.
23. Serva, Feinbiochemica GmbH and Co.690 Heidelberg 1, Römerstrasse 118, West Germany
24. Roumanian product
25. Veb Chemie Kombinat Bitterfeld, East Germany
26. The Permutit Co. Ltd., Gunnersbury Ave., London, England
27. United Water Softeners, Gunnersbury Ave., London, England
28. Jos. Crosfield and Sons Ltd., P.O. Box 26, Bank Quay, Warrington, Lancs., England
29. Nitrokémia Works, Balatonfüzfö, Hungary
30. Archer-Daniels Midland Co., U.S.A.
31. General Mills Ind., Minneapolis, Minn.
32. W. and R. Balston Ltd., Maidstone, Kent, England; H. Reeve Angel and Co. Ltd., 9 Bridewell Place, London, England
33. AB Pharmacia, Uppsala, Sweden
34. Asahi Glass Co. Ltd., Tokyo, Japan
35. Tokuyama Soda Ltd., Tokuyama City, Japan
36. Infilco Inc., P.O. Box 5033, Tucson, Ariz.
37. Cochrane Corp., 17th Allegheny Ave., Philadelphia, Pa.
38. National Aluminate Corp., 6216 West 66 Place, Chicago, Ill.
39. Ionic Inc., 152 6th St., Cambridge, Mass.
40. Chinese product
41. Varian Aerograph, 2700 Mitchel Dr., Walnut Creek, Calif.
42. Northgate Laboratories Inc., Hamden, Conn.
43. E. J. Du Pont de Nemours Co. Inc., 1007 Market St., Wilmington, Del.

44. Beckman Instruments Inc., 2500 Harbor Blvd., Fullerton, Calif.
45. Durrum Chemical Corp., 3950 Fabian Way, Palo Alto, Calif.
46. Hamilton Co., Representative: Micromesure b. v. P.O. Box 205, The Hague, The Netherlands
47. Zaklady Chemiczne Gamrat, Jaslo, Poland
48. Xenon, Poland
49. ZPA Kedzierzyn Poland
50. Carlo Erba, Divisione Chimica, Via C. Ibonati 24, Milano, Italy

Index

INDEX

A

Air, 230, 240
Aluminum, 21—35
　anion exchange resins, 21—22
　after anion exchange separation, 24—29
　in biological materials, 30—31
　cation exchange resins, 22—23
　after cation exchange separation, 30—34
　chelating resins, 23, 35
　in geological materials, 24
　in industrial products, 25—26, 29, 32—33
　microchemical detection, 23, 35
　in nuclear materials, 27—28
　in rocks, 24
　in steel, 25, 32, 33
Ammonia, 149, 150, 155, 160—162
Anion exchange resins
　aluminum, 21—22
　antimony, 205
　arsenic, 189—191
　bismuth, 211—212
　boron, 4
　bromine, 305
　chlorine, 293
　fluorine, 275—276
　gallium, 39—40
　germanium, 99
　indium, 59—60
　iodine, 311—312
　lead, 119—120
　nitrogen, 149
　phosphorus, 167—168
　polonium, 256
　selenium, 255
　silicon, 89
　sulfur, 227—228
　tellurium, 255—256
　thallium, 75—76
　tin, 107—108
Antimony, 205—208
　anion exchange resins, 205
　in biological material and synthetic mixture after cation exchange separation, 208
　cation exchange resins, 206
　chelating resins, 206
　in organic and inorganic materials after anion exchange separation, 207
Arsenic, 189—201, 333
　in alloys, 200
　anion exchange resins, 189—191
　after anion exchange separation, 193—196
　in biological materials, 193, 198—199
　cation exchange resins, 191
　after cation exchange separation, 195—200
　chelating resins, 191—192, 201
　in copper, 201
　in industrial products, 194, 200
　synthetic mixtures and, 195
　in waters, 193, 195—197, 201

B

Beer, 300
Bismuth, 211—222
　anion exchange resins, 211—212
　after anion exchange separation, 214—218
　in biological materials, 215
　cation exchange resins, 212—213
　after cation exchange separation, 219—221
　chelating resins, 222
　in geological materials, 214
　in industrial materials and products, 216—217, 219—220
　microchemical detection, 222
　synthetic mixtures and, 218, 221
Blood, 215, 260, 280, see also Plasma
Boron, 3—17
　anion exchange resins, 4
　after anion exchange separation, 13—17
　in biological materials, 7, 14
　in cast iron, 8
　cation exchange resins, 3—4
　after cation exchange separation, 5—15
　in electrolytes, 9
　in fertilizers, 12, 15
　in industrial products, 10, 15—17
　in nitrate solutions, 12
　in nuclear materials, 12
　in rocks, minerals, and glasses, 5—6, 13
　spectrophotometric determination, 11
　in steel, 8—9, 15
　in titanium alloys, 8
　titrimetric determination, 8—10
　in waters, 7, 14, 16—17
Brass, 139
Bromide, see Bromine
Bromine, 305—309
　anion exchange resins, 305
　after anion exchange separation, 298, 306—308
　cation exchange resins, 305
　after cation exchange separation, 309
　microchemical detection, 309
　in rocks, 306
　species in biological materials, 308
　in urine, 308, 309
　in waters, 307, 309
　in zinc sulfate, 309

C

Carboxylic acid resins, 326—327
Cation exchange resins
　aluminum, 22—23

antimony, 206
arsenic, 191
bismuth, 212—213
boron, 3—4
chlorine, 293—294
fluorine, 276
gallium, 40—41
germanium, 99—100
indium, 60—61
iodine, 312
lead, 120—121
nitrogen, 150
phosphorus, 168—169
polonium, 257—258
selenium, 257
silicon, 89—90
sulfur, 225—226
tellurium, 257
thallium, 76—77
tin, 108
Cement, 238
Cheese, 157, 172
Chelating resins, see Ion exchange resins; specific elements
Chloride, see Chlorine
Chlorine, 293—302
   anion exchange resins, 293
   after anion exchange separation, 295—299
   in beer, 300
   in biological materials, 295
   cation exchange resins, 293—294
   in geological materials, 295
   in industrial products, 296—297
   in plants, 300
   in seawater, 295, 300
   in urine, 295
Coal, 103, 229
Cyanide, 149, 154

D

Dental enamel, 266
Detergents, 181

F

Fertilizers, 12, 15, 158, 180
Fluoride, see Fluorine
Fluorine, 275—288
  in air, 278
  anion exchange resins, 275—276
  after anion exchange separation, 277—282
  in biological materials, 279—280, 284
  cation exchange resins, 276
  after cation exchange separation, 284—288
  in geological materials, 277, 284
  in industrial products, 281, 286—287
  synthetic mixtures and, 282, 288
  in urine, 279
  in waters, 278

G

Gallium, 39—56
  anion exchange resins, 39—40
  after anion exchange separation, 42—49
  in biological materials, 44
  in carbonate media, 49
  cation exchange resins, 40—41
  after cation exchange separation, 50—56
  chelating resins, 41
  in geological materials, 42—43
  in hydrochloric acid media, 54
  in hydrohalic acid media, 42, 47, 49
  in industrial products, 45—46, 50—51
  in meteorites, 42—43
  microchemical detection, 41, 56
  in mineral acid-acetone media, 52—53
  in neutral and alkaline media, 56
  in nitric acid media, 49, 54
  in presence of organic acids, 55
  in rocks, 42
  in thiocyanate and iodide media, 48
Germanium, 99—104
  anion exchange resins, 99
  after anion exchange separation, 101—103
  cation exchange resins, 99—100
  after cation exchange separation, 103—104
  chelating resins, 100
  in coal, 103
  in geological materials, 103
  in industrial products, 101
  synthetic mixtures and, 102, 104

H

Hair, 208
Halogens, 275—319, see also specific halogens
   in industrial materials, 301—302
   oxyanion separation, 299

I

Indium, 59—71
  in acetone-mineral acid media, 67—68
  in ammoniacal media, 65
  anion exchange resins, 59—60
  after anion exchange separation, 62—65
  cation exchange resins, 60—61
  after cation exchange separation, 66—69
  chelating resins, 61, 70
  in geological materials, 62
  in hydrochloric acid, 65
  in industrial products, 62—63, 66—67, 70
  in media containing organic complexing agents, 64
  microchemical detection, 61, 71
  in precipitation, 66
  in seawater, 70
  in sulfosalicylic acid and ethylenediamine media, 69
Iodide, see Iodine

Iodine, 311—319
  anion exchange resins, 311—312
  after anion exchange separation, 298, 313—319
  in biological fluids, 315—317
  cation exchange resins, 312
  in industrial products, 318
  in milk, 315
  in soils and rocks, 313
  in waters, 314
Ion exchange resins, 323—335, see also Anion exchange resins; Cation exchange resins
  chelating resins, 333
  manufacturers and distributors of, 334—335
  medium/weakly basic anion exchangers, 330—333
  strongly acidic cation exchangers, 323—326
  strongly basic anion exchangers, 327—330
  weakly acidic cation exchangers, 326—327

L

Lead, 119—145
  anion exchange resins, 119—120
  after anion exchange separation, 123—137
  in biological materials, 131, 138, 144—145
  cation exchange resins, 120-121
  after cation exchange separation, 138—143
  chelating resins, 121, 144—145
  in geological materials, 123—129, 144—145
  in industrial products, 132—135, 139—140, 144
  in organic solvent media, 141
  in presence of inorganic complexing agents, 143
  in presence of organic complexing agents, 142
  synthetic mixtures and, 137
  in waters, 130, 138

M

Meteorites, 42—43, 62, 79, 109, 214
Milk, 81, 315

N

Nitrate, 149—153, 156—159
Nitrite, 149—153, 156—160
Nitrogen, 149—162
  anion exchange resins, 149
  after anion exchange separation, 151—155
  in biological materials, 152, 157, 161—162
  cation exchange resins, 150
  after cation exchange separation, 154, 156—162
  chelating resins, 150
  in industrial products, 153, 158
  microchemical detection, 160
  in waters, 151, 156, 160

P

Perchlorate, 295
Phosphate, 174—178
Phosphorus, 167—183, 333
  anion exchange resins, 167—168
  after anion exchange separation, 170—175
  in biological materials, 172, 179
  cation exchange resins, 168—169
  after cation exchange separation, 176—182
  in geological materials, 170—171, 176—177
  gravimetric determination, 177, 180
  in industrial products, 173, 180—182
  optical methods, 178, 181—182
  in rocks, 176—178
  synthetic mixtures and, 175, 183
  titrimetric determination, 176—177, 180
Plants, 7, 81, 157, 199, 240, 300
Plasma, 161, 280
Polonium, 255—270
  anion exchange resins, 256
  anion exchange separation, 262
  cation exchange resins, 257—258
  cation exchange separation, 268
  chelating resins, 269
Pyrite, 263

Q

Quarternary ammonium type resins, 327—330

R

Rocks, see Silicates; specific elements

S

Selenium, 255—270
  anion exchange resins, 255
  after anion exchange separation, 259—262
  in biological materials, 260, 266
  cation exchange resins, 257
  after cation exchange separation, 263—266, 268
  chelating resins, 269—270
  in geological materials, 263—264
  in industrial materials, 261
  in rocks and meteorites, 259
  in urine, 266, 270
  in waters, 265, 270
Silicates, 5, 6, 24, 78, 94, 109, 126, 263, 295
Silicon, 89—96
  anion exchange resins, 89
  after anion exchange separation, 91—93
  cation exchange resins, 89—90
  after cation exchange separation, 94—96
  in geological materials, 94
  in industrial products, 92—93, 95—96
  in rocks, 94
  in waters, 91
Steel, 8—9, 11, 15, 25, 32, 112, 134, 139, 216
Sulfate, 230—233, 240, 241, 243, 244
Sulfide, 67, 231, 240, 243, 245
Sulfite, 233, 242—244
Sulfonic acid resins, 323—326
Sulfur, 225—245
  anion exchange resins, 227—228
  after anion exchange separation, 239—244

in biological materials, 233
cation exchange resins, 225—226
after cation exchange separation, 229—239
in environmental material, 240
in geological materials, 229
in industrial emissions and effluents, 234—235
in industrial products and materials, 238—239, 241—242
microchemical detection, 245
oxyanion separation, 243—244
in plutonium, 237
spectrophotometric determination, 231
in synthetic organic compounds, 236
titrimetric determination, 230
in uranyl salts, 237
in waters, 230—232

T

Tellurium, 255—270
anion exchange resins, 255—256
after anion exchange separation, 259, 261—262
cation exchange resins, 257
after cation exchange separation, 264, 267—268
chelating resins, 269
in geological materials, 264
in industrial products and materials, 261, 267
in rocks and meteorites, 259
Thallium, 75—86
anion exchange resins, 75—76
after anion exchange separation, 78—82
in biological materials, 81
cation exchange resins, 76—77
after cation exchange separation, 83—85
chelating resins, 77, 86
in geological materials, 78—79, 83
in industrial products, 84, 86

in meteorites, 79
microchemical detection, 77, 86
in rocks, 78, 83
synthetic mixtures and, 85
in waters, 80
Tin, 107—116
in alloys, 111—113
anion exchange resins, 107—108
after anion exchange separation, 109—114
cation exchange resins, 108
after cation exchange separation, 115—116
in geological materials, 109—110
in glass, 113
in hydrochloric acid, 114
in industrial products, 115
in iron, 112
in metals, 111—112
in meteorites, 109
in presence of organic complexing agents, 114
in rocks, 109
in salt, 113
in seawater, 110
in steel, 112
synthetic mixtures and, 116
Tobacco, 152, 158

U

Uranyl salts, 237
Urine, 138, 144, 162, 198, 215, 233, 266, 270, 279, 295, 308, 309, 317

W

Waters, see specific elements
Whey powder, 152
Wine, 138, 179